S0-DVE-326

THE COAL WAR

THE COAL WAR

A SEQUEL TO "KING COAL"

UPTON SINCLAIR

INTRODUCTION BY JOHN GRAHAM

COLORADO ASSOCIATED UNIVERSITY PRESS
BOULDER

This book was published
with the assistance
of a grant from the
E. Floyd Thompson Centennial Fund

INTRODUCTION by John Graham
THE COAL WAR by David Sinclair
International Standard Book Number: 0-87081-067-7
Library of Congress Catalog Card Number: 75-40885
Printed in the United States of America
Designed by Dave Comstock

CONTENTS

ACKNOWLEDGMENT

In preparing this text I have been indebted to many people and institutions, more than I can acknowledge. I wish to express my particular gratitude to Elizabeth Jameson, whose knowledge and encouragement have been of great value; David Sinclair, who has been uniformly helpful, and Pamela Graham, whose aid has been beyond summary description. I am grateful for the cooperation given by the librarians and archivists of both the Lilly Library, Indiana University, and the Denver Public Library Western History Department. *The Coal War* was published with the assistance of a grant from the E. Floyd Thompson Centennial Fund. My appreciation is also due the University of Colorado. A great deal of work was accomplished during a National Endowment for the Humanities Fellowship for Younger Humanists. I wish to acknowledge my appreciation.

INTRODUCTION

I

In early October, 1915, John D. Rockefeller, Jr., found himself in the unaccustomed position of addressing coal miners in Pueblo, Colorado. A long, bitterly fought strike had occurred, the miners had been defeated the year before, and Rockefeller spoke of a partnership between labor and capital. The keystone to this new partnership was the modern corporation. Rockefeller used a small table as illustration. Its four legs, none of which could stand independently, represented stockholders, directors, officers, and employees. Corporate earnings could be placed on a level table, he said, and he laid on it a handful of coins. Rockefeller showed that if one corporate leg reached up to take more than its share, the table tipped and the coins fell off. If another did not do enough, the corner dropped and the coins likewise fell away. Only when the four legs stood equally would the table remain level. Valueless alone, the legs were mutual partners. "Men," he said, "only when every man connected with that square corporation which is on the level, is interested unselfishly, not in what he can get out of the corporation, but what he can put into it for the benefit of every man in the concern, will that man himself get the most out of it."[1] The 1913-1914 Colorado coal strike, as Rockefeller understood, is a remarkable illustration of the transformation of the American political economy from laissez faire capitalism to the corporate capitalism of the twentieth century.

Laissez faire was the dominant ethos in post-Civil War America. As the country was industrialized with a speed and thoroughness before unknown, the doctrine of unrestricted freedom in the market place held virtually absolute control. By 1900 the development of the continent by railroads, banking, mining, oil, steel, and communication, for example, was fundamentally completed. Laissez faire advocates argued that the negative consequences of expansion, the graft and corruption which made Mark Twain prophetically call his first novel *The Gilded Age*, were minor when counted against the benefits. Any large industry not only gave work to laborers, but, in its use of raw materials and other manufactured products, created a systematic network of jobs and commodities for the American people. While it was true that industrialists amassed unheard of private wealth, the argument continued, employees' earnings increased in the same proportion. As technology and the factory system improved, more of the good things would be distributed throughout the land. Free competition left the field equally open to all, from the owner of a small refinery in Ohio to an immigrant Scottish bobbin boy. Nor was there need for any government interference, except, of course, for high protective tariffs. Adam Smith's invisible hand regulated and balanced competing forces in the market place through the law of supply and demand.

When reality intruded and it became clear that exploitation, not the good things, came to the working class, labor challenged the laissez faire economy. Battles with the Molly Maguires in the Pennsylvania coal fields were a portent for the future. Workers organized in protection associations recognized that merely trying to protect poorly paid, brutalizing jobs was an inadequate response and began to form unions. In 1877 a massive railway strike halted rail traffic in several states. Even more threatening to the new industrial order was the movement for an eight-hour day which culminated in Chicago in 1886. If the largest union of the time, the Knights of Labor, was in good part destroyed by its lack of support for the Haymarket defendants, it meant that labor was becoming more radical in its demands. In 1892 workers struck at Andrew Carnegie's steel plant, beat back a bargeload of Pinkertons with rifle fire, and had to be suppressed by the Pennsylvania National Guard. In Chicago again in 1894 President Cleveland was forced to issue a federal injunction and order in federal troops to defeat the Pullman strikers. Coal miners struck repeatedly in the mid-West and West. The radical Western Federation of Miners struck in the Coeur d'Alenes in 1892, at Cripple Creek in 1894, at Leadville in 1896, and again in the Coeur d'Alenes in 1899.

The federal government also became suspicious of the massive, growing combinations of capital. The Interstate Commerce Act was passed in 1887 and the Sherman Anti-Trust Act three years later. Although Sherman Anti-Trust became a dead letter to combinations after the Sugar Trust decision in 1895, and was applied against labor unions instead, the federal government's incipient interest in regulating capital was nevertheless apparent. More ominous was the organized appearance of socialism. Understanding that the operation and success of capitalism depended upon not paying labor for the full value of its productivity, workers became increasingly class conscious and militant. Other labor and middle class groups formed: the remarkably popular Edward Bellamy socialist Nationalist Clubs, the Christian Socialists, the Populists and Single Taxers. Industrial and commercial hegemony was under direct attack. William Dean Howells' traveler from Altruria put the question as directly as any: "Am I right in supposing that the effect of your economy is to establish insuperable inequalities among you, and to forbid the hope of the brotherhood which your polity proclaims?"[2]

American capital, which was far from a monolithic block, never satisfactorily answered Howells' question, but it did develop a variety of responses. On the right wing, unreconstructed laissez faire industrialists, local Citizens' Alliances, and the National Association of Manufacturers proved unable to develop any analysis beyond the simple reaction that unions and strikes were to be crushed, as brutally as necessary. John D. Rockefeller, Sr., and Andrew Carnegie argued as centrists that a peculiar form of Christian Stewardship was the solution to the conflict between labor and capital. Since the working class had little or no idea of what was in its interest, millionaires who did know were to serve as trustees for the poor and make the decisions which the laborers, intellectually ill-equipped and uneducated, could not be expected to make for themselves. If a few individual stewards doubted (as Carnegie never did) their absolute right to make those decisions, their fears were laid to rest by Herbert Spencer. A man largely without honor in his own country, a man, for that matter, whom the British correctly regarded as a fool, Spencer demonstrated "scientifically" that the principle of natural selection assured American industrialists the right to do whatever they thought best.

William Graham Sumner argued that the intense, free competition for the millionaire's station guaranteed its occupant's ability and wisdom, but far-sighted corporate leaders had written off unrestricted competition decades earlier. Although they continued to give public support to Social Darwinism, they recognized that market reality had demonstrated to any who cared to look that Adam Smith's hand was not only invisible but inoperative. James Logan, manager of the U.S. Envelope Company, was not alone in concluding, "Ignorant, unrestricted competition, carried to its logical conclusion, means death to some of the combatants and injury for all. Even the victor does not soon recover from the wounds received in the conflict."[3] While the ultimate purpose of trusts, established as early as the 1870s, was greater profits, only by the destruction of competition could the market be rationalized and those profits guaranteed. The merger fever which began in the late 1890s, and continued through 1903, was a consolidation by forces set in motion much earlier.

The liberal wing of capital recognized competition as economic chaos which demanded stabilization if the new industrial order was to be secured. As corporate leaders looked back on the past quarter-century they saw little fundamental difference between the Panic of 1873, the riots and strikes which followed it, and the social and economic consequences of the yet more severe double dip depression during the 1890s. By the turn of the century, as William Appleman Williams observes, they concluded that the "political economy had to be extensively planned, controlled, and co-ordinated through the institution of the large corporation if it was to function in any regular, routine, and profitable fashion."[4] They also perceived that Progressivism, regardless of what little power it had, was not going to disappear overnight. Ironically, the Progressives and their reformist ideology would be ultimately reduced to unwitting accomplices in the consolidation of the new order. The corporation would be the solution for the coming age, but as a solution it would have to rechannel and absorb Progressivism into its very basis. Otherwise labor could not be made to identify itself with corporate interest. Nothing else promised to guarantee capital's continued control of the political economy.

Led by the National Civic Federation, the liberal wing of capital set out to educate employers to the new reality. They had to be made to understand, for example, that status quo unions such as the AFL were useful both in stabilizing production and, as Samuel Gompers phrased it, in convincing the working class that there was "an identity of interests between labor and capital." The problem was not only the ignorance of employers but American radicalism, the only force which presented a serious challenge to capitalism. If such organizations as the Socialist Party (SP) and the Industrial Workers of the World (IWW) could win labor to their side the game would either be up or the rules considerably modified. Real concessions of power would have to be offered at the least. The Socialist Party not only understood business cycles to be a function of capitalism, but more importantly offered an alternative philosophy to capitalism, the labor theory of value and production for use, not profit. Like the business and political community, the Socialists also wanted industrial consolidation and a planned economy, but one in which control of the means of production would pass from private to public ownership, and labor's increased productivity would benefit the general welfare of society. Their electoral progress is instructive. When Eugene Debs first ran for the presidency in 1900 he received 95,000 votes. Four years later his vote more than quadrupled. Mark Hanna had correctly anticipated that by 1912 socialism would be the greatest issue facing the United States and, as Party membership tripled between 1908 and 1912, Theodore Roosevelt was forced to declare that socialism was "far more ominous than any populist or similar movement in time past." In 1911 there were seventy-three Socialist cities in America, including Milwaukee, Berkeley, Bridgeport, Butte, and Flint. A year later Debs polled 897,000 votes.[5]

Equally significant was the formation of the IWW in 1905. Unlike the accommodationist American Federation of Labor (AFL), which would be used repeatedly as a strike-breaking force against the IWW, and which anomalously concentrated on "skilled crafts" at the same time industrialism destroyed them, the Wobblies believed that all skills were necessary in a just society and began to organize the workers whom the craft unions avoided. The IWW agreed with the Socialists that reform would not alter materially the effects of capitalism. Arguing that the "working class and the employing class have nothing in common," that there "can be no peace so long as hunger and want are found among millions of working people and the few, who make up the employing class, have all the good things in life,"[6] the IWW set out to paralyze the economy by a bloodless general strike and obtain the means of production for the working class. By the conclusion of World War I and its aftermath, however, neither the Socialist Party nor the IWW was present as a vital force in America. While each organization and the left in general had its internal divisions and conflicts, external pressures were more crucial to their demise. Accusing radicals of lack of patriotism and "impeding the war effort," both business and government used World War I as an excuse to crush domestic radicalism. Understanding as well as Randolph Bourne that war was the health of the state, they made what seemed to many an obvious choice between war abroad and a feared class war at home. By the Chicago trial of 1918 and the Palmer Raids of a year later, anarchists were either jailed or, like Alexander Berkman and Emma Goldman, deported; the socialist movement was splintered into fragments and its leaders imprisoned; and the IWW was shattered and its leaders jailed. Even the legitimacy of providing a radical analysis of capitalism had become suspect, if not unAmerican. Corporate hegemony, although it would be challenged again in the 1930s, was fundamentally complete.

The 1913-1914 Colorado coal strike reflected this national pattern in fine detail. At the turn of the century the Rockefeller empire was faced with the not inconsiderable problem of reinvesting its oil profits and became a holding company, an organizational structure designed to dominate companies through ownership of a controlling proportion of their stocks and bonds. After examining the Colorado Fuel and Iron Company (CF&I), the state's largest producer of coal, the Rockefellers purchased forty percent of the company's preferred and common stock and forty-three percent of its bonds in 1902.[7] The purchase provided effectual control and a new majority on the Board of Directors was established. Primary among these appointments was that of Lamont M. Bowers as Chairman of the Board. An adherent of laissez faire capitalism in its most unrestricted form, Bowers, with the help of President Jesse F. Welborn, was given full administrative responsibility for CF&I's Pueblo steel works and multiple mining properties. When the coal strike began in September, 1913, Bowers reported to John D. Rockefeller, Jr., who had assumed direction of the Rockefeller holdings upon the semi-retirement of his father.

Although the younger Rockefeller accepted his responsibilities with some uneasiness, he believed with Bowers that private ownership of property gave him control over every machine or, interchangeable with those machines, every man who worked there. As late as July, 1914, Rockefeller roundly endorsed the sentiments of economist John J. Stevenson. "Unskilled labor," Stevenson wrote, "is merely animated machinery for rough work and adds very little value to the final product. One E. H. Harriman is of more lasting service to a nation than would be 1,000,000 of unskilled laborers, without a Harriman they would be a menace." Even more of a menace were labor unions: "their principles are no better than those of the India Thugs, who practiced robbery and murder in the name of the Goddess Cali." In a letter urging his publicity director to distribute Stevenson's article, Rockefeller described it as "one of the soundest, clearest, most forcible pronouncements on this subject [the relationship between capital and labor] I have ever read."[8]

In 1913 neither Rockefeller nor Bowers subscribed to the belief that Progressive measures were necessary for corporate control of labor. The hostility of both men toward unions was so thorough, in fact, that no attempt was made to distinguish radical from reformist labor. As a result, CF&I and the other Colorado operators flatly refused to discuss strike issues with United Mine Workers' (UMW) representatives. They judged the stability offered by a union contract to be undependable because the UMW could not be totally controlled; more importantly, unionization represented an unacceptable interference with the right of private property. The central issue of the strike, therefore, was capital's absolute control of the coal mines and its virtually complete control of the political apparatus supporting its power. The strike would be fought and broken on those terms. Not until late in 1914 was Rockefeller forced to conclude that a different method, a company union, would have to be devised to ensure continued control of the mines and the men who worked them. Although the Employees Representation Plan provided moderate concessions and the illusion of collective bargaining, it did not alter the power of private ownership in any fundamental way.

The 1913-1914 strike in the southern Colorado coal fields was unique in its intensity, duration, and solution, but it was only one of a series of coal strikes in the state which began in 1883 and continued at roughly ten-year intervals for the next thirty years. Since the majority of Colorado coal miners were native born Americans, and most early immigrants were of western and northern European origins, the mineowners deliberately used ethnic differences to subvert labor unity. After 1900 they increasingly recruited the new southern European immigrants to the coal fields, using their ignorance of English, American working conditions, and of one another to the benefit of management.[9] Many of the immigrants who first came to the mines as strikebreakers, however, remained to learn the English language and became strikers themselves.

The more far-sighted of the coal operators viewed the strikes as regrettable, but a cost which could be borne without great inconvenience. While strikes did temporarily halt production, the influx of foreign labor provided a ready supply of new strikebreakers and a labor surplus which could be used to depress wages. The United Mine Workers found organizing men of some twenty nationalities, who spoke even yet more dialects, a difficult obstacle to overcome. Moreover, the mining industry was the largest and most powerful in Colorado. With relatively few exceptions, such as during the term in office of Populist Governor Davis H. Waite (1892-1894), mining companies held the strings of control at the statehouse, the legislature, and at the county level whenever necessary. Except in rare instances, as strike after strike in both metal mines and coal mines proved, the Colorado National Guard could be depended upon to herd strikers into bullpens or expel them from the district. Secretary of State and ex officio Labor Commissioner James B. Pearce's statement in 1912 was hardly an exaggeration: "Certain interests have for so many years been accustomed to break strikes with the militia that it is a difficult thing to break them of the habit. It is much the cheapest and speediest method for them, as the taxpayers of the state pay the bills. A striking illustration of this is given in the bond issue of over $950,000 issued to settle the Cripple Creek war debt."[10]

Founded in 1890, the UMW was a reformist industrial union. Unlike its metal-mining counterpart, the Western Federation of Miners (which argued that since labor produces all wealth, all wealth belongs to the producers thereof), the UMW never fundamentally challenged capitalist ideology or the corporatization of America. John Mitchell, its president from 1898 to 1908, was firm in his belief that no irreconcilable conflict existed between labor and capital and, like Gompers, found his way to the National Civic Federation; Mitchell's replacement, Tom Lewis, became publisher of *Coal Mining Review*, the industry's trade sheet; and John White, president during the 1913-1914 strike, largely confined himself to an attempt to achieve improvements in the coal fields. In Colorado and other coal mining states the UMW worked to obtain the eight-hour day, union recognition and a check off, and better wages and working conditions. Defeated in West Virginia the year before, its finances drained, in 1913 the National Board of the UMW was in a certain sense reluctantly compelled to support its Colorado members. But once the decision was made, the National committed itself fully to the battle.

The union demands formulated in September, 1913, were moderate: (1) recognition of the union; (2) a ten percent increase in tonnage rates and a day scale to correspond with the rates in Wyoming; (3) an eight-hour day for all classes of labor in or around the coal mines and coke ovens; (4) payment for narrow work and dead work, such as brushing, timbering, removing falls, and handling impurities; (5) the election of checkweighmen without interference by company officials; (6) the right to trade at any store they pleased and the free choice of both boarding place and doctor; and (7) the enforcement of state mining laws and the abolition of the company guard system.[11] Although no less than four of these demands were guaranteed by Colorado law, the coal companies made little distinction between laws and demands. They regarded both as infringements upon their right to run their mines as they pleased. Still bound by nineteenth-century laissez faire assumptions, and no more aware than the union that the strike would last fifteen months, the operators had yet to understand that stability was a sine qua non for both a satisfactory and continuing rate of corporate profit.

The operators' belief in the power and privilege of private property showed itself in every aspect of the miners' working and social lives. In short, everything that was susceptible to company control was controlled. The southern Colorado mines were geographically isolated. Situated in foothills and remote canyons, the bulk of the coal camps were unincorporated, or "closed." They were owned outright by the companies, every square foot of land. A miner lived in a company house or boarding house. He or his wife bought food and supplies at the company store. Company money, scrip, was the common means of exchange. If his children did not work in the mines, they attended a company school, where the teacher was either selected or supervised by the company superintendent. If the miners drank, and many did, they did so at the company saloon or at one operated on concession by the company. If a miner went to church, he listened to a company-chosen minister. If a miner needed medical attention, he was treated by the company doctor. Miners received their mail at the company post office. If either the camp superintendent or company policy proscribed certain newspapers, *Harper's Weekly, The Rubaiyyat of Omar Khayyam*, or a mail order catalogue—and they did—the newspapers, magazines, books, and catalogues were destroyed or returned.

Closed camps were walled in by steep mountain cliffs and fences. A visitor or state mining official needed company permission to enter and was often refused. Since the companies controlled the local law officers, there was neither recourse when a miner's legal rights were abridged nor even a recognition of the existence of those rights. In cases of injury or death in the mines, regardless of company negligence, either no chance for damages existed or a minimal, informal settlement sufficed. The camp marshall, deputized by the company-controlled county sheriff, met any protest with brutality, immediate discharge, blacklisting, or all three. Dissatisfaction was spotted by company spies and the miner was fired, sent "down the canyon." Nor was there recourse at the polls. Company officials voted as many miners, or mules if they had names, as their Denver bosses instructed. Although the companies blamed the resulting strike on the even then timeworn claim of outside agitators, the level of industrial despotism made the miners recognize that their only hope in gaining a measure of control over their lives lay in unionization.

Company housing was initially a response to the absence of living quarters in remote mining areas. In the early days miners were permitted, if not encouraged, to construct their own houses as well. Thereafter, however, company housing served two other purposes, profit and intimidation. When a miner was discharged for pro-union sentiments, for example, he lost both his job and a place to live. Understandably men with families had a particularly close-lipped way of going about their business. Also, after J. F. Welborn assumed the presidency of Colorado Fuel and Iron, whenever a privately built miner's house was vacated, the company wrecked it. The companies rented out rooms at the standard rate of $2 per month. While some of the housing was in satisfactory repair, most was not. The Reverend Eugene Gaddis, the frustrated and impotent director of social work for the entire CF&I camp system, quoted the Sopris physician in his testimony before the Industrial Relations Commission: "Houses up the canyon, so called, of which 8 are habitable and 46 simply awful; they are disreputably disgraceful. I have had to remove a mother in labor from one part of the shack to another to keep dry." Gaddis then summarized the situation himself: "The C. F. & I. Co. now own and rent hovels, shacks, and dugouts that are unfit for the habitation of human beings and are little removed from the pigsty make of dwellings." The management of CF&I was well aware of these conditions; however, when the federal government sent a questionnaire regarding miners' living quarters, Gaddis testified that President Welborn directed the form be completed so as not to arouse suspicion or investigation.[12]

Sanitary conditions were, if anything, worse than the housing. During the year prior to the strike 151 cases of typhoid were reported at the CF&I camps. For over a year a cesspool next to a company store "was allowed to relieve itself by overflowing at the top and running down across the principal thoroughfare of the camp." Not surprisingly, both the store manager and his wife contracted typhoid. Drinking water at a number of camps was not fit for human consumption. Gaddis provided an instance: "Seepage water from a large mine at Walsen, with a distinctively dead-rat-essence flavor, was supplied to three camps, because it was cheaper than to tap the main pipe line, of one of the best water sources in the State, even though this water main ran d[i]rectly through one of the camps." "We do not believe," Gaddis testified of water closets and public washhouses, "more repulsive looking human rat holes can be found in America than those of Berwind Canyon before the strike."[13]

Unlike housing, the companies regarded schools as an expense which brought them no direct return on their investment. Some were nevertheless solid, substantial buildings while others were ill-equipped, unheated, and poorly lighted. At an ungraded school at Hastings, a Victor-American camp, one teacher was responsible for 120 students, ninety percent of whom were unable to speak English. One CF&I camp operated for ten years without a school. At another, a teacher who complained about the lack of heat was informed by the mine superintendent that had he known of her attitude earlier her contract for the following year would not have been approved. At the Morley camp, the daughter of the company store manager was made a teacher despite the objections of the county school superintendent and the fact that she was below legal age. When a camp superintendent did not himself select teachers, a company-controlled school board did.[14]

Religion, since it affected the miners directly, apparently posed a greater threat. The companies were fearful, in the opinion of the Reverend James McDonald, a mining camp clergyman, that church attendance would permit the miners to "become better educated and become better acquainted with their rights and grievances." CF&I was particularly vigorous in its harassment of ministers who felt a responsibility to combine social work with religious doctrine. Fifteen months at CF&I's camp at Sunrise, Wyoming, led the Reverend Daniel McCorkle to conclude "that the Colorado Fuel & Iron Co. endeavors to control the minister and his church absolutely. . . ." The independent McCorkle was in a position to know. In a sermon following the Ludlow Massacre he spoke of a citizen's duty to maintain the peace, but he also criticized the "hiring of private armies of so-called detectives by corporations for use in labor disputes." He was soon informed that "the company wants you in Sunrise to help keep the Greek and Italian people down, not to stir them up. If you do not help the company keep them down . . . it does not want you here at all." Chairman

Bowers angrily told McCorkle, "All this uplift and reform work is satanic."[15]

As a general rule the companies provided a physician for each of the camps. The companies argued that since they had hired the doctors at their own expense, the miners had the express duty to be treated by them. Victor-American also had an arrangement with a Trinidad hospital to care for seriously ill or injured miners, while CF&I maintained its own well-equipped hospital in Pueblo. Company medical expense existed only as a manner of speaking, however. The miners paid the cost both directly and indirectly. One dollar per month was deducted from every paycheck, a fee which entitled the miners (and their families, at half rate) to medical care except in cases of childbirth, venereal disease, and fight bruises. There is also ample evidence that miners were compelled to make additional payments for medical treatments. Gaddis brought out one reason for the extra charges: the company allotment for drugs and medicine (3¢ per man per month) was not nearly adequate to meet the actual expense. Also, more than one miner was able to make the comparison between men and machines. While the company stood the cost of maintenance and repair for the latter, if a man were injured in the course of his work, he paid the cost, not the company.[16]

The miners' demand that they be allowed to trade at whatever store they chose expressed a deeply felt grievance. Few aspects of daily camp life were as objectionable as the compulsion to shop at the "pluck me" stores. Wholly owned by CF&I, the Colorado Supply Company was established in twenty-two mining camps as a virtual monopoly. Like Victor-American's Western Stores, Colorado Supply netted an average profit of "around twenty percent per annum." The policies and prices of the company stores were such that, excluding all other causes, a CF&I physician of some eighteen years standing declared that "if a strike is called the Colorado Supply Co. will be responsible." Asked if prices were excessive, Mary Thomas, a miner's wife, answered, "Yes. In these company stores they would give you scrip and . . . it would say you could purchase anything in the company's stores in the mining camp; but not at the same store belonging to the same company at Trinidad, because in Trinidad they had to compete with other stores." Further, both miners and their wives were notified that jobs were dependent upon buying supplies at the company store. During the 1913-1914 strike one store took in forty-seven percent of the mine's payroll in receipts. After investigation the Colorado Bureau of Labor Statistics reported that "any attempt by miners to purchase supplies from any other than the company's store would mean instant dismissal to the purchaser."[17]

Although proscribed by law, scrip was the major form of currency in the coal camps. In 1899 the Colorado legislature passed a statute prohibiting "any agreement or understanding used by the employer, directly or indirectly, to require his employee to waive payment of his wages in money, and to take the same in goods of the employer. . . ." The statute forbade any understanding "that the wages of the employee shall be spent in any particular place or manner"; provided that any scrip issued in furtherance of such a system was void; and that any violation of the regulations "shall constitute a misdemeanor and cause forfeiture of the charter of a domestic corporation. . . ."[18] The coal operators flouted this anti-scrip law with impunity.

Testimony indicates that the smaller companies paid wages directly in scrip. Of the $30 Joseph Ray earned at the Broadhead mine, for example, $29 was given to him in scrip. The scrip system in force at CF&I and Victor-American, however, was indirect and thereby avoided the letter, although not the spirit, of the law. If a CF&I miner asked an advance on his wages due at the end of the month, he was given an order at the company store. Any unexpended amount was returned to him in scrip. Although the stores were wholly owned subsidiaries, the two companies themselves technically avoided legal sanctions. Further, while CF&I and Victor-American, for example, used scrip as a substitute for money, they refused to redeem scrip for dollars. Heavily discounted at non-company stores, nowhere in the camps could scrip be cashed at face value. Labor historian Donald J. McClurg comments, "The clear intent of the 1899 law was to prevent the employers from achieving the high degree of control over their employees' lives which a scrip system would make possible." It was no small irony that the power of the scrip system, direct or indirect, largely depended upon the open violation of another state law, a 1901 statute which provided for semi-monthly payment of wages. Had the operators obeyed it, of course, there would have been small necessity for advances in pay. As McClurg notes, CF&I's Bowers argued that the law was unconstitutional; unenforced by the state; unwanted by the miners; and in any event, that he and President Welborn were unaware of its existence.[19]

Coal mining was (and continues to be) a brutally difficult and hazardous job in the best of conditions. Since three and one-third more men than the national average were killed per tons of coal mined in Colorado during the period 1909 to 1913, it is clear that working and safety conditions in the mines were considerably less than adequate. Colorado's dry air was a contributing factor, but not one of major importance. While the companies (and company-controlled coroner's juries) asserted the miners' own negligence to be the cause of accidents, the evidence is indisputable that lives were sacrificed for increased profits. Deputy Labor Commissioner Edwin V. Brake's figures for 1913 are particularly telling. That year 8.31 per thousand employees were killed in Colorado coal mines. In the unionized mines of Illinois, Iowa, and Missouri, where the miners had some control over working conditions, the comparative figures were 2.06, 1.57, and .99 men per thousand respectively.

Company irresponsibility may be seen in the following examples. In October, 1913, a state mining inspector found the level of coal dust in a mine too dangerous and recommended that it be sprinkled daily. It was not. Two months later a dust explosion killed thirty-seven men. In another case a worker in CF&I's Primero mine informed his fire boss that the high concentration of gas made further work unsafe. He was ordered to keep his mouth shut and return to work. When he refused he was fired. A gas explosion occurred before he could pick up his tools and he was badly injured. Not so fortunate were the twenty-four men killed three days later in the same mine.

Except for one area where the dust was so thick that the mules had a hard time functioning, CF&I made no changes. Three years later another explosion in the same mine took the lives of seventy-five to one hundred more men. The state investigator not only judged CF&I antagonistic to safety regulations, but wrote: "To compel men who are working in a gaseous mine filled with dust, to work under the conditions imposed by the company at the time of the explosion was cold-blooded barbarism."[20]

There was normally no legal redress after mine disasters. So far as President Welborn could recall there had been no personal injury suits brought against CF&I in Huerfano County in the past twenty years. Almost invariably the hand-picked coroner's juries found mine accidents to be caused by the dead miners' "own negligence" or "by an unavoidable accident." The reason for this unanimity was explained by John McQuarrie, former undersheriff of Huerfano County. "I was always instructed," he testified, "when being called to a mine to investigate an accident, to take the coroner, proceed to the mine, go to the superintendent and find out who he wanted on the jury. That is the method that is employed in selecting a jury at any of the mines in Huerfano County." When Congressman John M. Evans found that of the thirty surviving lists of jurors (from 232 accidents) the name J. C. Baldwin appeared as foreman twenty-four times, he asked McQuarrie who this man was. Baldwin was a gambler and bartender, McQuarrie answered, the "secretary for the Republican county central committee."

As a "kind of pensioner," his duty was to serve as foreman on coroner's juries. In the ten years previous to the strike, coroner's records for Huerfano County show that only once in ninety inquests was the coal operator found negligent.[21]

The majority of miners were coal diggers, men who were paid by the ton. Preceded by a fire boss, a day man who checked the mine for gas content and general safety, the diggers entered the mines at daybreak or earlier. Unless the solid face of the coal was to be shot, the first operation was to undercut the seam. Lying on his side, a miner cut a space under the coal seam some three or four feet deep, or as far back as could be reached. An angled hole was then drilled by auger slanting upward to the roof. If the company did not employ shotfirers (men who prepared and fired charges after the miners had left at night) powder was placed in a cartridge, pushed deep into the hole and, the miners having safely retreated, touched off by battery. The amount of shot had to be precise: if too much, the roof or wall might collapse; if too little, the coal would still adhere to the roof. Once loaded the coal cars were pushed down the tunnel to the entry where mules pulled them to the main haulageways. The cars were then moved by trolley to the mine tipple, where the coal was weighed and the miners credited.

Coal diggers were assigned in teams of two to workplaces along the coal seam by the mine superintendent. The teams were composed either of two experienced miners, or a practical miner and an inexperienced buddy. The standard process of extracting coal in southern Colorado was the room and pillar method. As the miners followed the seam, thick pillars of coal were left standing to prevent cave-ins. After the seam was exhausted, or cut off by a fault or rock outcropping, the pillars were progressively pulled as the miners worked their way back out to the entry. The superintendent's assignment of a place was crucial. Some produced coal easily, some produced days of deadwork. Driving an entry, timbering, laying crosspieces and track for the cars, all were deadwork, for the greatest part unpaid by the companies. A miner worked in a crouch or on his side hour after hour. If no cars were available, his space was so cramped that he had to stop work. If no timbers were delivered to him, he had either to wait or work without them. If rock or slate were found in his coal, he could, and often did, lose credit for a thousand pounds of coal or even his whole car. His life depended upon his experience and good luck. Undercutting had to be done with skill; otherwise the coal could fall and crush the miner lying on the edge or under the face. He had to judge the safety of a roof correctly, know how to support it with timbers, and be able to recognize the sounds which often precede a fall. As pillars are pulled, for example, the roof gradually works loose from the overlying strata. The timbers begin to splinter slowly from the weight pressing downward as the roof works and moves. Only experience enabled a miner to tell from the condition of the timbers and the cracking sounds above him when the roof was ready to fall. However, many a miner died from falls, gas or dust explosions, which gave no warning at all.

In the best of places a digger might average $3.50 on the days he worked, less such charges as blacksmithing, powder, and medical care. In 1911, however, the average Colorado mine was operated only 174 days. In 1912 the Colorado Bureau of Labor Statistics computed the average daily wages in coal mines at $1.68 for an eight-hour day. Low pay combined with a certainty that the companies were not giving accurate weights made the union demand for a checkweighman, paid a small sum by each miner, a major grievance. Nor, despite an 1897 statute providing for a checkweighman, could the men as individuals insist upon their representative at the tipple without fear of being sent "down the canyon." When the companies permitted inspectors to examine the scales, state reports confirmed the miners' suspicions time after time. George S. McGovern, author of the most thorough examination of the 1913-1914 strike, cites a "typical" report, dated March 22, 1912.

> Colorado Fuel and Iron Company mine at Morley: Has no checkweighman. Find the miners complaining of weights. On inspecting two pair of scales, I find that neither will balance, and that the scales on the south tipple with 350 pounds increased the weight only fifty pounds. This is very unsatisfactory to the miners, who claimed that if they asked for a checkweighman they would be discharged. This is disputed by Mr. Harrington the company attorney at Denver.

A similar report gives substance to the fact that local officials in the coal counties were rarely if ever to be depended upon.

> March 21, 1912. Wooten Land & Fuel Co., two mines situated close to New Mexico line on the Santa Fe Railroad: I find the scales at the Wooten mine unbalanced, and scales at the Turner mine very unfair to miners. They could not be balanced, and, on weighing a car of coal, the weight of three men weighing 450 pounds only increased the weight of the car 50 pounds. Scales seem to be binding. They were inspected by the county inspector.

"Anticipating in time having to meet the demands of union labor," CF&I posted a circular in 1912 favoring checkweighmen at its mines and, the following year, established an eight-hour day and semi-monthly pay. All these concessions to Colorado law were primarily intended to undercut union organizing. However, Commissioner Brake testified in 1914 that he knew of only two mines in the entire state which had checkweighmen.[22]

Political control by the companies was crude but remarkably effective at all levels. In May, 1913, Bowers informed Rockefeller: "The Colorado Fuel & Iron Co. for many years were accused of being the political dictator of southern Colorado, and, in fact, were a mighty power in the entire State. When I came here it was said that the C. F. & I. Company voted every man and woman in their employ, without regard to their being naturalized or not, and even their mules, it used to be remarked, were registered, if they were fortunate enough to possess names. Anyhow, a political department was maintained at a heavy expense. I had before me the contributions of the C. F. & I. Co. for the campaign of 1904, amounting to $80,605. . . . "[23] Company influence at the party level may best be described by a speech given by Judge Jesse Northcutt in 1912, the year before his independence was hired away by CF&I.

> ". . . We will take for county clerk so-and-so; he is a good man for the purpose." Some other man says, "But still, I think probably sometime within the last 8 or 10 months he had some trouble with some pit boss," and there is just a suspicion if the company likes him. He isn't right with the company and they don't want him; he goes off the slate. And so it is from bottom to top. The candidates are selected, not with a view to their fitness, not with a view to their ability to discharge their duty, not with a view to their integrity, but "are they satisfactory to the company?" If they are, that settles it. And they have a majority of your conventions, and when they come to select delegates they select them in the same way. They send them in there to nominate, regardless of your wishes, for the office of district judge or State senator, the man whom the companies want, and if you don't like it you will have to take it.[24]

After Bowers was given charge of CF&I he assured Rockefeller, "Since I came here not a nickel has been paid to any politician or political party." While he may have drawn the line on direct payment, during the 1914 election CF&I turned out 150 employees to work, ostensibly, for the prohibition vote. Bowers was asked during the Industrial Relations Commission Hearings, "Didn't you use the prohibition sentiment that was strong in the State to get the support for what you called the law and order platform that was for the Colorado Fuel & Iron Co. and the others to aid in the ruthless prosecution of the strikers and the union officers and a relentless policy of suppressing those men?" "It was all interlocked together," Bowers agreed.[25]

In the mining camps themselves, the camp marshall, deputized by the sheriff who himself held office only by enforcing company policy, had legal control over the miners. He was further aided by company detectives whose methods are graphically illustrated in a letter written to W. H. Reno, head of CF&I's intelligence branch.

> Joe Mosco I drove out of town, also Rosario Dolce and his family. Nic Oddo refused to vacate and there was no way for me to get him out so I told Thompson to arrest him on the charge of vagrancy. That night he was taken before the justice of the peace and the case nolle prosequi. That was about 8 o'clock. I had Gordon, Barret, Smith, and King wait for him down by the bridge and they "kangarooded" him and the last I heard of him he was in the hospital, and he will not attempt to come back to Hastings.[26]

In 1913 a State Senate Committee investigating election irregularities in Las Animas County found, for example, that some one hundred Trinidad prostitutes, who had been forced by the local Republican machine to pay $6 to $7 per month to stay in business, were hauled by police to the polls and "assisted" in marking their ballots. The Committee concluded: "We express the hope that in no other place in the United States could be found a condition so degraded, so corrupt or so infamous as this."[27] As McGovern remarks, the Committee's hope, however strong, was fruitless. Barron B. Beshoar, who grew up in southern Colorado, writes: "The Huerfano County courthouse . . . was considered by many as little more than a branch office of the C.F.&I. Co. though citizens who held such opinions seldom voiced them in more than a whisper. The political boss and majordomo for the C.F.&I. Co. was Sheriff Jefferson Farr. . . ."[28] Indeed, Bowers himself testified that "the Colorado Fuel and Iron Co., in common with other coal companies in Colorado, worked jointly with Sheriff Farr with a view to controlling the political situation in that part of Colorado." Later, Bowers was asked:

> In the counties where these coal companies operate they have the judges and sheriffs, and through the sheriffs can select the jurors, as was testified to in the evidence before us of Sheriff Farr himself, and by others before us as to how the juries were appointed.... Under these circumstances I ask you, Mr. Bowers, as a man of great affairs and dealing with big business and big financial affairs, if you think a poor, humble miner, without any great amount of money, without any property behind him, or any influence, has any chance of getting justice in a situation of that kind?
>
> Mr. BOWERS. Why, no; no one need ask me that....[29]

Farr's re-election as Huerfano County Sheriff in 1914, when his opponent had to be escorted to the polls by federal troops, was contested as a direct result of the strike. Finding the election districts and public machinery to be under "the absolute domination and imperial control of private coal corporations, and used by them as absolutely and privately as were their mines," the Colorado Supreme Court overturned the election results. Precincts had been realigned to prevent access to polling places, company agents marked ballots for voters, and mine guards chose which citizens might vote and which might not. "There can be no free, open and fair election as contemplated by the constitution," the Court concluded, "where private industrial corporations so throttle public opinion, deny the free exercise of choice by sovereign electors, dictate and control all election officers, prohibit public discussion of public questions, and imperially command what citizens may and what citizens may not, peacefully and for lawful purposes, enter upon election or public territory...."[30]

Facing such corporate hegemony, the miners recognized that nothing could be done individually, that only through collective action could they hope to break operator control of their lives. The UMW had little enthusiasm for a strike. Not only was its national treasury low, but strikes in the Colorado coal fields in 1901 and 1903-1904 had been broken by the coal companies. In 1913 the operators were confident that if a strike came they could break it again.

With Rockefeller supporting the largest of them, they refused to negotiate or even meet informally with union representatives. The central issue was recognition, the union's first demand and the means of achieving all others. When the miners learned the companies were sending union sympathizers out of the coal camps, they devised an effective counter-tactic. Twenty-one pairs of men were sent to the camps, one an active organizer and the other passive. The active man moved in the open and made himself known to the superintendent. His companion worked secretly. He presented himself as strongly anti-union and obtained a job as a company spotter. When the active member of the team found a miner in favor of the union, he quietly signed him up. Any strikebreaker's name, however, was turned over to his passive counterpart, who reported to the company that "John Cotino" had just joined the union. "The result was always the same. The company sent John Cotino packing down the cañon.... In this manner a constant stream of anti-union and non-union men, the confirmed strikebreakers and scabs, were kept streaming down the cañons." Beshoar writes that this device was responsible for the replacement of more than 3,000 non-union men by union sympathizers.[31]

As a strike became more and more likely, Ethlebert Stewart was sent from the United States Department of Labor as an appointed mediator, first to confer with Rockefeller, who refused to see him, and then to Colorado in an unsuccessful attempt to bring about a meeting between the operators and the union. When Colorado Governor Elias M. Ammons also tried and failed to arrange a conference, he sent Labor Commissioner Brake to Trinidad for an investigation. Brake arrived on August 16, 1913, the same day Gerald Lippiatt, a UMW organizer, was taunted into a fatal gunfight by two coal company detectives, Walter Belk and George Belcher, and shot down in the street. Ten days later a miners' Policy Committee sent a letter to each of the southern operators:

> For many years the coal miners of Colorado have been desirous of working under union conditions and as you no doubt know have made this desire known on numerous occasions, a large number of them being discharged because of their wishes in this respect.
>
> While we know your past policy has been one of keen opposition to our union, we are hopeful at this time that you will look at this matter in a different way and will meet with us in joint conference for the purpose of amicably adjusting all points of issue in the present controversy. We are no more desirous of a strike than you are, and it seems to us that we owe it to our respective interests, as well as the general public, to make every honest endeavor to adjust our differences in an enlightened manner.[32]

Only two small companies responded.

The operators prepared for a strike. A committee comprised of J. F. Welborn, John C. Osgood, and D. M. Brown of CF&I, Victor-American, and Rocky Mountain Fuel, the three most powerful coal companies in the state, was formed to represent all the operators. Hundreds of men were sent into the coal districts and ordered deputized. Sheriff Farr swore in 259 deputies during the month of September alone. Although a deputy in Colorado had to reside at least a year in the county of his office, Farr made no examination of them whatever. Moreover, all were paid and armed by the operators.[33] Baldwin-Felts, a detective agency which specialized in breaking mining strikes, was engaged and Farr was directed to send deputized guards to the camps. The miners' Committee sent yet another letter to the operators advising them of a convention to be held in Trinidad on September 15. Company representatives were urged to attend. Not a single operator replied.

Commissioner Brake had returned to Denver convinced that unless the companies consented to meet with the union a strike was inevitable. He recommended to Governor Ammons that martial law be declared if necessary to disarm the Las Animas and Huerfano coal counties. When Ammons refused to act, Brake wrote to the three major coal operators:

"I shall ask you as the Chief Executive Officer of your company, to consent in writing, to submit your differences to a Board of Arbitration, as provided for in Section 9, page 302, Session Laws of Colorado, 1909." D. M. Brown's answer for Rocky Mountain Fuel typifies the other replies: "We know of no differences between our Company and its employes, and hence there is nothing to arbitrate."[34] Bowers' attitude was equally firm. CF&I, which would fundamentally dictate operator policy during the strike (Bowers wrote that Welborn was the recognized leader on the operators' committee), would flatly refuse to meet with union representatives or even indirectly discuss the issues with them. For that matter, in a long letter discounting the strikers' demands, Bowers reported to Rockefeller the belief that the strike could be avoided if the companies were willing to confer with the union: "it is thought on the part of a good many operators that the [union] officials, anticipating being whipped, will undertake to sneak out if they can secure even an interview with the operators. . . ."[35] Rockefeller soon replied, "We feel that what you have done is right and fair and that the position which you have taken in regard to the unionizing of the mines is in the interest of the employees of the company. Whatever the outcome may be, we will stand by you to the end."[36]

On September 15 the miners met in Trinidad. One after another they told of their grievances: short weights, exorbitant prices in the company stores, scrip, all the frustrations of years in the camps. Mother Jones, the eighty-year-old veteran of labor wars, addressed the miners at length:

> If you are too cowardly to fight for your rights there are enough women in this country to come in and beat hell out you. If it is slavery or strike, I say strike until the last one of you drop into your graves. Strike and stay with it as we did in West Virginia. We are going to stay here in Southern Colorado until the banner of industrial freedom floats over every coal mine. We are going to stand together and never surrender.[37]

The seven demands were unanimously approved and a strike was unanimously voted the following day. Commissioner Brake had written to the union as well asking its consent to arbitration. Although it continued to hope for a conference with the operators, the union declined.

The miners went out on September 23. The day was cold, rain mixed with snow, as the processions of strikers on foot and in wagons left the mines and slowly wound down the canyons. Don Macgregor, on whom Billy Keating is modeled in *The Coal War*, wrote:

> . . . They struggled along the roads interminably. In an hour's drive between Trinidad and Ludlow, 57 wagons were passed, and others seemed to be streaming down to the main road from every by-path.
>
> Every wagon was the same, with its high piled furniture, and its bewildered woebegone family perched atop. And the furniture! What a mockery to the state's boasted riches. Little piles of rickety chairs. Little piles of miserable looking straw bedding. Little piles of kitchen utensils. And all so worn and badly used they would have been the scorn of any second-hand dealer. . . . [38]

For the next few days the strikers were busy establishing tent colonies. Tents sprang up on rented land at various locations, some of which were frankly strategic. The large camp at Ludlow, containing some 1200 people, stood squarely in front of Delagua Canyon. Since state law prohibited picketing, the miners wanted to be in a position to inform and confront strikebreakers moving up the canyon into the mines. In response to the guards and deputies armed by the coal companies, the union also purchased rifles and distributed them to the strikers. George P. West, author of the Industrial Relations Commission's *Report on the Colorado Strike*, wrote: "In all discussion and thought regarding violence in connection with the strike, the seeker after truth must remember that government existed in southern Colorado only as an instrument of tyranny and oppression in the hands of the operators; that, once having dared to oppose that tyranny in a strike, the miners' only protection for themselves and their families lay in the physical force which they could muster."[39]

Violence occurred quickly. On September 24 a camp marshall at Segundo was killed in circumstances never made clear. Two days later began the first phase of the operators' campaign to have the state militia sent to the coal fields. On October 7 two strikers and a guard were killed at Ludlow. In another battle two days later a union man was killed. On October 17 CF&I's Death Special, an automobile covered by armor plating and equipped with a mounted machine gun, engaged the tent colony at Forbes. A striker was killed, a young girl shot in the face, and a boy hit nine times in the legs as he tried to crawl to safety. Numerous incidents occurred which did not result in death or serious injury.

The strike itself was successful. Of the 11,000 miners in the southern fields a conservatively estimated 9,000 went out, and others were held in the coal camps by armed guards. Those mines not shut down completely were working at severely limited capacity. The operators moved quickly. They imported out-of-state strikebreakers in massive numbers and publicly called for the state militia to suppress the disorder. At the same time, the coal companies exerted tremendous private pressure upon Governor Ammons to call out the Guard. Ammons and former Senator Thomas H. Patterson made continued attempts to arrange negotiations between the opposing sides, but operator rigidity had taken on a tactical purpose. Any concession would interfere with their plan to force the state to take over strikebreaking duties. As violence continued to mount, Governor Ammons ordered the militia to the coal fields on October 28. The operators left him with little choice. Any doubt about the coal companies' (and particularly CF&I's) power is laid to rest by Bowers' self-satisfied letter to Rockefeller explaining the situation:

> You will be interested to know that we have been able to secure the cooperation of all the bankers of the city [Bowers was writing from his Denver office], who have had three or four interviews with our little cowboy governor, agreeing to back the State and lend it all the funds necessary to maintain the militia .... Besides the bankers, the chamber of commerce, the real estate exchange, together with a great many of the best business men, have been urging the Governor to take steps to drive these vicious agitators out of the State. Another mighty power has been rounded up in behalf of the operators by the gathering together of fourteen of the editors of the most important newspapers in Denver, Pueblo, Trinidad, Walsenburg, Colorado Springs, and other of the larger places in the State.[40]

Governor Ammons' order to Adjutant General John Chase, as McClurg comments, "followed the usual form, by now familiar in the Colorado coal fields."

> I . . . direct you, in pursuance of the authority and power vested in me as governor by the constitution and laws of the State of Colorado to forthwith order out and assume command of such troops of the National Guard of Colorado as in your judgement may be necessary to maintain peace and order . . . and that you use such means as you may deem right and proper, acting in conjunction with or independent of the civil authorities . . . as in your judgement and discretion are demanded, to restore peace and good order in the communities affected and to enforce obedience to the constitution and laws of this state.[41]

The militia was directed, however, to disarm both mine guards and strikers; further, it was under orders not to escort strikebreakers into the mines.

When General Chase and his troops marched into Ludlow the road was lined with strikers, their wives, and children. Many wore their Greek or Bulgarian dress. The children, dressed in white, waved small American flags. All sang the union song.

> The union forever, hurrah! boys, hurrah!
> Down with the Baldwins, up with the law;
> For we're coming, Colorado, we're coming all the way,
> Shouting the battle cry of union.[42]

Chase's initial pleasure at the festive atmosphere was dampened by the turnover of guns, some thirty-seven in all. Quite a few were unserviceable and one was a child's toy popgun. The strikers held back their weapons for a number of reasons. They had no confidence whatever in General Chase, whose anti-union sentiments were notorious throughout the state, and who had helped General Sherman Bell and the militia break the Cripple Creek strike in 1904. Chase had stated then, "It is the intention of the mine owners to stamp out unionism. . . ." "The troops will remain there until this is done." More recent experiences also frightened the unionists. Miners from the tent camps at Sopris and Segundo, which had been disarmed before Ludlow, reported that their arms had been given to mine guards. After repeated attacks in the past, the strikers believed that they would be wiped out if they were unable to defend themselves, a fear which was to prove well-founded. The miners refused to surrender all their guns for yet another reason: a number of mine guards now wore militia uniforms.[43]

Relative calm prevailed during the first few days following the military occupation, but hostility lay just below the surface. On November 8 a strikebreaker threatened by union men asked company guards to drive him back to the mine. The car was ambushed by strikers and four guards were killed. In Aguilar a striker killed a non-union miner the same day. On Saturday night, November 20, George Belcher stood on the main street of Trinidad. As strikers and townspeople walked past, the crowd seemed to thicken around the Baldwin-Felts gunman. A shot was fired and Belcher fell with a bullet in his brain.[44]

As incidents in the strike district continued, renewed efforts to reconcile the two forces were made. President Woodrow Wilson had written Welborn on October 30 expressing his disappointment in CF&I's refusal to confer with the union. He also asked that the company report to him its reasons for rejecting "counsels of peace and accommodation in a matter now grown so critical." Bowers soon replied that recognition of the union was the issue at stake, and, since only 6 percent of CF&I's miners were union members, the vast majority were perfectly satisfied. On November 18 Wilson wrote Bowers in harsh terms: "I can only say this, that a word from you would bring the strike to an end, as all that is asked is that you agree to arbitration by an unbiased board. This is not only a reasonable request, conceived in the spirit of the times, but is one the rejection of which, I am sure, would be universally censured by public opinion."[45]

United States Secretary of Labor William B. Wilson then asked Rockefeller to use his influence to arrange a meeting between the operators and union representatives: "I will be at Albany Hotel, Denver, November 21, remaining indefinitely." Rockefeller's response reflects, in part, the information he was receiving from his Colorado management. "The failure of our men to remain at work is due simply to their fear of assault and assassination," he wrote. "The action of our officers in refusing to meet the strike leaders is quite as much in the interest of our employees as of any other element in the company. Their position meets with our cordial approval, and we shall support them to the end."[46] When Governor Ammons and Secretary Wilson were finally successful in arranging a conference between the operators and three selected strikers, it was on the condition that the men not be union representatives. Nor could the issue of union recognition even be raised. Bowers wrote J. H. McClement, CF&I Board member, "We reached no direct understanding; in fact, we wanted none, as we were almost sure that had an understanding between the miners and ourselves been reached it would have received the stamp of approval . . . and in that way been twisted into an arrangement between us and the [UMW]."[47]

When his final effort to bring the union and operators together failed, Governor Ammons again succumbed to coal company pressure. On November 28 he ordered General Chase to provide military protection to strikebreakers entering the mines. During the month previous Chase had given only marginal attention to Ammons' directive that the militia not escort strikebreakers into the mines, but the new order meant nothing less than the defeat of the strike and the operators knew it. Bowers wrote to Rockefeller: "If the governor had acted on September 23 as he has been forced to act during the past few weeks, the strike would have never existed 10 days. We used every possible weapon to drive him into action, but he was glove in hand with the labor leaders and is to-day, but the big men of affairs have helped the operators in whipping the agitators, including the governor." Bowers went on to remark that by "the number of miners we are getting in from the South and East, we will have all we can work in a week or so."[48]

For all Bowers' predilection for seeing enemies in every corner, it is still difficult to understand how he could have considered Ammons sympathetic to the union. Although it had taken two months, Governor Ammons and the State of Colorado had been "forced" to do the bidding of the coal companies. The 1913-1914 strike was defeated for the same reason strikes had been in the past and would again be in the future: state power served as the executive arm for private capital.

The strikers' suspicions of General Chase were confirmed when he established a military commission to try cases. While it did not supplant civil law in the region altogether, it did provide him with a rationale to imprison any striker he wanted, for as long as he wanted. During the military occupation Chase claimed "the right to arrest anyone at any place in the State at any time." As a result, strikers including Mother Jones were arrested for periods up to fifty-five days without a hearing or even having charges brought against them. As the weeks passed and businessmen and college students returned to their responsibilities, more and more mine guards were sworn into the militia. The symbiotic relationship between the operators and the troops was clear. Militiamen were quartered by the companies and furnished supplies by company stores. CF&I alone advanced $75,000 to $80,000 in certificates of indebtedness payable by the State.[49] Then, on January 27, a resolution introduced by Colorado Congressman Edward Keating to investigate the strike was passed by the House of Representatives.

On February 9, 1914, a Subcommittee of the Committee on Mines and Mining convened in Denver. It had been directed to investigate seven areas, including peonage, possible violation of immigration laws, whether constitutional guarantees had been abridged, and the causes behind the conditions in the Colorado coal fields. Over 2,000 pages of testimony were taken. But not until April 6, when the Committee met in Washington, did the witness appear whom everyone had awaited, John D. Rockefeller, Jr.

It soon became apparent that Rockefeller was either woefully uninformed about conditions in Colorado or that he chose to profess ignorance–if not both. He endorsed the principle of collective bargaining and was persistent in his claim that the strike was the exclusive responsibility of outside agitators. While he took a "profound" interest in the strike, he knew of no grievances that the miners might have. Rockefeller was asked by Congressman John Evans:

> Is it not a fact that your company refused to hear or meet with representatives of your employees at the time of the calling of this strike?
>
> Mr. ROCKEFELLER. Not that I know of.
>
> Mr. EVANS. You give that as your best judgment?
>
> Mr. ROCKEFELLER. I do.[50]

Bowers had been explicit in his letters to New York that CF&I refused to confer with union representatives and Rockefeller qualified his denial under repeated questioning, but he was adept in evading specific answers.

> The CHAIRMAN. The militia is supposed, is it not, to be sent into the field to preserve the peace?
>
> Mr. ROCKEFELLER. Yes.

The CHAIRMAN. And not to take part on either side?
Mr. ROCKEFELLER. Certainly.
The CHAIRMAN. Do you think that the swearing in of your guards and keeping them on the payroll as militiamen helps to do that?
Mr. ROCKEFELLER. Without having been on the spot and knowing the situation it would be difficult to say whether that was the best thing to do or not.[51]

Rockefeller's economic views may be seen in the following exchange.

The CHAIRMAN. And all this disturbance and loss of life, killing upon both sides out there, has not been of enough importance to you to cause you to say, Let us have a meeting of the directors, and find out more about it?
Mr. ROCKEFELLER. I have been so greatly interested in the matter, and have such a warm sympathy for this very large number of men that work for us, that I should be the last one to surrender the liberty under which they have been working and the conditions which to them have been entirely satisfactory, to give up that liberty and accept dictation from those outside who have no interest in them or in the company.
The CHAIRMAN. But the killing of these people, the shooting of children and all that that has been going on there for months has not been of enough importance to you for you to communicate with the other directors, and see if something might not be done to end that sort of thing.
Mr. ROCKEFELLER. We believe that the issue is not a local one in Colorado; it is a national issue, whether workers shall be allowed to work under such conditions as they may choose. And as part owners of the property, our interest in the laboring men in this country is so immense, so deep, so profound that we stand ready to lose every cent we put in that company rather than see the men we have employed thrown out of work and have imposed upon them conditions which are not of their seeking and which neither they nor we can see are in our interest.
The CHAIRMAN. And you are willing to go on and let these killings take place—men losing their lives on either side, the

> expenditure of large sums of money, and all this disturbance of labor—rather than to go out there and see if you might do something to settle those conditions?
>
> Mr. ROCKEFELLER. There is just one thing, Mr. Chairman, so far as I understand it, which can be done, as things are at present, to settle this strike, and that is to unionize the camps; and our interest in labor is so profound and we believe so sincerely that that interest demands that the camps shall be open camps, that we expect to stand by the officers at any cost. It is not an accident that this is our position—
>
> The CHAIRMAN. And you will do that if it costs all your property and kills all your employees?
>
> Mr. ROCKEFELLER. It is a great principle.[52]

Rockefeller went on to recognize the abstract right of men to bargain collectively, but his recognition was a transparent fiction. Both the strikers and the operators understood him to mean that the strike would be broken whatever the cost or consequences. Rockefeller's intransigent performance made it difficult for Bowers to put his satisfaction into words, but he nevertheless wrote to his employer expressing his "boundless delight with your magnificent and unshaken stand for principle, whatever the cost may be. Now for an aggressive warfare to 1916 and beyond for the open shop." John D. Rockefeller, Sr., who remained in the background during the strike, was so pleased with his son that he awarded him 10,000 shares of CF&I stock.[53] The operators' campaign would indeed be aggressive and the cost would be high.

Except for one incident, the strike district remained relatively quiet during the Congressional investigation. On March 10, however, the militia arrested sixteen strikers and then, for no apparent reason, razed the entire Forbes tent colony. Governor Ammons had begun ordering the gradual withdrawal of the militia in late February and, a month later, most units had been relieved of duty. Still in the field was Company B, some thirty-five to forty men under the command of Major Pat Hamrock, a Denver saloon-keeper, and Lieutenant Karl E. Linderfelt. Easily the most hated of the militiamen, Linderfelt had led mine guards in a battle with strikers as early as October 25. In mid-April Linderfelt's troops were reinforced by Company A, 130 men whom Bowers called "volunteers," but who were all mine personnel. Paid by the operators, not the state, Company A was such an informal private army that they had not had even a single drill.[54] Captain Edward F. Carson, commander of Company A, later stated that there were "not more than thirty" mine guards in the company, while the rest were "pit bosses, mine superintendents, mine clerks, and the like." State Senator Helen Ring Robinson wrote: "I was in those cities [Trinidad and Walsenberg] at the time and know they were recruited exclusively from mine guards, gunmen in the pay of the companies, and others of the same ilk. Husky young citizens of Colorado who were known to feel friendly to the strikers were refused enrollment because of 'physical disability.'" In his Industrial Relations *Report* West concluded: "Thus, by April 20th the Colorado National Guard no longer offered even a pretense of fairness or impartiality, and its units in the field had degenerated into a force of professional gunmen and adventurers who were economically dependent on and subservient to the will of the coal operators."[55]

On April 19 the strikers at Ludlow celebrated Greek Easter with festivities and a baseball game. When the game was interrupted by the militia a striker's wife made a comment and a trooper replied, "That is all right, girlie, you have your big Sunday to-day, but we will have the roast tomorrow."[56] The rumors that Ludlow, like the tent colony at Forbes, would be attacked were to be proven accurate the next day.

Early on the morning of April 20 the militia summoned Louis Tikas, the quiet Greek camp leader, to the train depot, and demanded that he produce a man thought to be held against his will in the Ludlow tent colony. Tikas determined that the man was not there and a brief argument ensued. As Tikas returned to the camp Major Hamrock ordered troops to Water Tank Hill—a strategic point which militarily commanded the area—where they set up a machine gun and exploded two bombs. Believing themselves under attack, perhaps by one of the militia's field guns, armed strikers ran for a protective sand cut at the same time. Tikas tried to stop them but soon realized that he could not. At this point in time, conflicting accounts make it impossible to state with certainty which side fired the first shot. Not surprisingly, each side accused the other. But the question itself is not of primary importance. More significant, as McClurg writes, military deployment occurred before the first sign of trouble. "This, plus the alacrity with which Hamrock decided that reinforcements were necessary and the singular ferocity with which the militia proceeded to pursue the battle, all lend strong support to the union claim that the colony was the subject of a premeditated movement and that the National Guard was the aggressor."[57] Gunfire soon blazed from both sides. The Ludlow Massacre had begun.

General Chase's constant and unexpected searches for weapons had depleted the number of guns at Ludlow. Pearl Jolly, a woman who was fired upon repeatedly despite the large red cross she wore, testified twice that there were only forty rifles and a few revolvers in the camp.[58] The militia was well-equipped with Springfield rifles and at least two of the operators' twelve machine guns. The battle lasted sporadically from morning to dusk. Strikers with weapons withdrew from the tents to draw gunfire away from the women and children, but the troopers poured a heavy stream of bullets into the camp. Many families were able to escape into the Black Hills, a small mountain range to the east. Others ran when an arriving train blocked the colony from gunfire. Some seventy-five to one hundred women and children climbed down a deep well. But only foresight saved the majority of strikers, wives, and children. Expecting an attack, they had dug cellars underneath their tent platforms weeks before.

Reinforcements sent by the Trinidad sheriff arrived in the early evening. Soon after their appearance Linderfelt led a charge on the colony, which was reached after encountering light resistance (just one militiaman was killed during the battle). The militia immediately set about torching the tents, and fires began to sweep the camp. Ludlow was in chaos as strikers and their families tried to escape the flames. While some troopers looted the tents, others helped women and children escape from the pits hollowed out beneath them. The Guard itself admitted in a subsequent investigation that the troops "had ceased to be an army and had become a mob." The report continued: "We find that the tents were not all of them destroyed by accidental fire. Men and soldiers swarmed into the colony and deliberately assisted the conflagration by spreading the fire from tent to tent. Beyond a doubt, it was seen to, intentionally, that the fire should destroy the whole of the colony. This, too, was accompanied by the usual loot."[59]

Before the fire became general, however, the militia captured Louis Tikas and two other union men who had remained in the camp to evacuate women and children. As the battle had begun it had been Tikas, waving a white handkerchief, who had tried to halt the running strikers. Linderfelt had threatened to get Tikas, had once beaten him, and now saw his chance. Swinging his Springfield rifle by the barrel, he hit Tikas on the head so hard that the stock snapped. The three defenceless strikers were then shot and killed.

Ludlow burned all night. By the time strike leader John Lawson was able to gather men and return from Trinidad, the colony was engulfed in flames and under the total control of the militia. Not until the following morning was the discovery made that thirteen people had failed to escape from a large pit dug below a tent. Barely recognizable, they lay huddled together in death, two women and eleven children. Union Secretary Ed Doyle telegraphed labor unions throughout the country: "Will you, for God's sake and in the name of humanity, call upon all your citizenship to demand of the President of the United States and both houses of Congress to leave Mexico alone and come into Colorado to relieve these miners, their wives and children, who are being slaughtered by the dozen by murderous mine guards?"[60] Bowers telegraphed Rockefeller that same day:

> Following withdrawal of troops by order of governor an unprovoked attack upon small force of militia yesterday by 200 strikers. Forced fight resulting in probable loss of 10 or 15 strikers. Only one militiam[a]n killed. Ludlow tent colony of strikers totally destroyed by burning; 200 tents; generally followed by explosions, showing ammunition and dynamite stored in them. Expect further fighting to-day. Militia being re[i]nforced. Suggest your giving this information to friendly papers.[61]

On April 22 an editorial appeared in the *Rocky Mountain News*, a Denver paper with state-wide circulation which had maintained neutrality toward the strike.

> The details of the massacre are horrible. Mexico offers no barbarity so base as that of the murder of defenseless women and children by the mine guards in soldiers' clothing. Like whitened sepulchres we boast of American civilization with this infamous thing at our very doors. Huerta murdered Madero, but even Huerta did not shoot an innocent little boy seeking water for his mother who lay ill. Villa is a barbarian, but in maddest excess Villa has not turned machine guns on imprisoned women and children. Where is the outlaw so far beyond the pale of human kind as to burn the tent over the heads of nursing mothers and helpless little babies?
>
> Out of this infamy one fact stands clear. Machine guns did the murder. The machine guns were in the hands of the mine guards, most of whom were also members of the militia. It was a private war, with the wealth of the richest man in the world behind the mine guards.[62]

Enraged by Ludlow, the strikers were certain they would be wiped out and issued a "Call to Arms." Since the state was furnishing no protection, the "Call" directed organized labor to gather "the men in your community in companies of volunteers to protect the workers of Colorado against the murder and cremation of men, women and children by armed assassins in the employ of the coal corporations, serving under the guise of state militiamen."[63] That same day, as more and more reports from Ludlow filtered in, the strikers attacked mine property after mine property. With red bandannas around their heads, shouting "Remember Ludlow," they burned mine buildings at Aguilar, at Hastings, Green Canyon, Delagua, Royal, Rouse, Primrose, and Broadhead. In Berwind Canyon they killed three members of Linderfelt's Company B in a pitched battle. The union song was different now.

> The union forever, hurrah, boys, hurrah!
> Down with the militia, to HELL with the LAW,
> We'll rally round the flag, boys, we'll rally round the flag,
> Shouting the battle cry of union.[64]

General Chase moved as many of the Guard back to the strike district as would go (after Ludlow some units simply refused), but they were no match for the furious strikers. It was a civil war which did not end until April 30, when President Wilson sent federal troops to Colorado. At least fifty people were killed during the ten days of battle, including the twenty to twenty-five (the number will never be known for certain) dead at Ludlow. George P. West concluded: "This rebellion constituted perhaps one of the nearest approaches to civil war and revolution ever known in this country in connection with an industrial conflict."[65]

As an enforced peace returned to the coal fields, the militia rapidly began court-martial proceedings against the soldiers involved at Ludlow. Beshoar quotes an officer: "We've got to take care of these men of ours. We've got to vaccinate them so no court in the land can touch them at some future date. A man's life can't be put in jeopardy twice in the courts and we'll take care of these boys."

Recognizing it as a charade, the union refused to take part in the proceedings, which ended with acquittals for all concerned. Meanwhile the Colorado courts returned some 400 indictments for crimes including murder, arson, and conspiracy to restrict trade against the union leaders and strikers.[66] Not until 1917, after an extensive series of legal battles, would all charges be dropped.

In May, 1914, Upton Sinclair brought the strike home to Rockefeller in New York City. Demonstrations at the Standard Oil building soon spread and Rockefeller was driven to the family estate in Tarrytown. But even Tarrytown proved no escape as anarchists and IWW members followed in close pursuit. While this publicity was bad enough, Rockefeller was also roasted by newspapers and magazines throughout the country as ultimately responsible for the massacre at Ludlow. In early June he countered by hiring Ivy Lee, publicity expert for the Pennsylvania Railroad. Lee was soon flooding the country with thousands of inaccurate pamphlets in support of the operators' position. Most notorious, perhaps, was Bulletin 14 in what was called *The Struggle in Colorado for Industrial Freedom*. Lee had discovered that UMW Vice-President Frank Hayes had been paid $4,062.92 for the twelve months preceding November 30, 1913. Of that total $2,395.72 was his salary and $1,677.20 his expenses. Lee took the figures, added the expenses again for good measure, and wrote that Hayes had received $5,720.12 for a nine-week period, "over $90 a day, or at the rate of over $32,000 a year." A similar procedure gave Mother Jones, who was not even paid by the union during the weeks she spent incarcerated by General Chase, $42 per day.[67]

Rockefeller's thirst for favorable publicity was so great that he became directly involved when Major E. J. Boughton, attorney for Colorado's metalliferous Mineowners Association, came to New York. Boughton and Ivy Lee had contrived a plan for Lee to compose a public statement on the strike for Governor Ammons to send out under his own name to President Wilson and every governor in the country.

Rockefeller had dictated a memorandum shortly before but, after a conference with his advisers, decided that the time was not ripe for him to make a personal statement. It was just the thing for Ammons to send along, however. "Several points in my memorandum," Rockefeller wrote Lee, "could well, even more appropriately, be used in the letter from Gov. Ammons to President Wilson which you are proposing to prepare as soon as the major's memorandum reaches you, which I hope will be very shortly."[68] The fact that Ammons did not send out the statement cannot be ascribed to Rockefeller's sense of the appropriate.

In September, after detailed investigation, President Wilson publicly proposed a plan for the settlement of the strike. Its provisions included a three-year truce, during which Colorado mining and labor laws were to be enforced. Wilson's plan provided that striking miners not found legally guilty be re-employed; that intimidation of union and non-union men was equally prohibited; that wages and regulations were to be posted; and that if grievances could not be resolved, they would be arbitrated by a commission of three, one member each from the miners and operators, and a third to be appointed by President Wilson himself.[69] Although the strikers were far from satisfied with the plan (union recognition and picketing were both proscribed), they had nothing better to hope for and ratified it almost immediately. The coal companies, led by CF&I, rejected it. Rockefeller's adviser Starr J. Murphy wrote to President Welborn: "The fact that the President of the United States has suggested a plan of settlement and given it out to the public produces a delicate situation which we have no doubt you gentlemen in the West will handle in the same careful and diplomatic way with which you have handled the whole situaion thus far, avoiding on the one hand any entanglement with the labor union, and, on the other, an attitude which would arouse a hostile public opinion." Murphy wrote a week later that "so far as we are concerned the strike is won; that the only remaining question is that of preserving law and order; and that if the authorities will do that it is a matter of indifference to us whether whether the strike is called off or not."[70]

At the same time, however, the New York office was aware that some alternative to President Wilson's plan had to be devised. Murphy wrote to Welborn: "I am impressed with the frequency with which they [newspaper editorials] make the point that the parties should either accept the President's plan or suggest some other. It seems to me clear that public opinion will demand either the acceptance of the President's proposition or some constructive suggestion from the operators. A mere refusal to do anything would be disastrous." With the help of Ivy Lee, Welborn accordingly wrote to Wilson that CF&I "will at an early date invite its employees to unite with it in creating within the company a permanent and impartial body which, while preserving to its lawful owners the control of the corporation, shall provide a mechanism for enabling the different elements in the company to present their views and suggestions to one another, for the peaceful adjustment of any differences that may arise. . . ."[71]

In the fall of 1914, as Welborn was writing, the strike was drawing to a close. The mines were operating at full capacity with imported labor and the UMW, its money running out and President Wilson's plan rejected, saw no chance of winning the strike. Nor were the state elections promising. Many Colorado newspapers had portrayed the strikers as criminals, and the two coal company candidates for governor and attorney general, George A. Carlson and Fred Farrar, both won on law-and-order platforms. On December 10, 1914, after nearly fifteen months of conflict, the miners formally voted to end the strike. Their defeat brought the birth of a company union.

Until the fall of 1914 Rockefeller had consistently deferred to the tactical judgment of his Colorado management. Bowers and Welborn had been trained and carefully selected for their positions; their attitudes coincided with their employers' views in New York; and the CF&I management was much more knowledgeable of the local situation than Rockefeller. He had permitted them to make decisions and, as a result, he was largely dependent upon them for his information about the strike. If much of it was self-serving distortion, it nevertheless fit Rockefeller's own prejudices so well that he made no attempt to determine its validity. But late in the year, as a consequence of the bad publicity, loss of life, and diminished profits, Rockefeller began to consider whether there might be a more efficient way to run CF&I. W. L. Mackenzie King, Canada's former minister of labor, was hired under the aegis of the Rockefeller Foundation to solve the problem. He began to construct an Employees Representation Plan which would improve working conditions and so remove some of the attraction of the UMW, establish stability at CF&I, yet keep meaningful authority away from labor.

Although the decision was made to institute the Plan after the strike ended, Rockefeller and his advisers began broaching it to their Colorado management in early October. "What would you think of the idea," Starr J. Murphy wrote Welborn, "of having in each mine a mine committee consisting of representatives of the operators and representatives of the miners employed in that mine chosen by the miners from their own number, which should be charged with the duty of enforcing the statutes of the State and also the regulations of the company looking to the safety and comfort of the miners and the protection of the company's property?"[72] After continued discussion Welborn was made amenable to the Plan, but Bowers, whose laissez faire attitudes had been formed to hard rigidity, was not. Unable to understand that a more sophisticated treatment of labor had to be developed, he was asked for his resignation. On January 5, 1915, announcements were posted at all CF&I mines informing the men of meetings "for the purpose of discussing matters of mutual concern and of considering means of more effective cooperation in maintaining fair and friendly relations."[73]

The central elements of the Plan were outlined in a letter from King to Rockefeller.

> A board on which both employers and employed are represented and before which at stated intervals questions affecting conditions of employment can be discussed and grievances examined would appear to constitute the necessary basis of such machinery. The size of this board and whether there should be one or many such boards would depend upon the numbers employed and the nature of the industry and whether or not the work is carried on in one or several localities. Where, for example, there are different mines . . . it might be that boards pertaining to each individual concern might be combined, with a provision for reference to a joint board covering the whole industry . . . to which matters not settled by smaller boards might be taken for further discussion and adjustment.

Employees would elect their own representatives and procedures would be developed to ensure that unresolved grievances could be carried to successively higher levels of appeal.[74] Industrial Relations investigator West was one of many who recognized that company control of the Plan provided only the illusion, not the reality, of collective bargaining: "The effectiveness of such a plan lies wholly in its tendency to deceive the public and lull criticism, while permitting the Company to maintain its absolute power."[75] Rockefeller, however, regarded the Plan as a solution to the conflict between labor and capital. In September, 1915, he traveled to CF&I's coal mines to sell it to the miners.

Rockefeller's performance at Pueblo, mentioned earlier, is worth consideration in more detail. After the parable of the corporate table, its four legs representing "stockholders, directors, officers, and employees," Rockefeller told the miners dramatically: "For fourteen years [of Rockefeller control] the common stockholder has seen your wages paid to you workers; has seen your salaries paid to you officers; has seen the directors draw their fees, and has not had one cent of return for the money that he has put into this company in order that you men night work and get your wages and salaries. How many men in this room ever heard that fact stated before?" he asked his startled audience. "What you have been told," he said, "is that those Rockefeller men in New York, the biggest scoundrels that ever lived, have taken millions of dollars out of this company on account of their stock ownership, have oppressed you men, have cheated you out of your wages. . . ." "Is it fair," he asked, "in this corporation where we are all partners, that three of the partners should get all of the earnings, be they large or small—all of them—and the fourth nothing?"[76]

The deliberate intent of this charade, in its shabby confusion of "stockholders," "common stockholder," and "stock ownership," was to make it appear as though (Rockefeller) capital had received "nothing" on its investment. None of the financially unsophisticated miners, men accustomed to dealing in a few dollars and cents, was aware that the Rockefellers, through their control of CF&I's Board of Directors, had ordered that common stockholders should be paid no dividends; that, instead, excess profits were to be plowed back into the company. Nor were the miners told anything about preferred stock or bonds. For all Rockefeller's attempts at evasion, that same year Industrial Relations Commission Chairman Frank P. Walsh pried the admission from him that, during the period 1902 through 1914, the Rockefellers had received over $9 million from their ownership of preferred stock and bonds in CF&I. For every $10 of employee wages, the Rockefellers received $1 in return; a percentage (on holdings of 40 percent) which represented a handsome surplus value realization. Walsh asked Rockefeller:

> My question is finally, would you consider it just and socially desirable that 15,000 employees who had worked for 12 years and many of them have been crippled and sacrificed their lives, should, as a matter of justice, receive 10 times as great a return as one man who had not visited the property—as a matter of social justice?

> Mr. ROCKEFELLER. I can not make any comparative statement. I think the employees should receive full wages, and I think they have. I think capital is entitled to a fair return [control of CF&I had cost the Rockefellers slightly more than $20 million]. There has not been a fair return. I think as between the two, the employees have fared better than the capital.[77]

Rockefeller's mission to Colorado was successful. Although some 2,000 employees apparently did not vote, the miners at CF&I approved the Representation Plan by a large majority less than a week after his speech. With the UMW no longer active in the state they had little alternative. In large part inspired by the Rockefeller Plan, variants of the company union became common not only in the Colorado coal fields but throughout American industry.

In 1919 the Russell Sage Foundation founded a lengthy and intensive investigation of the Plan. A number of desirable improvements had occurred at CF&I since the strike. The state mining code was enforced. Housing conditions, medical care, and schools were better, for example, and the miners were no longer compelled to trade at company stores. The Plan explicitly proscribed any discrimination against union membership. Also, CF&I had apparently withdrawn from state politics. But characteristic of employees in a company union, the miners were fundamentally powerless. Many were still fearful of the company. A miners' representative explained, "If I go to the men, they say they have grievances, but when I ask them to come and prove it . . . they say they are afraid." At Starkville, one of the mines where the investigators found the greatest number of grievances, not a single one had been formally reported. Of all the representatives questioned, only one who spoke unfavorably of the Plan would permit himself to be identified. A mine superintendent had told another representative, "if anybody comes and asks you about the Rockefeller Plan, you tell them it is all right."[78]

The investigators' conclusions are significant. Calling the Plan "an incomplete experiment," they found that "the enforcement of the agreement between the company and the employes, may be said to be, in the last analysis, almost exclusively a function of the managerial officials." While the employees' representatives might equally sit on boards during grievance hearings, they had no authority in determining the conditions that created them. Moreover, the representatives were themselves dependent upon CF&I for their jobs. Nor could the miners participate in any way in determining their wages. The Plan established competitive rates in other fields as the basis for wages, but nowhere defined "competitive." As a result, the rates agreed upon by the UMW in the Central Competitive Field were used until 1921, when they were suddenly reduced 30 percent to conform to those in unorganized districts in West Virginia and Pennsylvania. Not until after the nationwide 1922 strike did CF&I return to the Central Field as a basis.[79] Many miners recognized that only an independent union could represent their interests meaningfully, but they had been made powerless first by the defeat of the strike and then by the company union. What had been brutal autocracy was now corporate paternalism. Company unions would not be declared illegal until the National Labor Relations Act of 1935.

II

Shortly after the Ludlow Massacre, in an effort to bring the facts of the massacre to the public, the Colorado Strike Committee sent a small delegation to New York City. A mass meeting was arranged for the night of April 27, 1914, at Carnegie Hall and Upton Sinclair attended. Sinclair had discovered socialism in 1902, but he had not become fully engaged in working for fundamental change in the American economy until 1904. With $500 advanced by the *Appeal to Reason*, he went to Chicago to investigate the conditions in the packinghouses. Published first in installments in the Socialist *Appeal*, his indictment of the packers soon reached beyond the paper's half-million readers. In the furor created by the appearance of *The Jungle* in book form in January, 1906, Sinclair discovered the impact he could create. He concluded that if the American people were made aware of the inhuman conditions produced by capitalism, an aroused public opinion could force change. In 1914, by both art and action, he would try again to educate the people themselves to action.

Sinclair had read a variety of newspaper reports on the Colorado strike, but as he sat in the Carnegie Hall audience and heard descriptions of the killing of women and children, the burning and looting of the tent colony and its few possessions, he reacted with outrage and mounting frustration. Sinclair was certain that the Ludlow Massacre would not be fully reported by the press. He concluded that the most effective way to bring the events in Colorado to public awareness would be to conduct a mourning picket directly in front of the Standard Oil offices at 26 Broadway. He reasoned that such a demonstration would not only visibly link John D. Rockefeller, Jr., but capitalism as well, with the oppression in Colorado. Sinclair planned to organize small groups of people, all of whom would wear mourning bands of black crepe, to picket the Rockefeller offices. Regardless of the expected harassment, the marchers were to remain silent and nonviolent. Sinclair's wife, Craig, counseled caution: "We must give Mr. Rockefeller a chance. Let us go to his office with Mrs. [Laura] Cannon [a Colorado organizer who had spoken the previous evening at Carnegie Hall], and ask him to see her; if he does, that is a story, that will be publicity—without anybody going to jail. If he refuses, we have improved our position. People will see that we have been fair."[80] Although doubtful, Sinclair agreed.

The Sinclairs and Laura Cannon went to Standard Oil offices the next day. When Rockefeller refused to see them, the machinery for publicity was set in motion. Sinclair first arranged a public meeting at the Liberal Club that night. Then, with the help of a secretary, he sent off special delivery letters to the newspapers informing them of his attempt to speak with Rockefeller, together with notice of the meeting that evening. Consequently, when the Sinclairs arrived at the Liberal Club, there were not only radicals present but, perhaps more to his purpose, also a dozen reporters. When the author of *The Jungle* addressed a meeting about a labor battle, it was news. After describing the situation in Colorado, Sinclair announced his plan for picketing Standard Oil the next day and received strong support. The meeting closed with the resolution that "the American people must find some way to make clear their determination that the organized murdering of strikers by mine thugs and gunmen must cease."[81]

Sinclair arrived at 26 Broadway at 10 A.M. the following day. There he met not only those pledged to protest, but many others who had read the announcement of a demonstration in the morning newspapers. He gathered three supporters and began to walk slowly in front of the Standard Oil offices. Remarkable only for a black band of mourning, the pickets remained gravely silent as they marched. As a large crowd began to form, Sinclair and four women were arrested and charged with disorderly conduct. The police were impressed neither by Sinclair's contention that he and a few companions were entitled to walk quietly on a public sidewalk, nor by his explanation that the picket was a justifiable response to the violence in the Colorado coal fields. With reporters clustered around him, he described the massacre of miners and their families by the militia and explained the reasons for demonstrating against Standard Oil. Then he and the four women were taken to adjacent cells in the Tombs. They were arraigned that same afternoon, released on their own recognizance, and required to appear for trial the next morning, April 30. Sinclair left the Tombs and immediately returned to 26 Broadway, where the police were permitting the demonstrators to continue their picket. He promptly joined them.

Sinclair addressed a crowd of onlookers that afternoon. "I do this thing," he said, "because Mr. Rockefeller is here at the very headquarters of the invisible government with all his prestige and his power and his control of the press and the police."

> The public does not understand that when Mr. Rockefeller says that any miner can demand a check weigher, if he wants one, to see that he gets a fair deal, he merely means that the miner may do this, but that if he should do so he would be listed as a trouble maker and discharged and blacklisted, with never a trace on the records to show that asking for a weigher was wrong.
>
> The public does not understand that after posting promises to pay better wages and to grant shorter hours the company set about destroying all the leaders of the men and blackmailing them out of their jobs, and even hired gunmen to harass them, until the murder of a union organizer precipitated the present war.
>
> The public does not understand that the whole history of mining strikes is a history of promises and concessions made that were retracted and withdrawn as soon as the company found it could destroy the leaders of the men, one by one, after

> the peace agreement, instead of meeting them in open conflict in the strike itself.
>
> The public does not understand that the only reason the strikers want a contract with the union is that they have learned by bitter experience that no other kind of promise is any good.[82]

After Sinclair spoke, demonstration headquarters was established at 8 Trinity Place, a block away from 26 Broadway, and the Sinclairs returned to Morningside Heights: he to prepare for trial the following morning, Craig to rest for another day of picketing.

Early the next morning, with Craig already in front of the Standard Oil offices, Sinclair and the four women arrested with him returned to court. They refused legal aid and defended themselves. Sinclair described the atrocities in Colorado and explained the purposes of the demonstration. The magistrate did not find Sinclair's arguments relevant to the sidewalk. Instead, he found each of the defendants guilty and sentenced them either to pay $3, or to spend three days in jail. When the defendants refused to pay their fines, they were led away to their cells.[83]

Thousands of people gathered at the Standard Oil offices, asked questions and held discussions. When the news value of the mourning pickets threatened to wear thin from repetition, radicals from throughout New York City came to their aid. Leonard Abbott, Director of the Ferrer School, named in honor of the Spanish anarchist, temporarily took charge and effectively organized the forces at 8 Trinity Place. Members of the Rand School arrived to put their convictions and bodies on the line. Alexander Berkman, who had served fourteen years for his attempted assassination of Henry Clay Frick during the Homestead Strike, and was editing Emma Goldman's anarchist monthly, *Mother Earth*, arrived to increase the militancy of the demonstrations. Young Arthur Caron led anarchists and IWW members in a picket in front of the Rockefeller homes on West 45th Street, but the Rockefellers had left the city. John D. Rockefeller, Jr., had retreated to their four thousand acre

Pocantico Hills estate in North Tarrytown. Interviewed there, he stated, "To describe this condition as 'Rockefeller's war,' as has been done by certain of the sensational newspapers and speakers, is infamous. Our interest is solely in the Colorado Fuel and Iron Company, which is simply one of a large number of coal operating companies in the State of Colorado."[84] The publicist in Sinclair had chosen wisely: the Rockefeller name was now connected in the public mind with the Ludlow Massacre, battle lines were drawn, and Rockefeller, by virtue of a public denial, was responding. Also, the fact that the well-known author of *The Jungle* was in prison for peacefully protesting the Ludlow Massacre had sizable and symbolic effect. It was clear that oppression was not confined to anonymous miners in the distant mountains of Colorado.

Radicals were planning the first of multiple attacks against Rockefeller when Sinclair was released from the Tombs. On Sunday, May 3, demonstrations occurred both at the Calvary Baptist Church, where Rockefeller regularly taught Sunday School, and outside the Pocantico Hills estate. A well-attended memorial service for the Colorado dead was held on Bowling Green, and the militant Colorado War Committee met to endorse "every action and tactic" taken by the miners; collected money for food, arms, and ammunition; and resolved to open recruiting stations to gather "volunteers for the Colorado war."[85] These demonstrations also provided the first distinct sign of ideological conflict among the radicals. As early as April 30, Marie Gans, a young anarchist, had entered the Standard Oil offices and had verbally threatened to kill Rockefeller.[86] Sinclair, whose socialism precluded a personal vendetta against Rockefeller, had been clear from the beginning that the mourning picket demonstrations were to be nonviolent. He also feared that the public would not perceive the Ludlow Massacre as a direct consequence of capitalism if Rockefeller's personal life were permitted to confuse the issue. Sinclair argued against demonstrations at Rockefeller's church and home, but his arguments did not prevail. At the same time, however, he was well aware of the news value implicit in the Rockefeller name, and knew it could not be ignored. He recognized further that Rockefeller exerted more than nominal control over CF&I, and he knew that Rockefeller could change company policy. As a series of demonstrations began at Rockefeller's church and home, Sinclair followed his socialist convictions and continued the picket of Standard Oil. He wrote to Rockefeller:

> We are moving into a new era, Mr. Rockefeller, one which you as a member of the new generation must learn to understand. The older members of your family have by cunning and fraud, and sometimes by actual physical violence possessed themselves of hundreds of millions of dollars worth of natural wealth, which is necessary to the life of the people. The people know they have been robbed and are still being robbed. Inevitably they will hate the name of Rockefeller. You cannot divest yourself of that name; but you can proceed as rapidly as possible to socialize any of the stolen wealth which actually is under your control.[87]

On Monday, Sinclair attempted to extend the picketing of Standard Oil throughout the United States. Believing that the public would see that capitalism was the issue, not a single individual, Sinclair sent a telegram to Socialist leaders throughout the United States:

> Cannot the Socialist Party initiate movement in aid of the Colorado strikers to bring home to the masters of Standard Oil the intense abhorrence with which the American people regard their crime? Scores of telegrams have reached me suggesting this. There are branch offices of Standard Oil in every town. Cannot you or the National Executive Committee recommend that mourning pickets appear before these offices?[88]

Julius Gerber, secretary of the Socialist Party, rejected Sinclair's plan. "The Socialists here [in New York] are decidedly tired of this cheap clap-trap of Sinclair's," Gerber said. "They know it is the quiet work of organizing that counts and never this self-advertising noise. Sinclair's noise is his own personally organized affair, and we have nothing to do with it."[89] Upon Gerber's refusal, Sinclair decided to travel to Colorado. There he planned to examine conditions in the mining camps personally and gather information for a novel on the coal strike. As for Craig and the mourning pickets, she writes in her autobiography, "I went on for three weeks, spending every dollar left in our bank account."[90]

Sinclair arrived in Denver on Tuesday, May 12. With the help of union and radical contacts, he immediately set about investigating the strike situation and its political ramifications. The next day he spoke to a meeting of the Women's Peace Organization. After describing the mourning pickets and their effect in New York, Sinclair rhetorically asked why progress in settling the strike was deadlocked. He answered, "Simply because Rockefeller is ready to block everything, lose every dollar of his vast fortune and his reputation besides, to win this strike."[91] On May 14 Sinclair addressed a mass meeting of over 2,000 people assembled at the State Capitol.

> I say that the coal operators of your State have carried on a campaign of systematic and deliberate murder. Their purpose was robbery. None other purpose. They wanted the larger profits which they could get if they worked their miners as slaves instead of as free American citizens. They did the same thing in West Virginia. They have just finished doing it in Michigan, and in both cases they got away with the swag. They will continue to do it in this State and in other states, just so long as they can get away with the swag. I say that to permit them to win this strike by the methods they have used is to encourage the systematic and wholesale murder of working men everywhere throughout the United States. I say that to force a just settlement of this strike is to serve notice upon corporations everywhere throughout the United States that government by mine thugs and gunmen must cease. I say that the issue before us is now one simple issue of fundamental morality. Rockefeller has murdered labor. Shall he be permitted to rob the corpse?[92]

The Colorado legislature was preparing to adjourn as Sinclair spoke. President Wilson, angered by the prospect of leaving federal troops in the coal fields indefinitely, recognized that the state government intended to take no remedial action and telegraphed Governor Ammons that maintaining order in the strike district was Colorado's responsibility, not the federal government's. Wilson saw no reason for the "inaction of the state legislature" and flatly said that Colorado had no constitutional right to rely on federal troops "when it is within the power of her legislature to take effective action."[93] Ammons was in a difficult position. State monies were low and impartial investigation—with its attendant publicity—had shown the state militia to be heavily corrupted by the coal companies. It would be exceedingly convenient for both the state and the operators if federal troops could be counted upon to take over strike-breaking duties. Ammons accordingly withheld Wilson's telegram from the nearly adjourned legislature. Not only did the governor enlist "the leaders of the hand-picked machine majority in the state legislature" in his scheme, Sinclair charged, but Ammons also sent the president's telegram to the coal operators, who replied under the governor's signature that a legislative "committee on mediation on the present strike has been provided for and appointed."[94]

Sinclair examined a verbatim record of the legislative proceedings for May 15 and found the resolution to be an investigatory measure (when a number of investigations were already underway) providing that "remedial legislation may be enacted at the next General Assembly which will tend to prevent a recurrence of insurrection and public disorder."[95] The bill was a sop, in no way directed at mediating the strike. Sinclair believed Ammons and the coal operators deliberately misinformed President Wilson in order that he would have no viable alternative to keeping federal troops in the coal fields—

which is precisely what was to occur. Sinclair wired President Wilson that his telegram had been withheld from the legislature the day it adjourned. "All newspaper men know that during that time your telegram was in the hands of all coal-operators in this city, and they know the men who took it to them." Governor Ammons' telegram to you contains a falsehood," Sinclair wrote, "the word 'mediation' did not appear in the measure referred to, which provides for investigation only."[96]

The next day an interview with Governor Ammons appeared in the *Denver Post*. Ammons said that Wilson "was misled when he sent me that first wire Saturday, complaining that the legislature was about to adjourn without doing anything," and described Sinclair as an itinerant investigator interested in his own advertising and aggrandizement. "It is true," explained Ammons, "that I did not send the telegrams [to the legislature], but their contents were known to almost half the body and there was no attempt at covering up." Ammons denied Sinclair's charge that Wilson's telegram was put into the hands of the coal operators. Then he directly responded to Sinclair's statement that "mediation" was beyond the scope and powers of the appointed committee: "Probably that particular word does not occur, but a reading of the resolution will show that it gives the legislative committee power 'to assist in settling the strike.' If that isn't mediation I'd like to know the true meaning of the word."[97]

Once again Sinclair reread the resolution. The significant words, "to assist in settling the strike," did not appear in it. In a final telegram to President Wilson, Sinclair urged him "to get the full text of this resolution and realize what it means that the Governor of this State is wilfully and deliberately endeavoring to deceive you and the public in this crisis."[98] Sinclair tried in every way to publicize the machinations of Ammons and the coal operators. When the Denver Bureau of the Associated Press refused to carry any word of the controversy, Sinclair telegraphed a story collect to some forty newspapers, the majority of which accepted and printed it. Sinclair saw the situation in Colorado as nothing less than class warfare. On one side were thousands of wage slaves. On the other were some hundred million dollars of capital which controlled, directly or indirectly, the governor, legislature, and the Associated Press. "The directors and managers of the Associated Press were as directly responsible for the subsequent starvation of these thousands of Colorado mine-slaves as if they had taken them and strangled them with their naked fingers."[99]

The remainder of Sinclair's stay in Colorado was comparatively quiet. With the state legislature adjourned for the year and Governor Ammons hostile toward him and the union cause, Sinclair had small opportunity to accomplish anything further. Most of his time was given over to personal investigation and documentation of mining conditions. He traveled to the strike zone, visited the embattled camps, and interviewed miners and their families. As he prepared to return to New York, Sinclair wrote again to Rockefeller.

> The strike has now reached a critical point. The Federal troops are preventing picketing. They are admitting strikebreakers, provided they come individually and not under the superintendence of the companies; and, of course, the companies can arrange to send in as many as necessary under these terms. So the strikers, who have endured many years of oppression and torture, who have been driven from their homes to hide and fight in the mountains like wild animals, are now to be slowly starved into subjection, and to return to work under the old conditions of slavery. The American people are to see constitutional government permanently suppressed in several counties of Colorado, and are to be prepared for an even more terrible outbreak of civil war in another decade.

> I have seen a good many strikes, and I know the symptoms of them. I know that period of slow strangling, which is the most heart-breaking and terrifying of all—the more so because it is a silent process, because it happens after the excitement has died down and the country has forgotten. . . . It is enough, more than enough, that for nine months your agents have hounded these ten thousand wretched slaves; if now they are permitted to reap the profits, in the form of larger dividends, it becomes incitement to every criminal corporation in the country. I know the feelings of hundreds of people with whom I have talked upon this subject; so I am writing this last appeal, and I beg you to listen. . . . [100]

Rockefeller had pledged himself to stand by his management whatever the outcome and he would not alter his policy in the spring of 1914.

The New York demonstrations against Rockefeller and capitalism had not halted with Sinclair in Colorado. On Sunday, May 10, the Reverend Bouck White and followers from the Church of the Social Revolution interrupted services at Rockefeller's church and were forcibly thrown out and arrested by police. The mourning pickets continued their vigil in front of the Standard Oil offices and Mother Jones, only lately imprisoned in Colorado, arrived in New York in an unsuccessful effort to confront Rockefeller. Arthur Caron and Leonard Abbot twice asked Tarrytown authorities for permission to hold public meetings and were twice refused.[101] With peaceful overtures rejected, direct action became the only alternative.

On May 30 the *Mother Earth* group, a loose association of anarchists and IWW members organized by Alexander Berkman, returned to Tarrytown. When they found the Rockefeller estate well-secured, they gathered at Fountain Square, where Salvation Army and even socialist speakers had been heard in the past. It soon became apparent, however, that anarchists and IWWs were not to be accorded an equal right. The first speaker was arrested immediately and, as others followed, they were arrested in rapid succession.

The radicals showed little dismay. The tactic of mass arrests had been successfully used in free speech fights in the West, and there was little doubt that the White Plains jail could soon be filled to overflowing. The police evidently reached the same conclusion. They halted after twelve arrests and began clubbing the remaining demonstrators. That evening and the next ended with the police and townspeople clubbing and shoving demonstrators aboard trains for Manhattan. [102] When Sinclair returned from Colorado, he found himself progressively drawn into the battle for free speech in Tarrytown.

Sinclair met with Tarrytown trustees on June 6 and again on June 8, when an ostensible compromise was worked out. Tarrytown was an incorporated village which apparently contained no public land. But the trustees, convinced that an absolute prohibition of free speech would prove unwise, agreed to permit use of a large hall for a public meeting. The villagers did not support the trustees' decision. Inflamed by the local newspapers, which proposed that townspeople wear small American flags on their lapels as a sign of their opposition to dissident notions of free speech, feeling in Tarrytown grew more hostile. When it became apparent that no meeting hall would be made available, Sinclair began searching for a landowner who would permit a meeting on private property. His inquiries led to Annie Gould, a North Tarrytown resident who offered use of her open-air theatre for a free speech meeting on the condition that anarchist and IWW spokesmen would be excluded.[103] Upon Sinclair's agreement, the conflict that had split the radicals earlier became crystallized.

Sinclair had attempted throughout to focus the demonstrations specifically on the issue of capitalism and to restrict them to the Standard Oil offices. The *Mother Earth* group, however, had assumed direction of the demonstrations and saw no reason to confine them to any institution, even Standard Oil. Men stood behind capitalist policy and should not be protected by corporate anonymity—particularly when those men were the Rockefellers. The demonstrations had taken on a life of their own during Sinclair's absence, and Tarrytown's prohibition of public meetings had led necessarily to a battle for free speech. While Sinclair's agreement to a free speech meeting that explicitly excluded anarchist and IWW speakers arose from his conviction that the Colorado strike had been permitted to fall into the background, in no fundamental sense, even had anarchist spokesmen been included, could a meeting on private property have resolved the right of free speech. Sinclair tried to act as an intermediary during the Tarrytown battle, but he compromised what the anarchists temporarily regarded as the primary issue—free speech for all, regardless of political beliefs. Their decision either to speak, or to break up the meeting, comes as little surprise.

On Sunday, June 14, some 300 people came to Mrs. Gould's estate. Sinclair spoke first. For an hour or more he discussed the meaning and implications of the Colorado strike. He described the social unrest in America as the most serious since the Civil War: "Between 600,000 and 700,000 men of the United Mine Workers will go out on a strike all over the country if they have to." "If the Federal troops continue to be used as strike-breakers as they are in Colorado now," he said, "there will be revolution all over the country." Sinclair attacked Rockefeller: "We must establish the principle . . . that a capitalist is guilty of murder if he hires gunmen to go out and kill people to increase his dividends."[104] John W. Brown, a UMW organizer, related the history of the Colorado strike, described what had occurred in the past year, and explained the union position. A resolution was offered declaring that Rockefeller's treatment of the miners was sufficient for President Wilson to nationalize the mines, and, after considerable argument and discussion, it was passed unanimously. The question of whether there should be a permanent public forum for free speech established in Tarrytown was raised, and this too passed without a dissenting vote. The meeting was running smoothly when Adolf Wolff, an anarchist sculptor and part of the *Mother*

Earth group, gained the floor and denounced Sinclair. The right of free speech was not subject to compromise and negotiation. "We shouldn't plead, we should take!" Wolff shouted. Then he angrily called the people of Tarrytown "cowards, curs, and traitors."[105] The meeting collapsed into chaos.

Annie Gould was not alarmed that the forum had almost literally become a free speech fight and she offered the use of a large meadow for any future meetings. It was learned, however, that New York City owned the aqueduct that ran through Tarrytown. To those prepared for confrontation, not compromise, this narrow strip of public land was preferable to any piece of private property. To Sinclair neither of the future sites was satisfactory. Each was too far removed from the working class area of the village.

With the discovery of the public aqueduct in Tarrytown, it was only a matter of time until the *Mother Earth* group returned to take advantage of it. On Monday evening, June 22, Berkman led some forty anarchists to Tarrytown. The villagers were taken by surprise. After marching past City Hall and up Main Street, Berkman led the way to a small fenced area on the aqueduct strip. As he began to speak, the Tarrytown police chief suddenly arrived and wrestled him to the ground before he had completed a sentence. It developed that in order to speak publicly on New York City property a permit had to be obtained. This Berkman had failed to do. As he argued with police, hundreds of local citizens arrived at the aqueduct to reinforce the outnumbered police. Many villagers brought eggs which they hurled at the radicals. Others threw rocks. When the anarchists refused to leave the aqueduct, the villagers, their ammunition exhausted, became furious. They burst through the fence and charged the demonstrators, forcing them to retreat. The radicals fled through the street with the villagers in pursuit.

The Tarrytown police had earlier called a unit of the New York City mounted police, which held jurisdiction over the aqueduct. When they appeared the radicals were run down by horses and clubbed to the ground. The villagers then pulled and dragged them to the railroad station. The demonstrators were lined up, beaten further, and, when a train for New York appeared, pushed into the cars. Becky Edelsohn, who had stood up to the mob at the aqueduct for nearly an hour, received two black eyes. A number of radicals, including Arthur Caron, had their heads smashed open by clubs.[106]

The day after the aqueduct riot John D. Rockefeller, Jr., decided to take his family for a lengthy vacation to Seal Harbor, Maine. John D. Rockefeller, Sr., stayed on at Pocantico Hills unruffled. The *Times* reported: "The I.W.W. is not worrying John D. Rockefeller these days. Mr. Rockefeller feels safe with his sixty guards and the twenty-four Deputies under Sheriff Doyle, and is spending his mornings golfing and his afternoons hearing organ recitals on the big organ in his home."[107] Had the elder Rockefeller been aware of all the radical plans, he might have been more concerned—his eighty-four guards notwithstanding. At 1626 Lexington Avenue, Arthur Caron and at least three others were making a bomb to destroy whichever Rockefeller they encountered first. Early on the morning of July 4, Independence Day, the bomb exploded prematurely. It killed Caron, Carl Hanson, Charles Berg, a woman in a neighboring apartment, and blew the upper three floors of a seven-story tenement to ruins.

After the disagreement with Berkman over the wisdom of confrontation and violence, Sinclair had become more removed from radical councils. The *Mother Earth* group had chosen to escalate its demands and actions and viewed Sinclair as too much given to negotiation and conciliation. For his part he saw no purpose to any more violence. The Sinclairs had spent every dollar they had or could borrow, and Craig was on the edge of nervous exhaustion. They learned of the explosion and deaths when the police asked her to identify Arthur Caron's body.

Although one man, Mike Murphy, who had been in the far end of the flat, had escaped, the exploding dynamite had blasted Caron's body downward through two floors of shattering concrete and rubble, where it was caught, torn and dead, on a fire escape. He was nineteen years old. They were all young, none over twenty-four. Berg's body, its face and feet gone, was found pinned among debris against the remains of a wall. Hanson was blown into bloody pieces, some of which fell and lodged on a nearby Lutheran church.

A great deal of uncertainty surrounded the explosion. Mike Murphy, who had miraculously escaped death, proved equally successful in eluding the police. Dazed and in shock, he had been stopped on the street by a policeman, but then had staggered away to telephone and meet Berkman. Nor were the various radicals themselves certain of what had occurred on the morning of July 4. Many thought that an agent of the Rockefellers may have planted the dynamite in the building. The police believed the bomb to have been intended either for the Rockefeller estate or for the Rockefellers themselves. Detectives reportedly found batteries, wire, IWW literature, and a loaded revolver in the debris of the apartment. The police concluded that the men had been working on the bomb when it exploded.[108]

On July 5 Berkman announced that a public funeral was planned for Saturday, July 11. The three bodies would be taken from the funeral parlor by sympathizers, and a procession would travel by automobile to Union Square. The hearses would stop at a temporary platform, the caskets would be taken out and laid upon it, and eulogies given. The proceedings were intended to be grave and very public. As plans for the demonstration developed, General Sessions Judge Thomas Crain upheld Sinclair's conviction of disorderly conduct. After this last legal flurry, Sinclair's activist response in support of the Colorado miners ended. While he understood the frustration of Caron, Hanson, and Berg he believed that bombs and violence would only beget harsher oppression. Sinclair wrote in the *Appeal*: "I know that the working class has terrible wrongs to endure;

I fear that it will have still more terrible wrongs to endure before it reaches the goal of the co-operative commonwealth; I know that no man can face these wrongs and endure the horror of the battle against them . . . without wishing that he could blow his enemies out of existence. But that temptation is one that must be resisted at all hazards. For our enemies are not men . . . ." Sinclair included a stern judgment: "If Caron was himself the bomb maker, then it is better that he died as he did. It would be better that all bomb makers should blow up themselves, for they can do nothing but harm to the cause they would serve."[109]

With the possibility of police retaliation in the air, the IWW also publicly repudiated Caron and disclaimed any connection with the explosion. Joe Ettor, who with Arthur Giovannetti had led and been imprisoned during the 1912 Lawrence Strike, declared that Caron was not a member of the IWW and, further, that he had been refused membership because he was out of work. "The I.W.W.," Ettor said, "does not approve of dynamiting or setting off bombs. We have been accused of violence, but the charges are false."[110] Italian anarchist and IWW leader Carlo Tresca, however, would have none of repudiation and called Ettor's comments "false, entirely uncalled for and cowardly." Caron had been a member of the IWW's Unemployed Local, which had been specifically organized for unemployed workers. Nor did the fact that Caron had lately called himself an anarchist preclude membership in the IWW, which accepted working men irrespective of their politics. "When people get 'cold feet,'" Tresca said, "and rush into print at the least sign of danger and repudiate violence, like Ettor, then I want to go on record—like my comrade Alexander Berkman—that under certain circumstances I favor violence."[111]

There was no question of disavowal by the *Mother Earth* group. Berkman was clear that violence was both justified and necessary in the struggle of labor against capital. To think otherwise was to be bound by a bourgeois morality which refused to admit the relationship between terrorism and its cause. Caron, Hanson, and Berg, Berkman wrote in the July issue of *Mother Earth*, "have taught the country that there is a class war . . . and that all and every means are justified in the defense and offense of labor against its Ludlow masters." The significance of the Lexington Avenue explosion was two-fold: first, it taught the enemy to fear and respect the power of the working-class; second, that as labor gathers power "its success will be hastened, its courage strengthened by tempering oppression with dynamite."[112]

The radical demonstrations set in motion by the Colorado strike were coming to an end. The police had successfully prevented the potentially uncontrollable aspects of the public funeral. There would be neither bodies nor a parade. Berkman was far from happy, but he realized that while he could counsel politics of confrontation in Tarrytown, any duplication in New York City would be suicidal. After the bomb explosion of the previous Saturday, public opinion would countenance virtually any action taken by the police.

The massive meeting was set for Saturday, July 11, at 2 P.M. in Union Square. In the early afternoon people in twos and threes and in large groups began their journey to the small park at the upper end of Greenwich Village. The day was clear, hot even for New York in July. The crowds moved slowly, solemnly, gathering force, converging at corners and then moving on. Men wore red roses in their lapels, the women red ribbons in their hair. Many wore black. Over 10,000 people came to Union Square. They were surrounded by 800 police.

As the crowd waited, red and black banners were unfurled. "Capitalism the Evil, Anarchism the Remedy." "Those Who Die for a Cause Never Die—Their Spirit Walks Abroad." "You Want to Do Away with Violence? Do Away with Capital and Government that Provoke and Breed Violence." The speakers' platform was constructed from two large dry goods boxes, draped in red and black cloth. On the platform stood a great triangle of red roses and green leaves. Worked out in the red flowers were the words, "Caron, Hanson, Berg, Soldiers of the Revolution." At 2 P.M. Berkman mounted the platform. He wore a red necktie, around his right arm was a black band of mourning. He stood quietly and the band began to play the *Marseillaise*. As the singing died away, Berkman began to speak. "We hold that our Comrades Arthur Caron, Charles Berg and Carl Hanson died either martyrs to the cause of labor, or victims of the capitalist class." It is possible, Berkman said, that the three "were directly murdered by the enemy, perhaps by agents of the Rockefellers." But, he declared, "I want to go on record as saying that I hope our comrades had themselves prepared the bomb, intending to use it upon the enemy."

Leonard Abbott spoke of the three dead men in moving terms. Both Berg and Hanson were Lithuanians, and both had participated in the great social upheaval that shook Russia in 1905. Upon arriving in America, they showed themselves "as zealous as they were in their native land in their devotion to social ideals and to the cause of working class emancipation." "As I speak of [Caron] ," Abbot continued, "on one side, I see a young working man, the champion of the exploited and the disinherited. . . . And on the other side, I see the richest man in the world passive while hired gunmen and soldiers train cannon on a tent colony of his striking miners, massacre their wives and children, and set fire to the tents."

Elizabeth Gurley Flynn spoke not as a representative of the IWW, but as a member of the working class. She developed the possibility that the bomb had been planted by agents of capitalism. But to the Tarrytown veterans, for whom oppression had foreclosed any possibility of civilized response, dynamite held an altogether different purpose. It was a weapon, as Becky Edelsohn paraphrased Albert Parsons, that equalized all mankind, that made the poor as powerful as the rich. The kept press might write of the violence committed by the working class, but "all the violence that has been committed by the labor movement since the dawn of history wouldn't equal one day of violence committed by the capitalist class to keep itself in power."[113] Berkman concluded by giving the simple inscription on the urn containing the ashes of Arthur Caron, Carl Hanson, and Charles Berg: "Killed July 4, 1914–Caron, Hanson, Berg." In the silence that followed, the speakers left the platform and the thousands of people departed as quietly as they had come.

The Union Square meeting ended with sober grace. With its end radical demonstrations in support of the striking miners drew to a close. On July 28 the much delayed trial of the original Tarrytown demonstrators, their ranks diminished by death, took place. The trial was brief. With the exception of one who recanted, and Becky Edelsohn, who had been previously sentenced to ninety days, they were all found guilty and sentenced to sixty days in prison.

In February, 1915, an unexpected symmetry imposed itself. The aftermath of the coal strike concluded in Colorado. In Denver to found a Ferrer School, Alexander Berkman spoke candidly about the explosion. "The plot against the lives of the Rockefellers," he said, "was originated in a spirit of vengeance for the brutality visited upon the I.W.W. workers at Tarrytown last summer, when they went up there to protest against the outrages that brought on the battle of Ludlow." The bomb had been nearly assembled when it exploded. "They had no definite idea when they would use the missile," Berkman said, "but it was their plan to wait for an opportunity and hurl it into the carriage or automobile of the Rockefellers when they were leaving the Tarrytown estate. They wanted to get both of them together, if possible, but would have taken the life of either one."[114] As Berkman spoke, the coal strike had been broken and the union defeated two months before.

III

With the public battle for justice in Colorado and Tarrytown at an end, Sinclair and Craig moved to Croton-on-Hudson. *Sylvia's Marriage*, a novel which was to sell some 100,000 copies in England, had recently been published in New York, but its early sales in America did not provide even a temporary income. Sinclair had been too involved in the coal strike and its aftermath not to write about it, but lack of money dictated an interim project. Accordingly, he proposed an anthology of social protest literature to the Winston Company and received a $1,000 advance. *The Cry for Justice* unexpectedly grew to nearly 900 pages and occupied him for nearly a year.

Sinclair's stay at Croton was temporary. As the spring of 1915 approached, Craig's parents invited them to the family home in Gulfport, Mississippi. Sinclair went primarily to write and lost little time in beginning *King Coal*, initially conceived as "a big novel dealing with the Colorado strike and the solution of the labor problem along the lines of syndicalism."[115] *King Coal* was first rejected and then, in revised form, accepted and published by the Macmillan Company. This initial rejection, coupled with Sinclair's lengthy correspondence with George P. Brett, president of Macmillan, significantly affected not only the shape of *King Coal*, but determined the existence of *The Coal War* as well.

The evidence, letters and the manuscripts of *King Coal* and *The Coal War*, is conclusive that Sinclair originally planned one novel which incorporated both stories. When Macmillan first rejected *King Coal* because it was too much a historical reconstruction of the Colorado strike, Sinclair basically cut his large manuscript in half. He rewrote the first part with Brett's suggestions in mind. As a climax, in place of the Ludlow Massacre, he used a mine disaster modeled on the 1909 Cherry, Illinois, explosion which claimed some three hundred lives. With the story of the Colorado strike still untold, Sinclair then wrote a documentary sequel to *King Coal*, *The Coal War*. It is doubtful if he would have rewritten *King Coal* without the alternative of *The Coal War* firmly fixed in his mind.

When Sinclair began work on *King Coal* in May, 1915, he had had ample time for the story to develop. As a novelist, he considered the actual events of the Colorado strike so compelling as to make major replotting unnecessary. As a socialist, he believed that a novel constructed from substantiated fact would be more effective than one built upon fictional circumstance. Readers of *King Coal* were to be made well aware (principally by means of a factual appendix) that the action of the novel was not imagined but, instead, historically true. Sinclair was more concerned with making the history of the coal strike widely known than with writing a novel which merely used the strike as its general background. At the same time, however, he knew that a novel would find one hundred readers where an exclusively historical treatment might reach one.

In barest outline, Sinclair planned to create a small number of fictional characters, place them in the context of the Colorado strike, and let their fate be determined by history. The events in *King Coal* were to be seen primarily through the son of a coal operator, Hal Warner, who takes an assumed name to learn about the mining camps firsthand. Known as Joe Smith, Hal is introduced to mine work and what passes for life in an isolated Colorado coal camp. Both his split existence and the brutality of mining conditions serve to educate him to an economic reality which he has never known and he comes to learn that his attitudes and beliefs are largely class prejudices. Engaged to a banker's daughter in Western City, Hal's dilemma is made more acute by a growing attraction for Mary Burke, a young woman in the mining camp. He joins the struggle to improve conditions, helps to unseal a mine filled with trapped men after an explosion, and is then a participant in Sinclair's reconstruction of the Colorado strike. However, Hal is trapped by his old allegiances. When he finds himself about to shoot his brother, a member of the militia, he gives up his gun and surrenders. "At the conclusion of the story," Sinclair writes in this early, but later altered synopsis, "the miners' leader is in jail for life, the master of the mine is introducing his new paternalism, a tame union, fathered by the companies, and carefully prevented from obtaining power. Hal and his sister go out to appeal to the people for industrial democracy and self-government."[116] Conflicted and unable to declass himself, Hal can do little but return to his own world.

The writing of *King Coal* began well, but it was soon interrupted by an emotionally draining custody suit filed by Sinclair's first wife. Sinclair found himself unable to work productively and, shortly after being awarded custody of his son for six months each year, moved to Coronado, California. He had the first chapters of his novel printed, attached a synopsis of the remainder, and sent them to Macmillan. George Brett, who had earlier refused *The Jungle*, responded on December 7. "I consider that you have a great public for the book which you are now writing," Brett wrote, "and I believe further that you are probably the only writer in America who could do this great theme full justice." However, Brett emphasized that Macmillan wanted a novel, not history: "I am very glad indeed to know that you are making the incidents of history . . . supplementary to the story itself, i.e., the story interest and the story climax. This, as you know, is most necessary if the novel is, as I hope, to have a very large sale and exert great influence on our times."[117]

Sinclair, still plagued by financial problems, asked ten days later if a contract could be signed and an advance given before completion of the book. Brett replied that while he "was much interested in" *King Coal*, and "meant to publish it," he nevertheless preferred to wait until the novel was finished. Brett feared that Sinclair would give more emphasis to the strike than to fictional structure.

> You speak in your letter of making the thing literary, i.e., giving a literary view of the whole story, and this I greatly hope for but your letter also refers to your attempt to make it an historical narrative. Pray don't make the mistake of falling between two stools in this matter. This book professes to be, and is, and ought to be, a novel. As a novel it will reach a thousand readers where as an historical narrative it would reach one, and you will find it very dangerous to your aim to get a large circulation for this book, as you should; if you attempt to depart from your original plan of telling the whole story as a novel.[118]

Sinclair soon responded in full agreement with the editorial advice, but it is apparent that he refused to swerve from his original conception of the book. The resulting conflict between the opposing demands of fiction and history was not only to be the central issue in the composition of *King Coal*, but was also to create the necessity for *The Coal War*.

Sinclair sent off *King Coal* at the end of March and Brett rejected it in late April. On May 1, Sinclair offered to rewrite his manuscript but said that "his furnace fires were dead within him."[119] On May 7, disappointment and fatigue gone, he asked Brett what needed to be done to the book. Brett answered in plain terms. "The manuscript as I see it now will not in my opinion live beyond its first furor or sale," he wrote. He made three interrelated suggestions. First, the Colorado coal fields should be largely forgotten during revision. *King Coal* "should be a novel of the coal world for the world and not merely for the Colorado strike." Brett then urged Sinclair to create a plot worthwhile in itself and to place it in "the world of King Coal, using such of the incidents from the Colorado strike as were of universal interest and omitting others, thus giving a true picture of the mining kingdom of coal." Last, Brett argued that much of "the incident and document of the present last half of the MS will have to be sacrificed and in place of this a story invented which will carry the thing to its natural and proper climax."[120] Brett concluded with the hope that his opinions would encourage Sinclair to revise *King Coal*.

When Sinclair began to rework the novel, it soon became evident that minimal revisions would not suffice: his manuscript would have to be recast and rewritten—a task which was not completed for some six months. He was not happy about it. Not only had Brett rejected the novel, but the story of the Colorado strike, nothing less than the informing purpose of the book, would apparently have to be cut. As Sinclair considered the problem a deceptively simple solution came to mind. He would rewrite *King Coal* as Brett suggested, but he would also write *The Coal War*, the story of the coal strike.

Sinclair began rewriting *King Coal* in May and finished October 29. "I imagine you will think it my very best," he wrote to the Dutch critic, Frederick van Eeden, "because I have taken pains with every little detail, as you wanted me to. It is all thanks to Craig. She has worn herself to a skeleton."[121] Brett accepted the novel on November 29, 1916. Macmillan planned to issue *King Coal* the next spring, but the effect of war conditions in the publishing industry prevented publication until September, 1917. Sinclair received $500 on account and royalties of fifteen percent on all copies sold.[122]

Sinclair started *The Coal War* shortly after *King Coal* was accepted and finished it ten months later. A documentary novel of the Colorado strike, it was to be his book this time, not Brett's. The social purpose of *The Coal War* is made clear in his Postscript: "This book goes out as an appeal to the conscience of the American people: an appeal for millions of men, women and children who are practically voiceless—not merely in Colorado, but in West Virginia, Pennsylvania, Michigan, Alabama, a score of states in which miners and steel-workers have been unable to organize and protect themselves."

Edward C. Marsh rejected *The Coal War* for Macmillan on November 7, 1917, in terms Sinclair had heard before. While the new novel was more important than *King Coal* in rounding out the story of the Colorado strike, Marsh wrote, *The Coal War* "is deficient in story interest." "Perhaps the book tells too much about the strike." Marsh wanted Sinclair to revise *The Coal War* in the same fashion as he had *King Coal*. "Can you not go over the manuscript of 'The Coal War' once more," Marsh asked, "and strengthen the story interest by every means in your power?" "You will understand, of course, that we have every wish to go on with the publication of this volume as a sequel to 'King Coal.'"[123] Sinclair considered the history and intent of *The Coal War* and refused. Other factors contributed to Macmillan's rejection as well. Although *King Coal* had more than met costs, it was not a big seller; the Colorado strike was no longer topical, but a relatively stale issue; and the American public was daily concerned with a much larger war in Europe.

Sinclair viewed the Colorado coal strike as a battle in the class war, an economic lesson which demonstrated the brutal, often bloody, costs inherent in the American economy. As a result, *The Coal War* is a thesis novel which shows capitalism to be viciously destructive. Recalling the impact of *The Jungle*, Sinclair again wanted to arouse his readers so strongly that public outcry would force change. *The Coal War* is a hybrid, an amalgam of fiction, history, and propaganda.

To Sinclair his books, fiction and non-fiction alike, were primarily significant to the degree that they exerted social influence. The concluding pages of his autobiography reveal this overriding purpose clearly. He asks himself, "Just what do you think you have accomplished in your long lifetime?"[124] and then provides ten answers. All involve social change and, with two exceptions (his role in founding the American Civil Liberties Union and the League of Industrial Democracy), his books were instrumental in that change. Nowhere in this list of accomplishments is there a judgment that any of his novels represent an exclusively literary achievement. Art which was not at the same time socially beneficial was largely irrelevant.

Social effect was at the heart of Sinclair's aesthetic. Choosing not to turn his back on his times, and recognizing neutrality as a human impossibility, Sinclair scorned the illusion of a disinterested posture. He had come to understand that he was not an isolated individual at odds with the world, but a man who shared a common condition with other men. He also understood that his life was interdependently affected by the suffering or liberty of others. Freedom for all could come only through the efforts of all. In neither his life nor his writing did he wish to evade his socialist commitment. It is not surprising, then, that Sinclair describes art as "a representation of life, modified by the personality of the artist, for the purpose of modifying other personalities, inciting them to changes of feeling, belief and action." His definition of art is an aesthetic justification of propaganda. He faced this issue squarely. "Great art," he writes, "is produced when propaganda of vitality and importance is put across with technical competence in terms of the art selected."[125] Art had to address itself to the material problems which men faced in their daily lives, not avoid them. With misery abroad, Sinclair was not content to treat fiction as entertainment.

Sinclair was not a writer whose socialism was incidental to his art: as Granville Hicks comments, "what interested him as a socialist interested him as a novelist."[126] It cannot be too strongly emphasized that for Sinclair socialism became an aesthetic, one that explains flaws which other writers easily avoid. He perceived character, for example, as determined by social and economic relationships and he constructed it that way in his fiction. As his novels show, however, the demands of fiction and ideology are often in opposition. Literature does not easily lend itself to the conceptual forms necessary for political understanding. Irving Howe's judgment is a good one: "Ideology . . . is abstract, as it must be, and therefore likely to be recalcitrant whenever an attempt is made to incorporate it into the novel's stream of sensuous impression. The conflict is inescapable: the novel tries to confront experience in its immediacy and closeness, while ideology is by its nature general and inclusive."[127] Sinclair developed an aesthetic which theoretically combines art and ideology, but *The Coal War* is not an example of their successful fusion. Ideology either grows organically out of the action of a novel, is fully absorbed in the flow of experience, or it interrupts the fictional process itself.

The lack of inner complexity of Sinclair's characters has yet another explanation, however. It is not merely that Sinclair is a novelist of ideas, but *The Coal War* is a nonfiction novel. In no significant way could he responsibly diverge from history. *The Coal War* takes the action of the Colorado strike as its plot, and nearly all of its characters are as closely modeled upon actual participants as Sinclair's information permitted. Only Mary Burke, Hal, Jessie, and to differing extents their families do not have direct counterparts in the strike. Sinclair had traveled to Colorado three times during and just after the strike, obtained information from interviews, newspaper and dictated accounts, and when the federal government printed the results of two congressional investigations—over 6000 closely packed pages—Sinclair examined the hearing records in detail. He bought George West's report written for the United States Commission on Industrial Relations and read every article on the strike he could find. As Sinclair writes in his Postscript to *The Coal War*, "episode after episode has been made out of the sworn testimony of actual participants or witnesses. Every case of injustice, every practice named among the causes of the strike, has been sworn to by witness after witness. The speeches made at the convention which called the strike are from stenographic records. The statements as to political conditions were sworn to by actual participants in these practices. The statements as to violation of the state mining-laws, and the maintaining of peonage during the strike, have been sworn to by state officials." His reliance upon physical evidence was great and exacting.

As Sinclair is simultaneously novelist and historian, *The Coal War* is "a novel of contemporary life." Sinclair did not believe there should be a fundamental difference between art and the life it described. He judged the truth of a book more by its accuracy of observation than by its imaginative grasp of character or motivation, and he equated factual truth with aesthetic truth. For example, Sinclair had originally prepared a collection of documents demonstrating the historical accuracy of *King Coal*, but his plan to append them as a postscript to the novel was superceded by the Colorado Supreme Court decision on the Farr election. After quoting from this decision at length, Sinclair comments, "It is not often that the writer of a novel of contemporary life is so fortunate as to have the truth of his work passed upon and established by the highest judicial tribunal of the community!"[128] For the reader, however, the fact that incidents in a novel actually happened is not a sufficient register of truth. The remaining dimension must always be how those incidents are rendered, whether they have both the complexity and the aesthetic coherence to make them credible. Ironically, Sinclair had a keener political and historical awareness than our finest historical novelist, James Fennimore Cooper, but Cooper's fiction benefits from his lack of systematic analysis. He knew that for the novelist history succumbed to the richness of metaphor and symbol, not the finally delimiting accumulation of facts.

Unlike Cooper, Sinclair resisted the temptation to transmute history imaginatively. Writing out of a realistic tradition antithetical to the romance, he wished instead to reconstruct the strike through close, accurate detail. His writing would be functional and reportorial. Sinclair knew how *The Coal War* would conclude: historically it had already happened. Not even Hal Warner, the novel's central character, can in any way alter the events of the strike. Similarly, when Floyd Dell, who read *King Coal* in manuscript, suggested that the characterization of a labor leader as a self-seeking scoundrel would add dramatic impact, Sinclair refused because such an addition would distort history. He was bound by his decision that imaginative effect must be sacrificed to historical realism.

This emphasis upon historicity fundamentally affects the texture of the novel. Most importantly, it prohibits the free play of imagination, a force which remains subordinated to fact. Accepting the role of historian, Sinclair did not so much create his characters as replicate them from life or his research. With the great majority of them either alive or known to others, he was not free to enter their interior lives as intimately as a novelist must. Apprehended from the outside, developed from the general to the particular, and tightly linked to the historical event, their actions necessarily take on a predetermined quality which often drains vitality from the novel. Their autonomy subsumed by historical perspective, they too rarely have the potentiality of controlling their lives or events. Further, the protagonist of *The Coal War* is capitalism. As a consequence, Sinclair was not primarily concerned with the inner world of his characters. While those dimensions had meaning, they were excluded as largely secondary to his reconstruction of the Colorado strike.

Sinclair's characters are not free agents in the fictional sense. While character and action serve to further meaning in any novel, in *The Coal War* they do not further personal so much as historical and ideological meaning. Superimposed purpose tends to reduce action to illustration and character to social abstraction. In his Postscript to *The Coal War* however, Sinclair makes it clear that this result is in great part deliberate: "all the characters having social significance are real persons, and every detail of the events of the strike is both true and typical." Instead of an emphasis upon individualized qualities, Sinclair purposely concentrated on the generalized and the typical. A reader would not be able to discount the book on the basis of unique characters involved in unique incidents and so unrelated to his or her life.

An understanding of Sinclair's aesthetic can explain why literary faults occur, but it cannot altogether justify them. *The Coal War* is after all a novel, an art form in which we expect to encounter human life in all its complexity. The force of love aside, Sinclair developed his characters as solely determined by social and economic conditions. Close description of the concrete social milieu, class analysis, takes the place of interior definition. But while Marxian theory is satisfactory as an explanation of the economic forces and relationships evident in the Colorado strike, it is not (nor was it represented by Marx to be) adequate to explain the intricate dynamics of human psychology. As Freudian excursions into history can explore only one dimension of action, Marxism is analogously limited when it is used to confront the dense tissue of personal motivation. Sinclair was so much concerned that his characters not be seen apart from economic forces that he chose not merely to emphasize a Marxian interpretation of *The Coal War*, but, by not providing any other, to argue by implication that no other was necessary for a full understanding of behavior. His belief that any loss in the internal dimensions of his characters would be more than compensated for by their ideological significance is fictionally viable only if his characters are made credible as a first priority, otherwise the potential impact of a novel can neither be achieved nor the importance of ideology be fully recognized.

Inadequate characterization, then, is an important flaw in *The Coal War*. The strain of combining fiction, history, and polemics was too great: few writers would have risked it. But the novel is hardly without value. It provides a graphic rendering of an all too representative battle in the war between labor and capital during a crucially formative period of American history. Sinclair's value was never as a novelist of character but rather as a writer who chronicled a part of our national life which we have gone to great trouble to ignore, if not expunge. His reconstruction of the strike is compelling and his concentration on the power of capitalism, of class and ideology, be their effects direct or indirect, is nothing short of necessary if understanding our past is to become a meaningful phrase. *The Coal War* makes visible, as few American novels do, the human costs unheard and lost in the roar of fortunes being made.

The Coal War completes the story, begun in *King Coal*, of the incomplete conversion of Hal Warner, a wealthy young man who entered the coal fields as much on an adolescent lark as to learn directly about mining. Sinclair knew as a novelist that he had to develop a unity between individual and collective history. Hal's introduction to economic reality and the personal dynamics of his search provide the additional dimension of fiction. He finds himself in love with two women, Jessie Arthur and Mary Burke, the one from his own class and the other a miner's daughter. Each offers and dramatizes an opposing way of life. Mary is the New Woman or, in Joe Hill's words, the Rebel Girl of the early 1900s. She comes from straitened, brutalizing circumstances, but her intelligence and increasing political awareness make her a woman whom any radical would be happy to have as a partner. Jessie's class background, on the other hand, has unfitted her for any other role than an ornament. She is the classic prototype of the inflexible "rich girl" who under no circumstances, apparently, will ever understand social reality.

Hal's journey, as it is in *King Coal*, is one of education. His responsibility is to make sense of the strike, sort out clear perception from class prejudice, and to try to translate that perception into a basis for choice and action. His recognition early in the novel that he must somehow find a way to bring the owners and workers together, that both are impoverished by the boundaries cutting them off from the other, is likewise a model for the conflict within himself. As he tries to reconcile the two parties long after the possibility of reconciliation is gone, so also does he refuse to make a full personal commitment. Somewhere he knows that the reward—his own liberation—is proportionate to risk, but, with no guarantee of success and no way to control the consequences, he ultimately retreats into compromise.

Likeable, well-meaning, and initially without a coherent political analysis, Hal is a trustworthy interpreter of the strike. Sinclair develops him carefully so as not to alienate the reader. Hal believes in the efficacy of moral force and is repelled by violence. When he is forced to pick up a gun immediately after the Horton Massacre, his internal struggle against precisely such an eventuality makes his decision sympathetically perceived and the strikers' position understood. However, his dilemma is not proportionate in its personal significance to the strike itself. Hal is too callow, too desensitized by his background to understand his situation fully or to develop a consistent pattern of action. Sinclair shows that Hal is not only in a position where he can choose, but that he must choose. Unlike the strikers, his birth has given him options. Aware of what he has to lose, he must consciously step over the line beyond the point of return. Instead, he reaches the danger point and then both scrambles and is hauled back to safety.

Sinclair's partial success in characterizing his major figures is reflected in the hope that Mary and Hal will not marry at the conclusion of the novel. Mary is drawn in direct contrast to Jessie throughout *King Coal* and *The Coal War*. While Jessie has been pampered all her life, Mary has had to struggle: while Jessie is primarily concerned with caste and appearance, Mary is warm, open, and passionate. A battle is waged between them, with Hal as the dubious prize. Jessie has more power and she wins, as the operators win, but Mary is better off without Hal and his divided allegiances. All three major characters are given equal opportunities to learn from the strike. While Jessie rejects hers out of hand, Hal attempts to join the working class and is accepted on his own terms, even at times treated as someone from another world. In an effective parallel, Mary is given as much exposure to the class structure as Hal. For her, however, there is only one way she can enter the upper class–as a servant. Her journey, like Hal's, is one of education; unlike Hal she succeeds. Both return to their classes, but Mary has an unconflicted depth and solidity which Hal never gains.

Sinclair's description of the daily life at Horton is ably done. His eye for detail serves him well although early in Book III, for example, the density of background information drags the story to a halt. For many readers *The Coal War* was presumed to be their first knowledge of the hard work and unromantic routine behind an effective strike. Sinclair shows the mechanical problems of paying strike benefits, organizing men of some twenty nationalities, and the efforts of the leaders to prevent bloody revenge on the mine guards. The strikers are frequently referred to as colonists. Their self-government, and particularly their struggle, make inevitable their connection with earlier American colonists who went to war to gain their liberty.

Sinclair presents the moral problems which any political movement must face, the polarities of pacifism and violence and the multiple positions between them. The pacifists, Louie the Greek and John Edstrom, are casually murdered (as Louis Tikas and James Fyler were in the actual strike) and those who see no alternative to violence are mercilessly crushed. The coal companies and the military were simply too powerful. Although Sinclair did not have an answer, he did for a time consider "a syndicalist solution" both to *The Coal War* and to the pitched confrontation between labor and capital in America. In one of the earlier drafts of *King Coal*, before it was separated in two, Hal's father asks what the strikers had hoped to accomplish:

"When I started out a week ago," Hal answered, "I didn't know what I wanted. But it became quite clear to me as I went on—I want to take the mines, I want to take them and hold them against all comers."

He paused, but his father had no words.

"I'll try to explain it to you, dad. When I started out on this adventure, a year and a half ago, I was a nice, respectable Socialist. I believed in the law and the constitution, and all the old religions. I thought we were going to build a political party, carry elections, put our candidates in office, and then proceed to tax incomes, and by means of bond issues and such respectable devices, gradually take over the industries of the country. But after I'd lived in a closed coal-camp, I saw that wouldn't do. If you didn't vote the way you were told, down the canyon you went. Why, you couldn't even be naturalized unless you stood in with the machine. When certain camps didn't vote right, they re-districted the neighborhood, making it eighteen miles long and a half a mile wide, so that you had to travel over several mountains to register and vote. I saw a hundred things like that—I could spend all night telling them to you. The point is that politics are a farce—it's like telling a man to break out of jail with a toothpick. Can't you see that, dad?"

"Yes, I suppose I can see that," admitted the other.

"Well, then I thought I would become a Syndicalist. I would advocate industrial solidarity, I would tell the workers to use their economic power; to form themselves into a big union, strike all together, paralyze the industry, and get justice that way. An[d] we were on the way to do that—when they brought in the gun-men and militia. And then you see what happened; they drove men and women to desperation—so suddenly I discovered the way out. It isn't any invention of mine, it isn't any theory that I got out of books; it's something I saw happening before my eyes. We went in and took the mines—we just took them, that was all. The mines for the miners; the railways for the railway-men; our sugar refineries for the sugar-slaves! You see, dad, how simple it is!"

"My poor boy," said the other, shaking his head. "Why the fools hadn't any idea but to burn the mines or to blow them up with dynamite!"

Hal smiled, a bitter smile. "That same remark was made to me by a camp-marshal; poor Jeff Cotton, the ex-train-robber! It's a matter that any military man could explain, dad. When an army burns up buildings, it's to keep them from falling into the hands of the enemy. They never burn them if they are sure they can hold them."

"And you imagine these foreigners have intelligence enough to run a coal-mine?"

"Jeff Cotton asked me that," said Hal. "I make you the same answer—they haven't enough sense to want them yet. They're like a giant, lying in slumber. He's only beginning to stir; but some day he'll open his eyes and sit up; and then he'll know that he wants the mines, and he'll take them. He'll offer jobs in them to all who wish to do honest labor—even those who used to be bankers and stockholders."

By 1917, however, both Hal and Sinclair saw little hope for the American syndicalism of the Industrial Workers of the World. For all the accuracy of its basic analysis and its real social effects, the IWW never adequately treated the problem of how state power was to be assumed, nor, as a serious force, was it able to survive the political repression during and after World War I. Sinclair's comments at the conclusion of *The Coal War* are prophetic: "The Syndicalist might wish ever so hard to ignore the state—but the state would not let itself be ignored. It would come in and smash your labor organization . . . before you had a chance to become strong!" Hal is left with a task as difficult as any he faced in the strike—the fusion of syndicalism and socialism into a practical ideology in order that he can face the future effectively.

In the end Hal's class assures him of lawful liberty, regardless of his crimes, but there can be no personal freedom for him. He is encumbered by his past and by Jessie Arthur. As vapid as she is pretty, it is impossible to remember anything she says, but her face, golden-brown hair, and appearance are easily recalled. "The smile upon her lips, the light in her eyes, the very odor of the perfume she used, the touch of her soft garments—all these were intoxicating to [Hal's] senses, and threw his mind into confusion." This passage characterizes Jessie well, for it is precisely these externals which both attract and finally envelop Hal Warner. When Jessie deliberately compromises herself by publicly coming to his rooms, the values which he has so vainly tried to deny inexorably force him to marry her. He is recaptured more by his class than by Jessie. Taken only so far as his background will realistically permit, Hal is more ambivalent reformer than committed radical. As the miners cannot win their strike, Hal cannot win his freedom. Speak for their cause as he will, he is trapped in his own contradictions.

The Coal War is characteristic of much of Sinclair's fiction in both purpose and execution. Not surprisingly his comment that his World's End series of novels can be read as history, politics, and art is apt for *The Coal War* as well.[129] Sinclair may have set himself a task so difficult as to preclude harmonious resolution, but if his fiction is found deficient by standard artistic criteria, the exclusive application of those criteria is in itself deficient. His literary faults constitute his strength as a social novelist. Alfred Kazin described Sinclair as "one of the great social historians of the modern era,"[130] and Edmund Wilson wrote: "Practically alone among the American writers of his generation, he put to the American public the fundamental questions raised by capitalism in such a way that they could not escape them."[131] Sinclair set an opposing truth before readers besieged daily with misinformation and outright lies. He wrote some eighty books and, when publishers refused either to publish them or to keep them in print, Sinclair published them himself at prices that all could afford. He is one of the very few American novelists who had not only an impact on the ideas of his time, but on the social and economic conditions as well. Sinclair fought a hard, honest fight for over sixty years. These are not inconsiderable achievements.

BOOK ONE

THE SOCIAL CHASM

[1]

It was the last afternoon of the year, and in the sunlight the distant peaks of the mountains shone dazzling white. The houses of Western City made a frame for this snow-picture, and the young man who was walking down the street kept his eyes upon it so continually that he was hardly aware of the brown slush under his feet, nor of the unlovely neighborhood about him. This was characteristic of the young man, whose preoccupation with distant loveliness sometimes got him into immediate difficulties. He was twenty-two years of age, erect and keen-looking, with wavy brown hair and sensitive features, generally serious, but capable of lighting up with sudden humor. He was well-dressed, but in an inconspicuous way, as if it had happened by accident.

He came to the number he was seeking, and rang the doorbell of a cheap lodging-house. Of the woman who opened the door he inquired, "Does Mrs. Minetti live here?"

The reply was, Third floor, straight ahead at the back. Evidently the etiquette of the place did not provide for visiting-cards, so the young man climbed the stairs and knocked. Quick steps came, and the door was opened by a small boy, who gave one glance, and then a shout: "Joe Smith! Joe!" He made a leap, and the young man caught him and tossed him up so high that he almost bumped his head on the ceiling.

"Hello, Little Jerry! How's the boy?"

The boy's mother, a black-eyed Sicilian girl, had started from her chair. It was plain that she too was glad to see the distinguished-looking caller, but her shy welcome was eclipsed by the eagerness of the child. "Say! How you know we was here?" And then, "You heard about my father?"

"The union wrote me," said the visitor.

"Say, ain't it rotten?" cried the youngster. "Say, I wisht I was a man! Wouldn't I go for them mine-guards!" Little Jerry added exclamations of a kind which would look disturbing in print.

"But my father go back!" he declared. "He get into them camps again!"

Hal asked for particulars of the elder Minetti's fate, and Rosa sat with her hands clenched in her lap and a look of distress on her face, while he spelled out her husband's Italian letter. Big Jerry had been doing organizing work in one of the "closed camps" of the coal-country, and the company guards had caught him, beaten him unconscious, and then, to get rid of him, had thrown him on top of an outgoing coal-car. It had been pretty bad, said the letter, but now it was all right, for he had been found by a section-man who was a union sympathizer, and while this man's home was only a box-car on a siding, there were flowers in the windows, and a woman to take care of a broken head.

Little Jerry took up the story in his eager, high voice. He and his mother and the baby had been living in the coal-town of Pedro, and at night three strange men had broken into their lodgings, and tumbled them out of bed. They had torn everything to pieces, searching for letters. From their remarks it was plain that Big Jerry had been caught at his perilous work; but that was all the family had known for more than a week. The union people had advised them to move up to Western City, where they would be out of danger. So now they were all right, said Rosa, except that it was so lonely.

Hal looked about him at the cheerless lodging-house room, a hall-room with only one chair in it, upon which he himself sat. Rosa sat on the bed, which was hardly more than a couch, so that he wondered how she kept herself and Little Jerry and the baby from rolling off at night. There was a small chest of drawers, and a wooden box with a gas stove on it, and a sauce-pan boiling. One did not have to lift the cover to know what this sauce-pan contained.

During his three months sojourn in the coal-camps, the young man had learned to put up with odors, so now he wrinkled his nose, and grinned at Little Jerry and said, "Um! Um!"

"Um!" said Little Jerry, and grinned back.

Rosa added, "You stay supper with us?"

"Sure, Joe!" cried the child.

The mother put in quickly, "You say Mister Warner!"

And the young man laughed. "Let him call me Joe. And you call me that, too." Then, seeing Rosa look embarrassed, "I'm mighty proud of having been a miner. I still have my union-card, you know." He was looking at the girl-wife as he spoke, and noted that she had lost some of her pretty color. It occurred to him that boiled cabbage is not a sustaining diet, especially when flavored with terror.

It was the Christmas season, and the young man had witnessed many festivities. But here, it seemed, was a family which had been overlooked by Santa Claus. He made inquiry and learned that Little Jerry had never made the acquaintance of the benevolent old gentleman of the rein-deer; perhaps North Valley had been too high up in the mountains for these creatures to climb. Had there never been a Christmas tree at the North Valley church? Yes, but Little Jerry did not go to it. His father had no use for churches.

So Hal recalled that Minetti was a Socialist, and of the Italian variety; he spoke of priests as "black beetles". Since he had come to America, and earned his living as a shot-firer in the feudal fortresses of the General Fuel Company, he had seen nothing to disabuse his mind of this hatred of religion. The General Fuel Company had taken fifty cents a month from his wages for the maintenance of the Reverend Spragg, and Jerry had paid this, as he paid all other charges, under protest; he had kept his family away, considering it an effort to steal their minds by the agency of General Fuel Company theology, baited with Christmas trees and Sunday school prizes.

But now Big Jerry was far away, and had a broken head, and could not interfere; and the idea possessed Hal Warner—what a shame this Christmas-tide should pass entirely over the head of a Dago mine-urchin! He recalled the parties he had seen, the preparations for parties, the remains of parties. So many bright and shining faces, so many bright and happy homes, full of gifts and laughter and song; and all of it passing undreamed of over the head of a Dago mine-urchin!

Ever since Hal had gone to North Valley, and made a practical test of the life of a coal-miner, his thoughts were continually being lured into experiments in social amalgamation. What a cruel thing was this chasm between the classes! Cruel to both classes—not merely to those who had too little, but also to those who had too much! To the smooth, comfortable, kindhearted, generous, blind people, who went to church on Christmas morning, and sang carols, and went home and ate turkey and plum-pudding, really believing that God was in his heaven and all was right with the world! What an education for these people, if one could bring them to this lodging-house room, full of the odor of boiling cabbage, and let them hear the story of this child-wife from Sicily, with her two babies, and another soon to be born, and a husband lying with a broken head in a far-distant box-car!

Hal suddenly thought of one party that had still to be: a New Year's day party at the home of Robert Arthur, the banker, a party for the old gentleman's eleven grandchildren, and at least twice as many of their friends. An inspiration flashed over him. The Dago mine-urchin should go to that party!

It was such a thrilling idea that he could not wait—not even till he had secured an invitation. "Little Jerry," he said, "do you remember Miss Arthur, the pretty young lady who came to your house at North Valley?"

"Sure I remember!" said Little Jerry. "Your girl!"

"So you said," laughed Hal. "Well, you know, her father is Santa Claus."

"Go on!" said the mine-urchin, who had learned the American way of speech.

"Honest!" said the other. "He has a big house near here, and he's going to give a party tomorrow afternoon, a New Year's party. I'm going to get him to invite you."

"Aw!" said Little Jerry. "He wouldn't let no Dagos come!" Nevertheless, the black Dago eyes began to shine; and when Hal insisted that his prestige as "best feller" of the pretty Miss Jessie would enable him to get an invitation, a fountain of Dago questions was set flowing. What was a New Year's party like? What did they do at it? Did they have grub? Plenty? Ice-cream? *Santa Maria!* All one could eat! *Santissima!* And a Christmas tree? And Santa Claus? How many children would be there? Girls too? Were they pretty, like Hal's girl? Dressed up fine, like her?

So came an important matter; Little Jerry must be dressed for this party—dressed as never a Dago mine-urchin had been dressed in history before. In the first place, he had to be scrubbed. Perhaps Rosa could borrow a wash-tub from the landlady; also soap, and a towel, and plenty of hot water. Little Jerry listened in dismay. Yes, *hot* water! Not too hot, not hot enough to imperil the skin, but hot enough to take off the dirt. There must be no mistake about it, every particle of dirt must come off; the whole body, even the feet; the backs of the hands, as well as the palms. The coal-dust must be mined from under the Dago finger-nails, and from out the ears, and from behind them. The Dago hair must be scrubbed—yes, even the hair; it would shine in lovely wavy black curls—but no oil, or pomade, or anything to make it smell good. Just soap and hot water! And then Hal would send a lot of new clothes, in which Little Jerry would be arrayed. What size suit did he wear? What was his chest-measure? What was the size of his foot? And of his hand? Yes, even his hand! He was going to have a pair of new kid gloves. "*Jesus!*" cried Little Jerry.

And he danced about, clamoring; his eyes shining, ready to pop out of his head. What time would the party be? And how long would it last? Would they eat all the time, or what? They played games? What sort of games? One might have to kiss the little girls? *Santa Maria!* No doubt that explained why you had to be so careful to be washed; but for God's sake, why did you have to wash your feet?

[2]

Hal took his departure, and hastened to "Perham's Emporium", where he purchased a complete party outfit for a boy of seven, to be delivered at the Minetti lodgings that evening. And having thus committed himself, Hal called up the pretty Miss Jessie on the telephone and told her about it. He did not say he was preparing an experiment in social amalgamation; he merely reminded her of the cute little Dago mine-urchin, to whose home he had taken her when she had visited North Valley. Now the Minetti family was in Western City, in miserable lodgings, with the father away, ill; and they had had no Christmas at all, and it was a shame; could not Hal bring Little Jerry to the party—especially as he had already invited him, and bought his clothes, and arranged to have him made clean?

Jessie was in a state of dismay. It was difficult to imagine what Little Jerry would look like made clean; it was difficult to imagine him among party-children. "Hal," she said, "you know Papa's so fussy. And he's terribly cross with you! He talks about it all the time, the way you've behaved, getting mixed up with strikers, and all that!"

"I know, dear," said Hal. "But it's going to be all right now, for I'm coming to the party, and I'll bring Little Jerry with me, and he's a winner, and your father will be taken captive completely. When he hears what happened to Big Jerry, he'll want to call a strike himself."

But Jessie was not to be drawn into jesting. "Hal! The child's language!"

"I'm going to see to that, dear. You leave it to me! And by the way, there's somebody else I want to interest in my Dago family—that's Uncle Will. I wonder if you'd ask him to the party?"

"Why, of course, if he'd come."

"Well, ask him. Tell him to drop in for a few minutes, anyhow. You see, the Minettis are in trouble, and that's the sort of thing Uncle Will lives on."

So this remarkable experiment in social amalgamation came to a climax. Promptly at two o'clock next day the big maroon-colored touring-car of Hal's father drew up in front of the Minetti home, and Hal climbed the stairs of the lodging-house again, and found Little Jerry waiting at the top, in excitement so intense that it was almost painful to witness. He had wanted to wait at the front door, but Rosa had held him back, for fear of some accident to his wonderful clothes. For two hours the little chap had been arrayed in all his glory, unable to sit still for anxiety. Was it sure Joe Smith was coming? What would they do if he failed to come?

But here at last he was! And he drew the trembling Little Jerry into the room and inspected him; he inspected hands, wrists, ears, eyes, hair. Yes, they were all right! And the clothes were all right, the new grey suit without a crease in it, the snowy collar and silk tie, the black stockings and solid shiny shoes, the thick warm overcoat with cap to match, the shiny brown kid gloves, each with a round gold button! Yes, he was a regular little swell! He would pass at the party for the crown-prince of Italy!

Rosa stood beaming with pride at Hal's praise; sure, he was a fine kid! She came down to the front door with the baby in her arms, and stared in wonder at the monster automobile; the neighbors were staring, also—the windows of the other lodging-houses filled with faces. It was not often that an equipage of that splendor condescended to stop in their street.

Little Jerry got in. He said a weak fare-well to his mother, and then, while the car rolled away, he sat in silence, awe-stricken. He was overwhelmed by that man sitting up so solemn in front, wearing a big coat of fur. Was that the owner of this car? Or was it the gentleman who was giving the party?

[3]

Hal began telling about the place they were going to, and how Little Jerry must behave. He would be polite to Miss Arthur, of course, and to all the ladies at the party; and he must be especially nice to two gentlemen he would meet—one of them the old gentleman, Mr. Arthur, who was giving the party, and the other a middle-aged gentleman, Mr. Wilmerding, who was a particular friend of Hal's. Little Jerry, who was used to strange names among the polyglot hordes at North Valley, pronounced the name very carefully—"Mis-ter Wil-mer-ding". This gentleman was a priest, Hal went on to explain, but of a sort they did not have either in Italy or in the coal-camps. He was called an Episcopalian; Little Jerry said this name very carefully—"An-ny-pis-co-pa-ling".

He was a good man, who had been Hal's friend since Hal was a boy like Little Jerry. Hal called him "Uncle Will", though he was not really Hal's uncle. Now he was distressed because Hal went away and lived with working-people, and got mixed up in strikes like other working people. It was not that Mr. Wilmerding didn't care about such people—it was that he didn't know about them. He had never met a miner in his life, and he used coal to keep his house warm without thinking of the sufferings of those who toiled to dig it out of the ground. Some day Hal intended to take Mr. Wilmerding to meet Big Jerry, and hear what had happened to him—not merely since he had become an organizer for the union, but since he had come to this country, which Mr. Wilmerding thought was a good country, a land of freedom. Little Jerry must watch out, and if he got a chance, must tell this gentleman something about the lives of coal miners.

It was almost the same with Mr. Arthur, Hal went on. Mr. Arthur was a very rich man, and a lot of people thought he must be a selfish man, but it wasn't really so—it was just that he didn't know. He was cross with Hal because he wanted Hal to stay at home like other rich men's sons, and not go off stirring up the working-people and making trouble for the owners of mines. Mr. Arthur was a friend of Mr. Peter Harrigan, who owned the General Fuel Company, and was making fortunes out of the misery of the people in the coal-camps; but Mr. Arthur had no real idea about the way Mr. Harrigan ran his mines—he actually thought that miners were men who wanted to loaf and get drunk, and had to be driven and made to obey their masters. Little Jerry would be helping the union if he would tell this old gentleman a little of the truth. "You know what I mean," said Hal.

"Sure, I know!" replied Little Jerry. And a calm, firm resolution took possession of his bosom. In early life he had meant to be a shot-firer, that had seemed to him the highest destiny of man; but recently a new vista had opened, to be a union organizer, a teacher of working-class solidarity—and here was the first step to that thrilling career!

There was one thing more, Hal said; Little Jerry must be extra careful not to swear. The people he was to meet were different from mining-camp people in this respect, they had peculiar notions about the most every-day cuss-words. Had Little Jerry ever heard the story of the beautiful young lady who was bewitched, so that every time she opened her mouth there hopped out a toad or a snake? Well, that was the way these people at the party would feel about the simplest "damn". Little Jerry must be very, very careful.

"You won't forget?" said Hal.

"Hell, no!" answered Little Jerry.

So Hal had to explain in detail just what he meant by "cuss-words". Perhaps the safest way would be for Little Jerry to say his exclamations in Italian; then nobody at the party would understand. Of course, he must shake hands politely with everyone he was introduced to; and when he was given things to eat, he must eat carefully, and not spill things on his fine new suit, nor on his host's carpets or chairs.

"Sure, I know that!" said Little Jerry, reassuringly.

"And you'll remember the old gentleman's name—Mr. Arthur."

"Sure, I got it. Mr. Otter."

And then, Hal's instructions being completed, the flood-gates of Dago questions rolled open! Was it a very big house? As big as the superintendent's at North Valley? And the pretty Miss Otter lived there? When she and Hal got married, would *he* live there? And this ottermobile? Did it belong to Hal? How much did it cost? Would it go very fast? Could Hal ride in it all he wanted to? Holy Smoke! (That wasn't swearing, was it?)

[4]

The car had passed the limits of the city, and following a great boulevard along the slope of the mountains, came at last to the Arthur home. A blanket of snow lay over the grounds, but one could see that they were vast and amazing. There were whole rows of ottermobiles drawn up along the drive. Did everybody come to this party in an ottermobile? Why did they have so many different kinds?

The car stopped in front of the door; and there was a man in short pants to open the door for them. Little Jerry had never seen a grown man in short pants before; he would have thought it was a play—only he had never been to a play!

He got out, holding tightly to Hal's hand, as they approached a bronze-barred doorway which was like a jail. Little Jerry knew about jails; they had one at North Valley, and Little Jerry had seen his friend, "Joe Smith", looking out through the bars of it.

But these impressions came so quickly, that the Dago mine-urchin had scarcely time to realize them. The bronze doors swung back—and there was a vast, mysterious apartment, with something that smote Little Jerry's eyes, so that he stopped and stared, paralyzed. The Christmas tree!

The entrance hall of the Arthur home went up all the way to the roof; and in this tall place the Christmas tree towered, enormous, awe-inspiring as cathedral arches. It gleamed with a thousand tiny electric lights—green, red, blue, yellow, purple. It shone with something like snow, it sparkled with something like jewels. A score of kinds of fruits grew red and yellow on it, fairies and goblins and angels danced about in its branches, mysteries beyond counting peered from its coverts. And beneath it stood a strange and awful figure, a man in a long gown of white furry stuff trimmed with red, with a white furry hat trimmed with red, and a white beard all the way to his waist. There was only one thing Little Jerry could think of—Hal had brought him to heaven, and this mysterious personage was God!

The Dago mine-urchin had been able to take care of himself in all emergencies which had hitherto arisen in his life. Standing upon the cinder-heaps of the village of North Valley, his voice had been as loud, and his fists as hard, and his cuss-words as prompt and powerful as any boy's. But here was something new and appalling, depriving him utterly of his *savoir faire*. He forgot his wonderful clothes, he forgot his role as heir apparent to the throne of Italy, he even forgot his duties as organizer of the United Mine Workers. He stood, clutching Hal's hand, his mouth open.

When Hal started him across the room, he walked like a mechanical toy. Everything about this place was bewildering: the carpets under his feet, which were like soft grass; the chairs, which were like feather-beds when you sat in them; the dim lights, the softly shining furniture, the pictures, which were not like other pictures, but were realities magically brought and transferred to the walls. In one corner stood a giant of a man, all steel, with a steel head and a spike on top, and a steel hand with a long and deadly spear. Yet Hal seemed to go near this creature without heeding him. If the old man under the Christmas tree was God, Little Jerry could only think that the one with the spear must be the devil!

There came the pretty young lady to welcome them—the one who had been in North Valley. Her eyes were soft brown, and her hair was the color of molasses-taffy when you've pulled it—but all fluffy and wonderful, with star-dust in it. She was clad in something soft and filmy, snow-white, with ribbons of olive-green, and a dream-like scarf of olive-green about her shoulders. She shook Little Jerry's hand, but his lips were stiff, and his tongue was a lump in his mouth—he could not speak to such a vision of loveliness!

There came children to be introduced; the room was full of children, Little Jerry began to realize, and all having been scrubbed like him. Yes, Joe Smith was right, they had sweet smells, they looked like fairy-tale children! The little girls had bright red cheeks, and hair all curls, and dresses of white and pink and blue and yellow—like they had just stepped out of store-windows. When one was introduced to them, they bobbed down on one knee in a funny way; Little Jerry felt like a lost soul, he did not know anything about knee-bobbing. He recalled Joe's statement that he might have to kiss these little girls, and consternation possessed him.

Miss Otter was very nice to him. She introduced him to another Miss Otter, who was her elder sister, and to several more Miss Otters, who were her cousins, and to a nice, stout, smiling old Mrs. Otter, her mother. All these people had heard that a little boy from a mining-camp was coming to the party, and they did everything they could to make him at home; they asked him questions, and took him about and showed him the Christmas tree at close range, and explained to him that the old gentleman with the white beard was not God, but that other important personage who brought presents to little boys at Christmas time. When the ladies learned that he had not brought anything to Little Jerry, they were surprised, and said that they would call his attention to the oversight. They gave Little Jerry some fine candy, which made him feel more at ease, and they introduced him to other little boys, who said Hello, just like boys who did not live in heaven. So gradually the Dago mine-urchin found himself, and recalled his duties as an organizer!

[5]

Yes, for the time was at hand for the organizer to get on his job. Joe Smith came up, and pointed out a round-faced old gentleman with flat white side-whiskers, sitting in a big black leather chair; that was Mr. Otter, and Little Jerry must be introduced to him. They walked over; and the old gentleman smiled, and took Little Jerry's hard, rough hand in his soft, pink hand, and said, "So you live in a mining-camp, little man!"

Which, obviously, made it easy for the organizer to get down to business. "Naw," said Little Jerry. "I used ter, but they throwed us out."

"Is that so?" said Mr. Otter.

"Sure thing!" said Little Jerry. "They don't let no union men stay in them camps, you bet!"

"Are you a union man?"

"My father is, you bet! And when there was a strike, they found it out, an' there was nothin' doin' no more. They got a black-list up there, an' if you don't watch out, you get on it an' you don't get no job in none o' them mines."

"Indeed!" said Mr. Otter—what else could he say?

"After that my father was a Norganizer for the union, an' he was in East Creek, an' they found him out, an' they busted his head an' throwed him onto a coal-car. An' that's the way they do you in them camps. If they find you're a Norganizer, maybe they kill you an' stick you under the dirt right where you lie."

"Is that possible, little boy?"

"Sure thing, it is—you ask Joe Smith here. He was in North Valley, an' he seen it. There was some men wanted him to be check-weighman—'cause you see, the men don't never get their weights. You dig and dig, an' load up a car, an' then the weigh-boss gives you anything he pleases. Maybe some feller steals your cars, but you can't do nothin', 'cause he stands in with the boss, he's one of the fellers that gives him drinks an' keeps on the good side of him, an' maybe tells him lies about the other fellers, an' so he gets a good place, an' he can get out lots o' coal. Maybe the other feller gets a place that's no good a tall, an' he works like the—that is, he works awful hard, an' it's all dead work, gettin' out rock and stuff, an' they don't pay for that a tall. An' when it comes to the end o' the month, the feller gets his pay-check, an' it ain't enough to pay his bill at the store, 'cause prices is high—you have to pay twict what you pay down to Pedro, but you can't go down there to buy nothin', 'cause if the boss finds that out, it's down the canyon with you. They don't let you trade outside, 'cause they want to hog all the money for themselves."

The round-faced old gentleman's eyes had become as round as his face. He had adjusted his spectacles, in order to stare more closely at this Dago mine-urchin; and when at last there came a pause in this torrent of information, he exclaimed, "Well, well, little man! You seem to know a lot about mining-camps!"

"Sure! I know about them. Why shouldn't I? I lived in 'em all my life."

"But your life hasn't been so long!"

"I'm seven years old, an' that's enough. I hear all what the men say. When they can't get out o' debt, then they're kickin' all the time. An' they have to pay their rent, an' that's the company's too, an' comes out of your pay! An' if your roof leaks, you can kick, but they tell you to go to—that is, they don't fix it for you. An' if you try to take in a gentleman to board to help pay, then they fire the gentleman, 'cause maybe he left the company boardin'-house, an' they was takin' his board out of his pay, an' no matter how bad the grub was, he hadda stay there. An' they have accidents all the time, they kill you in the mine, but you don't get no damage. They don't sprinkle the mine like they should—my father knows that, 'cause he was a shot-firer an' that's a mighty dangerous job. He was all the time sayin' how he hadda risk his life. But he dassn't say nothin' to the camp-marshal, 'cause if you don't like it, you can go—that is, you go down the canyon. I tell you, that feller Jeff Cotton, that's camp-marshal up to North Valley, he's a terrible feller; he beats the men all the time, an' when he's drunk, he'll shoot you right dead. He's a Norful bad man, Jeff Cotton."

Mr. Otter was continuing to gaze at the Dago mine-urchin, with a kind of dismayed fascination in his eyes. "Joe Smith", ex-check-weighman and miner's "buddy", decided that it was a good time for him to retire into the background. His presence might possibly make the old gentleman self-conscious; distrustful of these torrents of information, so generously poured out from little Dago lips!

But the ex-buddy hovered in the vicinity, keeping watch over the situation. He could see his victim going deeper and deeper into the snare. Having got over his first surprise, the old gentleman began to ask questions. Several times he was observed to laugh; and once he looked horribly shocked, and put his finger to his lips; Hal wondered what particular toad or snake had happened to jump from the Dago lips! But apparently no serious harm had been done; the Dago lips went on moving, and Joe Smith saw the triumph of his dream of social amalgamation. He would make a union propagandist out of the founder and head of the banking-house of Robert Arthur and Sons!

[6]

The time for ice-cream came; and this of course broke up the interview. The old gentleman, wheezing slightly, lifted himself from his big leather arm-chair, and patted Little Jerry on the head; Little Jerry, thus dismissed, looked about for his chief, to make his first report as a Norganizer. But before anything could be said, the second victim made his appearance. Will Wilmerding came in, beaming like Santa Claus, rosy-faced from a ride behind rein-deer. He shook hands with his host and hostess, and with the young ladies, and with all the little boys and girls within reach. He wore a black clerical suit, and a white clerical collar; the latter being hidden by a reddish brown beard, very bushy and stickery, as you discovered when he kissed you, which he frequently did if you were a child. He had blue eyes, a kind of knobby, rough complexion, and a benevolent laugh which had been trained to take in a whole roomful of children.

Presently he espied Hal, and came over to him. "Well, boy! I've been hearing wild tales about your doings. I want to hear about them. Maybe I want to scold you."

"I'm coming to see you," said Hal. "It's too long a story for a party."

The other assented to this. He could not stay very long, anyway; and Hal made note of this remark, and put his wits to work. Somehow, no matter how short his stay, the clergyman must be got into touch with the organizer!

The problem, as it happened, found its own solution. Mr. Wilmerding's profession made it impossible for him to stay even for a few minutes without trying to do something for his fellow-man; and just then ice-cream was being passed, and this is one of the recognized functions of the modern clergy. So here was "Uncle Will", with a heaping saucer of chocolate and vanilla in each hand, headed straight towards the Dago mine-urchin!

"Will you have some of this, little boy?" he inquired, with his best brotherhood-of-man smile.

Hal was only a few feet away, watching with his heart in his mouth. He saw Little Jerry staring with shining eyes. The crown prince of Italy was forgotten; the union organizer was forgotten; the plain boy was on top. "*Jesus*!" cried Little Jerry.

The two heaping saucers trembled perilously in Mr. Wilmerding's apostolic hands. But then the owner of the hands recovered his self-possession. In his capacity as advocate of high church ritual, he had a rule that no matter where he was, or under what circumstances, when that sacred name was pronounced he made an inclination of the head. So now he inclined his head over the two saucers of chocolate and vanilla.

After which, his obvious duty was to make inquiry. "Where do you come from, little boy?"

"From North Valley."

"North Valley? Where is that?"

"That's a coal-camp."

"*A coal-camp!*"

"G.F.C. camp," said Little Jerry. "My father was a shot-firer. Only they made him a Norganizer for the union. He went into East Creek, an' they caught him at it, the guards did, an' they beat him up an' smashed his head. They come to our house, an' smashed it up, an' they drove us out of

Pedro, an' now we're in Western City, an' my father's in a box-car till his head gets well. An' maybe you don't know how them fellers treat you if they hear you tryin' to talk about a union! They raise—that is, they treat you rough, maybe they beat the face off'n you."

"Dear me!" said Mr. Wilmerding. "But how did you come to be at this party?"

"Joe Smith brung me," said Little Jerry—and then, pointing, "Him!"

"Oh!" said the clergyman, and light dawned.

"Yes," said Hal, coming up. "And I told Little Jerry that you were a man who visited the sick and afflicted, and that you would come to see his mother, because his father can't work, and they've got a little baby, and not much money, and Santa Claus didn't come to see them."

So of course Will Wilmerding had to say he would come. He gave Little Jerry one of the heaping saucers, and put off his apostolic admonitions until a later occasion. After he had gone, Hal gave his fellow-conspirator a nudge of delight, and saw that he got a second saucer of ice-cream.

[7]

Then came the games. Little Jerry was kissed on the cheek by a white fairy and a pink one, an experience never to be forgotten; and after that, amid endless chatter and confusion, the children were bundled into their wraps, and taken for a tour of the estate.

Old Mr. Arthur was blessed with four grown sons, and so he had withdrawn from business, and devoted his time to pottering around this country-place. He had a passion for building things, and was forever getting ideas of new things to build. He subscribed to several "country-life" magazines, which made a specialty of inventing outdoor foolishness for the diversion of old gentlemen in his position; and he would read of these inventions and set to work to realize them, quite regardless of congruity. He had begun with an Elizabethan palace of dark red brick and marble trimmings; to this he had added Italian gardens and Greek pergolas, a Dutch tea-house and a Colonial ice-house. His place was a series of history-lessons, a regular trip around the world; it was almost a Noah's ark—there was a deer park, and peacocks and lyre-birds, and ducks from China, and pheasants from Thibet, and chickens from a score of places which had to be looked up in the atlas. There was a green-house with no end of fascinating things—bananas, oranges and lemons gleaming on tropical Christmas trees, vanilla and chocolate beans, papyrus-plants, custard-apples, sapodillos. And the stout, round-faced old gentleman with flat white side-whiskers would follow you about, showing these treasures; it was his favorite form of exercise, his substitute for a golf-game.

All this, of course, was glorious to a Dago mine-urchin, whose mind was untroubled by considerations of architectural congruity. With the rest of the children he tumbled in the snow, and threw snow-balls, very gently, and laughed and shouted, not too loud, with glee. He went to the white marble swimming-pool, which was now a sliding-pond; he saw a Dutch tea-house and a Japanese pagoda, which went perfectly together when covered with snow; he walked through the hot-houses, and was given a strange fruit, white and sweet and creamy inside; he inspected the Chinese ducks and the Thibetan pheasants and the Annamese rooster, all of whom were housed more luxuriously than any mine-worker he had ever known.

At last, all too soon, this miraculous party came to an end. Hal and Little Jerry said good-bye to all the children, and to Mr. Otter and Mrs. Otter and the young lady Otters; and so they drove off.

And all the way home Hal questioned the union organizer about the success of his propaganda. Just what had he said, and what had the old gentleman said? Had he told about the companies not obeying the law? How the men were not allowed to have unions, although the law gave them this right? Yes, said Little Jerry, he had told that; he had told how the men were robbed, how the bosses were in on the graft, how you had to pay for any sort of chance in the mine. He had told how the company doctor was drunk half the time, but you had to pay what he charged, it was taken off'n your account, 'cause he was a cousin of the super. He had told how the shacks wouldn't keep out the cold, but you could never get the company to make them tight, 'cause Jeff Cotton would tell you to shut your face. All that he had told–

"And what did Mr. Otter say?" asked Hal.

"He didn't say much. He jest asked about it."

"Didn't he say if he believed you, or anything like that?"

"No, he didn't say nothin' like that. He said it was too bad. He said it hadn't ought to be that way. Then he said–oh, yes, I remember one thing he said–that I hadn't ought to call Jeff Cotton a son-of-a-------!"

[8]

Before going back to college, there was one other experiment in social amalgamation which Hal wanted to get under way. To that end he went to call upon his friend Adelaide Wyatt.

Adelaide was one of those people you read about in the society columns as a "young matron of Western City". She was the wife of a famous polo-player, who spent his time between New York and Southern California, and never failed to stop off to see her when crossing the continent. That is to say, they did not get along together, but managed to keep friendly, and to avoid too much gossip. It was possible that Adelaide was a little in love with Hal, but if so, he had no idea of it, and looked upon her as a good pal to whom he could tell his troubles. She sympathized with his attitude upon social questions, making it clear that a woman who had luxury need not be walled up in caste-prejudice.

Hal had not seen Adelaide since his expedition to the coal-country. Now he spent part of a day telling her his adventures: a story difficult to make real to a lady in a blue silk morning-gown, reclining at ease upon a brocaded couch. These coal-camps were places of terror such as one read of in Russia; situated as they were in remote mountain recesses, everything in them belonged to the company—the stores, the saloons, the schools, the churches, the homes of the miners. They were "closed" camps—that is, no one could enter them without a pass from the company, not even a doctor or a priest. Sometimes they kept out the state mine-inspector and his deputies.

Hal told how he had got a job at North Valley, and of the mine explosion which he had witnessed. The company had sealed the pit-mouth, meaning to sacrifice the lives of the men to save its coal from catching fire. It had chanced that young Percy Harrigan had been at Pedro, with his private train, and a party of friends, and Hal had gone to him and forced him to go up and have the mine opened. Adelaide, of course, had heard about that; now Hal was interested to hear what she had to contribute to the story—what this person had said and that, the terrible rage of Old Peter Harrigan, his threats against the Warner family. He would doubtless take it out on Hal's brother Edward, the business-man of the family; he could not very well compel the faculty of Harrigan College to "flunk" the amateur sociologist!

Hal came to the matter which was on his mind. There was a young Irish girl, Mary Burke, who had lived in these camps all her life, with a younger brother and sister to take care of, and no one but a drunken father to help. When Hal had first met her, she had been desperate, without hope either for herself or for the miners; but when the disaster had come, she had flamed out in a way that had amazed him. She had shown courage and devotion, the stuff out of which a leader of her people might be made. Hal was trying to figure out some way for her to take care of herself and the young people. Could not Adelaide give her a place as a servant in her home? She was young and handsome, with a treasure of auburn hair; also she would surely be what the advertisements described as "willing and obliging". She could be made into a parlor-maid or a waitress, and would have a new start in life. She would save part of her wages; also she would improve her mind. It would be an adventure for a young matron of society to have a revolutionary parlor-maid!

Then Hal stopped; his friend was eyeing him closely. "Hal Warner," she said, "what's this you're letting me in for?"

"How do you mean?"

"You go and live in a mining-camp—a rich young scapegrace—and you come out with a handsome girl with a treasure of what you politely term auburn hair. And Jessie Arthur has seen this treasure, your brother Edward has seen it—so we can't keep the secret."

"Why should we, Adelaide?"

"Don't you know what people are going to say?"

"Of course—they have said it already. But I can't refuse to help the girl for that, can I?"

"No," was the reply. But if *I* am to help her, at least I have to know how matters stand. Is the girl in love with you?"

Hal hesitated. He had not intended to go into that aspect of the matter. "Why, I don't know."

"You'll have to deal with me frankly."

"Well—" And he laughed. "Possibly she is, a little. You understand, she's had no chance—"

"I can understand quite easily. You're a young god out of the skies to her, she's ready to follow you to the ends of the earth. And you, with the usual stupidity of a man, haven't realized what she means."

Hal was silent. That was not precisely the situation, but it would do for the present, he thought.

"Having been brought up under the clerical supervision of Will Wilmerding, you have treated her in saintly fashion, so you have a good conscience, and don't care what the world may say about me or my parlor-maids. Is that it?"

He laughed again. "Yes, that's it."

"You're so naive, Hal!"

Again he let it go at that. He was not so naive as his friend thought; there had been some encroachments upon his naiveté—one of them right here in Adelaide's own home. Three years ago, a freshman in college, he had sat at a dinner-party in her beautiful dining-room, with its painted panels of scenes from the "Lady of the Lake", next to another "young matron" of Western City society, a brunette beauty in one of the newest disappearing gowns, eked out with a necklace of rubies. It was Mrs. "Pattie" Perham, wife of the "Emporium", from which Little Jerry's outfit of princely raiment had come. The "Emporium" himself was old and fat, while Hal was young and engaging; so before the meal had passed the sherbet in the middle, Mrs. Pattie had slipped her little foot out of its little red silk receptacle, and was reposing it gently on top of his. A problem in etiquette for a young man making his first appearance in "society"!

Adelaide was going on with her cross-examination. "Are you the least bit in love with this girl, Hal?"

"You know I'm engaged to Jessie," he replied.

"Yes, but answer my question." And, as he did not answer with alacrity: "I suppose you weren't altogether pleased with Jessie in North Valley?"

It would have been a relief to talk that out with Adelaide, but Hal could not think it quite loyal to analyze the girl he loved to another woman—especially one who was so matter-of-fact. "You know," he argued, "Jessie's very young! She's had no experience of life at all."

"Yes, I know that."

"And she loves her mother and father. You can hardly blame her for believing what they tell her."

"I'm not blaming her," said Adelaide. "I'm asking if you blame her."

"Well, I don't," said Hal. "I'm hoping to teach her. The trouble is that my own experiences were so maddening—I got drawn in further than I intended. And of course that's made Jessie's people angry, and it's hard for her."

Adelaide sat in silence—passing before her mind a procession of the members of the extensive Arthur family. There was Jessie's oldest sister, who was married to Percy Harrigan's brother, a vice-president of the "G.F.C." There was Garret Arthur, Jessie's oldest brother, acting head of the banking-house; pale, prematurely bald, silent and methodical—Hal had referred to him as a "bond-worm". Yes, truly, Hal would have a hard time imparting his revolutionary fervor to the Arthur family!

[9]

Adelaide went on at last. "Let me ask you one thing; did Jessie notice your interest in Mary Burke?"

"No," he said, "I don't think so."

"Why don't you think so?"

"Well, she didn't say anything."

The other could not help laughing. "Let me tell you, Hal—she noticed it! She's going to continue to notice it!"

"But—she has no reason to!"

"I'm not so sure about that. But anyway, when she hears that Mary has come to this city, she's bound to be troubled. And when she hears the girl is in my home, she'll think I'm acting—well, in a way not friendly to her. Her mother will think it, her sisters and her friends will have their eyes on this place. You may be as naive as you please where you're acting alone, but where I am concerned, Hal, you have to have your eyes open. You must understand what people would say about me if I took this rose of a mining-camp into my home, and it turned out to be a stage in the development of a romance."

"It's not going to be anything like that," declared Hal. "But the girl must have a chance!"

"What chance can I give her? To marry the milkman or the gardener?"

"Is it necessary to think about that aspect of it?"

"It certainly is, Hal. Anyone but you would know it. And the opportunities of parlor-maids are restricted."

"You ought to meet her, Adelaide! Then you'd understand better. She has a mind, and she's going to use it."

"To what end?"

"To lead her people."

"A girl labor leader?"

"Why not? The miners can't win without the help of their women, and there have to be some with ideas and understanding."

"But I can't help her in that way, Hal."

"In the first place, she will help you. She'll teach you something about working-girls—just as she taught me when she took hold of that crowd at North Valley with a speech. If you liked her, and took a real interest in her, there's so much you could show her—human things, social things. The care of her person, for instance."

"*Oh!*" said the other—and there came a trace of dismay into her voice. "*Such* things!"

"Yes, *such* things! Think about a girl's life, in a three-room cabin that you can't keep the rain out of, nor the storms in winter! There's only the kitchen stove, so in cold weather if she wants to take a bath she has to wait till her father and brother have gone to bed, and then take it in a wash-basin on a draughty floor."

"So she doesn't take a bath often enough, you mean?"

"Of course she doesn't! I didn't myself, when I was there. The girl has never been able to afford a decent dress in her whole life-time; she can't even afford a clean one, because she has to do other people's dirty work."

"Dear me!" exclaimed the woman. "I hadn't realized how bad it is."

"No," Hal answered. "That's the hell of it! Nobody ever realizes how bad it is."

Before such actualities the idea of a "romance" faded suddenly in Adelaide Wyatt's mind. Of course it had been impossible for a youth like Hal to love such a girl! Impossible, even if he had not been one of the Reverend Wilmerding's Protestant-Episcopal saints. "Will she take such instruction from me?"

"You're a woman," said Hal. "You'll know how to give her hints. And if you show her that you really respect her, you can tell her anything. She has read some, she has her mind made up to learn. You'll see a miracle in a year."

The other pondered for a moment. "Will you come to see her, Hal?"

"Of course." Then he laughed. "You are wondering about the etiquette? How you'll make out with a revolutionary parlor-maid!"

"I'm thinking about my other servants, Hal—and my neighbor's servants. If you are going to set out to make war on the exploiters of this state, you must realize that a slander bureau will be set after you, and after everyone who gives you aid. If I'm going to the trouble of pretending to find work for Mary Burke in my home, I've got to have the say as to how it's done."

"I'll do anything you ask," said Hal, soberly. "You are the only person I know who might help me."

"Well," said the other, "you can always come to see me, and I can call Mary into the room. If you're not going to make love to her, I suppose it wouldn't do any harm for me to be with you."

"Delighted!" said Hal.

He could not help laughing. But Adelaide persisted. "I think you ought not to be seen on the streets with her while she's in my employ."

"It's a bargain," said he. "If you think it's necessary, we'll have a stenographer take down our conversations, and offer the enemy a transcript of them."

"No need to do that," was Adelaide's reply. "If the enemy decided it would pay them, they would find a way to get a transcript for themselves."

[10]

Hal Warner went back to finish his senior year at Harrigan. He had given his word to Jessie that he would "behave himself"; but alas, the story of his sojourn in the coal-camps was known, and there was no way he could avoid being a storm-center. Discussions would start in the most unforeseen manner; a perfectly innocent conversation would take a turn, perhaps through some "joshing" remark; and the first thing one knew, there on the campus would be a group of vehement young men and women, determining the basis of property rights and the moral validity of watered securities. "Cut it out! Cut it out, you fellows!" the crowd would say; but all the crowd could accomplish was to drive the disputants to their own rooms, where they would go on, perhaps until midnight.

There were some who took Hal's side: a Russian Jew named Lipinsky, whose father had escaped from Siberia; a tow-headed ranch-boy who was earning his way tending furnaces; a son of a "copper-man" who had been put out of business by the trust. On the other hand there developed a party of angry conservatives, gathered round Percy Harrigan, defending him in spite of himself. One of these was Laurence Arthur, Jessie's youngest brother, and another was "Dicky" Everson, who had been in Percy's car at North Valley. These men, Hal's boyhood friends, were not slow to tell him what they thought of him for casting in his lot with "roughnecks" and "goats". If he chose to mix with that element it was his right, but he could hardly expect his old friends to follow him. —So the class-war crept into the classic halls of Harrigan, to the great distress of the chancellor and the faculty's wives.

There were even two or three of the professors who showed traces of leaning to Hal's side. The professor of economics, instead of lecturing on the labor problem, spent a whole hour arguing with the ex-miner. The news of this made a sensation, even a scandal. Imagine Percy Harrigan having to sit in a classroom and listen to talk about conditions in the properties of the "G.F.C."! What would the newspapers say to a thing like that? What would Old Peter say?

It was amazing, the rate at which radical ideas were making headway in American colleges; most disconcerting to men who had the financing of the institutions to provide for. The students of Harrigan of course knew upon whose bounty their culture was nourished. They had a song about it, which they had sung for years—a typical college-song, full of gay nonsense, mingled with good-natured "joshing". They had been wont to sing it even in Old Peter's presence, and he had laughed, and thought it as funny as anyone—

"Old King Coal was a merry old soul,
And a merry old soul was he;
He made him a college all full of knowledge—
Hurrah for you and me!

"Oh Liza-Ann, come out with me,
The moon is a-shinin' in the monkey-puzzle tree;
Oh Liza-Ann, I have began
To sing you the song of Harrigan!

"He keeps them a-roll, this merry old soul—
The wheels of industree;
A-roll and a-roll, for his pipe and his bowl
And his college facultee!

"Oh, Mary-Jane, come out in the lane,
The moon is a-shinin' in the old pecan;
Oh, Mary-Jane, don't you hear me a-sayin'
I'll sing you the song of Harrigan!

"So hurrah for King Coal, and his fat pay-roll,
And his wheels of industree!
Hurrah for his pipe, and hurrah for his bowl—
And hurrah for you and me!
"Oh, Liza-Ann, come out with me—"

And so on—like Tennyson's brook, this song went on forever. Tradition had it that on one June evening it had been sung until midnight without a halt, inspired composers springing up and "spelling" one another with new verses. By now such happy days were gone; a serpent had crept into this Eden—a serpent of radicalism, which gave a sinister meaning to harmless phrases, and turned the Harrigan song into a taunt. Now when men heard its strains they looked about to see if the Coal King's son were near; and often he was not near—he had sought refuge in his rooms, to escape some labor agitator denouncing his father!

[II]

The original seed of this evil had been Morris Lipinsky, the Russian Jew. Prior to coming to Harrigan the only Jews Hal had known had been Jesus and Heinrich Heine; and somehow he had not connected them with the members of their race in Western City—the Abrahams of the clothing-trade and the Isaacs of the "show business". Lipinsky was under-sized, sharp-faced, and painfully apologetic; it was rumored that his father, the escaped Siberian exile, kept a stationery shop in some obscure quarter. The son labored under a constitutional inability to pronounce the letter w, and without reflecting about the matter especially, Hal had set him down as "impossible".

But early in the sophomore year Lipinsky came to Hal's rooms, and timidly explained that he was making an effort to interest some of the students in the subject of Socialism.

Hal was obliged to admit that his ideas of the subject were of the vaguest. Lipinsky began to explain; and Hal was surprised to see a shrinking and apparently inferior little Hebrew become suddenly lighted up with enthusiasm. He was a person with a secret religion, it appeared; and Hal, who was used to this in Judea, hardly knew what to make of it in Harrigan!

It happened that Hal himself was in a state of transition at this time. He had been brought up a member of St. George's, but he had ceased to believe the creed of his church, and what was more important, he had come to doubt its practical efficiency. Under Will Wilmerding's guidance he had taught Sunday-school to a class of urchins, gathered with considerable difficulty from the poorer quarters of Western City. He knew that these urchins were going out to face the temptations of a community in which the vice-interests were in alliance with police and politicians, a "machine" subsidized by the great public service corporations; could he feel that he was giving an adequate equipment for life, teaching about Moses in the bulrushes and Jonah in the whale? On the other hand, he could not very well teach them about graft in Western City, because the heads of the interests which subsidized this graft were the pillars of the church in which he taught!

Morris Lipinsky had faced such facts as these, and he had a remedy which he was willing to stay up all hours of the night explaining. He mentioned incidentally that there was an organization of college men interested in spreading such ideas, and might it not be a good thing to have a lecturer come to Harrigan? Hal offered the use of his rooms, and there was a meeting attended by some twenty students, who had brought to their attention the neglect of a most fascinating subject in the curricula of American colleges. The lecturer suggested the disturbing possibility that there might be some connection between this neglect, and the source from which the funds of American colleges were derived. And when some of the students wished to go on with this fascinating subject, their project met with a reception which caused them to recall this suggestion of the lecturer. They proposed to organize a "study chapter", but the chancellor of Harrigan put his foot down; he would not have the name of Harrigan associated with Socialism!

So throughout the precincts of the college broke out a lively controversy. Young men and women who had had no remotest desire to study Socialism took part in vehement discussions as to whether they had a right to study Socialism if they wished. Even one or two of the professors spoke up. Surely it was an unusual thing to forbid college students to study! Between athletics and fraternity life, it was so difficult to get them to study anything!

Hal and Lipinsky took counsel, and called on the chancellor with a new and subtle proposition. The authorities would not permit a "Socialist" organization; but what if the students should meet to discuss, not Socialism, but social problems in general? Surely the authorities would not forbid that! Seeing the authorities begin to weaken, Hal pressed his advantage, stating that he intended to invite friends to his rooms to discuss social problems, unless this procedure was expressly forbidden.

So once a week there was a gathering of students, and a discussion so exciting that it was sometimes difficult to get rid of the discussers. Even the professors took to coming, and taking part in the proceedings on equal terms with their students. A brand new experience to these gentlemen—to see American college boys taking ideas seriously enough to get angry over them! The time came when Hal's rooms would hardly hold these gatherings. The most aloof and fashionable men in the student-body, pillars of fraternity-life like "Bob" Creston and Laurence Arthur, were first drawn in by curiosity and then drawn out by arguments, until you might hear them defending themselves and their privileges under the very thinnest veils of economic formula!

[12]

The climax came near the end of the year; when Dan Hogan, the notorious "I.W.W." leader, happened to be in Western City. The brilliant idea occurred to these young collegiate revolutionists that the students might be interested to know what the "I.W.W." had to report about the state of the country. So first they invited him—and then they announced that he was coming.

Never would Hal forget the scene in his rooms that night; a cultured leisure-class audience, packed like sardines in a box, and a one-eyed, battle-scarred old veteran of the class-war, backed against the wall facing them. He had not wanted to come, he told them; he had no interest in leisure-class audiences, no faith in them; he was only wasting his breath, talking to them about things they could never understand. But the boys had insisted, and so here he was; he stood like a mangy old bear at bay, growling at them for their blindness and indifference to the horror of the life of millions upon whose toil they fed. He told them what he had seen and known—not because it was any use to tell them, but because his mind was full to bursting with it, because his soul was shaken to the deeps with it.

It was facts that "Big Dan" told; no one could doubt that they were facts, no human imagination could have invented such things! He told of starvation and oppression; he pictured masses of men and women driven to desperation, flaming out in blind revolt, crushed into submission by club and bayonet and machine-gun. He told of prison-hells where men were driven insane, of bull-pens where men, and women too, were beaten and starved and frozen, or left to die of loathsome diseases. He told of the brutality of police and soldiery, of the corruption of courts and juries—the whole enormous, relentless machine of oppression which was "government" to the man underneath. He pictured the long agonies, the sacrifices and martyrdoms of labor's nameless and forgotten heroes. He did not plead for them–he was too powerful for that; in the midst of his most moving cry of despair, of yearning for deliverance for his people, his scorn and fury would flash out again. He would hail the mighty hosts of labor, marching to their final and inevitable triumph! His voice would rise like a trumpet-call to this battle of the morrow.

The news of this meeting spread like wild-fire through the college; it spread farther–to the trustees, to the politicians, to the Chamber of Commerce. The chancellor sent for Hal, and in a state of the wildest agitation denounced this outrage on college dignity. The college had barely escaped a hideous calamity; the Western City "Herald" had got hold of the story, and it had required the influence of one of the wealthiest of the trustees, exercised over the telephone at midnight, to avert this horror from the classic shades of Harrigan. There would be no more stuff about "free speech" in colleges; this undignified and unacademic propaganda would come to an end forthwith!

Nor was this all for Hal. He was summoned to Western City by telephone, and made to listen to a discourse from his brother Edward.

Two years previously Hal's father had suffered a paralytic stroke, and his doctors declared that excitement of any sort might cost him his life; so Edward had taken over the affairs of the Warner Company–and incidentally, the duty of lecturing his younger brother. Hal was free; he had all the money he wanted, and only one thing in the world was asked of him, that he should not bring disgrace upon the family name. Did he realize that the one-eyed old ruffian whom he had received in his rooms and invited to supper in a college dining-hall had barely escaped hanging for a cowardly assassination?

"I know that," said Hal. "I know also that a jury found him not guilty."

"Stuff and nonsense!" cried Edward. "Let's talk straight

to each other. Are there any of Hogan's own followers who don't believe that he did the murder?"

"I don't know as to that," said Hal. "I haven't had a chance to meet his followers. Have you?"

Edward scorned this attempt at repartee. All the world knew that Hogan's organization was carrying on a campaign of terrorism and blackmail, and that Hogan was the driving will of it. Hal might talk about the sufferings of miners, but Edward could answer by picturing the desolation in one American home. He knew the family of Dan Hogan's victim; the son was a member of their college fraternity, a class-mate of their cousin, Appleton Harding. How would Edward feel, how would Harding feel, having to meet this young man in business and social life—and he knowing that Hal Warner had been publicly giving aid and support to the man who had blown up his father with a dynamite bomb?

Hal thought about these complicated problems when he should have been thinking about Greek and trigonometry. In his mind the matter boiled itself down to one question—was modern industry lifting the worker, or was it degrading him? Was it disciplining him and fitting him for wider responsibilities, or was it beating him down, making him unfit for citizenship? There was the heart of the controversy, and upon the answer depended the attitude one should take.

And Hal saw that however much both sides might differ, they were in agreement on the point which concerned *him*. Dan Hogan's scornful words were burned into his brain: "You, living easy lives, getting the pretty thing you call culture—what do you know about the slaves of your mills and mines? You belong to some higher order of beings, who have the right to eat your fellows!" And on the other hand were the men of affairs—his brother, Jessie Arthur's brothers, his cousin, "Appie" Harding—all calling him a theorist and a dreamer, taunting him because he was raised on books. What did he know about the cares of employers, the difficulty of handling large bodies of ignorant and jealous men?

Hal saw that whatever position he took, he could not maintain it until he could say that he knew something of his own knowledge. And so little by little there formed itself in his mind the concept of his Great Adventure. He would devote his summer's vacation to a course in practical sociology—"field-work", as it was called. He would put behind him his comfortable home and his leisure-class friends; he would take the clothes of a workingman, the name, the manner of speech, above all the purse of a workingman, and go into one of the centres of industry and see the labor problem for himself.

Because of the arguments he had had with Percy Harrigan, he chose for his experiment one of Old Peter's mountain fortresses. He saw all the evils of which Dan Hogan had told; and when the mine disaster took place, and he saw men penned up and left to die, he set out to rescue them—and so quite suddenly, what had begun as a sociological experiment was turned into a battle of the class-war. Before that battle was over, Hal had made pledges to his fellow-workers which he knew would change the whole future of his life.

[13]

Back in college again, Hal was trying in vain to persuade himself that he was interested in the Greek enclitic, in the membership of fraternities and the "line up" of a football team. There came letters from the coal-country—from Mary Burke, giving details of the war of bludgeon and revolver which the companies were waging against organizers; from Mrs. Jack David, telling how union literature in half a dozen languages had been smuggled into North Valley and distributed; from Mike Sikoria, the old Slovak whose "buddy" Hal had been, telling how he had sought a job in four camps in succession, and been met with a fist in his face; from Johann Hartman, secretary of the union local in Sheridan, telling of the "kangarooing" of Little Jerry's father.

How was it possible to preserve the academic ideal, the "passionless pursuit of passionless intelligence", with such images as this in a man's mind? Hal became more than ever a disturber of the classic halls of Harrigan. He got some of the rebels together, and forced the issue of free speech, proposing the organization of a "Social Study Club". When this project was vetoed by the chancellor, there was another controversy, more vehement than before. Somehow the facts were whispered to Hal's friend Billy Keating, a reporter on the Western City "Gazette", which spread the whole story upon its front page, and called for a mass-meeting of citizens to protest against the state appropriation of money for coal-company colleges! So the chancellor weakened again, and there was a gathering in Hal's rooms every Friday night—not a "Socialist" gathering, but one which discussed unacademic questions, and in a manner far from passionless. It was pledged to have no outside speakers, but there was no way to keep the spirit of a one-eyed old labor-agitator from coming to cast its shadow over the proceedings!

In the Easter holidays Hal went home, to get the reaction of these events upon his relatives and friends: more especially upon the Arthur family.

He was very much under the spell of Robert Arthur's beautiful daughter. She was the thrill and rapture of first love to him; mysteries lurked in her fair hair, swift emotions chased one another across her features, like shadows of April clouds across a mountain lake. In chaste and secret whispers he heard in her presence the voice of new life, craving to be; as for Jessie, when she saw him after long absence, she went faint with excess of happiness.

He came to her as a young hero, robed in light. But now this light was dimmed, and the April clouds threatened showers of tears. Laurence Arthur had "told on" Hal, how he was making himself president of a club of "rough-necks" and "goats", leading a revolt against good society at Harrigan!

Hal tried to explain, but there was no making Jessie see that a Socialist "goat" was different from any other kind of "goat". A cheap and obvious device for people to attain prominence in college-life! Hal himself was generous, naive; he had no conception of the horrid insincerities of some people, their craving to thrust themselves forward, to make the acquaintance of their betters. Why, there was a Jew in this "Study Club" of Hal's! A fellow who, at the last commencement, had pushed himself into a group of Jessie's friends without being introduced! No, said Hal, that was another Jew. But Jessie was not pacified—all Jews looked alike to her.

What Hal was doing would ultimately cut him off entirely from social life. How could he expect people to tolerate him, if he persisted in discussing their private affairs? And so, of course, came an argument—what are a man's private affairs? His coal-mines where he works thousands of other men as serfs? If Percy Harrigan chose to take that attitude, and to "cut" Hal because he sympathized with the employes of the "G.F.C."—why, Hal would simply forget that he knew the Coal King's son. Percy was a good fellow, but he wasn't the brainiest man in the college, by any means, nor the one Hal would prefer as an intimate. Had not Jessie herself expressed the opinion that he would never be able to get away with his important airs, if it were not for his father's millions?

Jessie admitted that Percy was not the greatest of her troubles. There was her brother, who was terribly stirred up; and more serious yet, there was her father. Someone had teased Mr. Arthur at his club, mentioning that he was to have a Socialist for a son-in-law, and the old gentleman talked about it all the time—he was threatening what he called "a serious talk". What would Hal do about that?

Here indeed, was cause for uneasiness. Hal had never failed in tact in his dealings with Mr. Arthur; he had taken the proper interest in the country-place, he had duly inquired as to new developments, and gone promptly to see them—finding his consolation in the fact that Jessie went along, and that he could hold her hand while inspecting the Annamese rooster and sampling the Siberian strawberries. But there was one thing he could not pretend, no matter how much he might be in love with the lovely Jessie—that was, that he was impressed by the old gentleman's views on social questions. After once he had heard these views, he sought tactfully to shift the conversation, pleading an engagement on the tennis-court. But now the tennis-court was covered with snow; and before he and Jessie could think of another excuse, here came the old gentleman puffing in!

[14]

The head of the banking-house of Robert Arthur and Sons was a self-made man, and saw nothing to be ashamed of in the fact. He had begun life as a clerk of the mining-exchange, way back in the early days; he had been sober, plodding, and acquisitive, and now in his late years he was one of the financial powers of the city. All that he had, he had earned, and he was satisfied of his right to it. That jealous persons should dispute his right—the unsuccessful whom he had watched and condemned for their lack of sobriety and ploddingness and acquistiveness—that anyone could understand. That such persons should get together and found a school of thought; that they should start a political party and take to making speeches on street-corners—that too was comprehensible. The one incomprehensible thing was that these jealous and greedy spirits should receive the backing of men of education and responsibility—for example, a young man who had been permitted to become engaged to Robert Arthur's favorite daughter!

The old gentleman invited Hal into the library, and invited Jessie out; having got himself into his big leather arm-chair, and Hal in a smaller chair opposite to him, he began what was clearly to be the "serious talk". He had recently heard about certain ideas adopted by his future son-in-law, and he desired to make an effort to understand these ideas.

Hal began very tactfully, taking his start from Little Jerry. And that was all very well—Little Jerry was a clever child, and if conditions in North Valley were as he described them, the matter ought to be brought to the attention of the executive officers of the General Fuel Company, who would correct the evils. And so at once came controversy; "Joe Smith", ex-miner's "buddy", revealed his notion that the heads of the General Fuel Company, gentlemen with whom Mr. Arthur did business every day, knew about these conditions, and deliberately maintained them! Such a notion revealed a disordered state in a young man's mind—it revealed that he had been listening to agitators!

And what was the young man's program—what did he expect to do about the wrongs he described? When the young man started to explain, his future father-in-law was unable to hear more than three sentences. Why, that was Socialism! And did not the young man realize that Socialists wanted to divide things up and start over again? Could not he see that if this were done, it would be only a short time before the capable men would have everything again? Socialists wanted to have everything owned and run by the state; and could not anyone see that that must lead to the building up of a political machine, worse than any ever known before?

Hal was so indiscreet as to attempt to point out to Mr. Arthur that these two ideas about Socialism were diametrically opposed, and so at least one of them must be incorrect; but at this the old gentleman became still more irritated;

he made clear that his ideas were his ideas, and were to be treated respectfully, like his roosters. The old gentleman's face was round, and his eyes were round, and his two fists were round; he had a way of lifting both fists and pounding down with them to impress his arguments upon you. It was as if he were working a heavy churn, and at each thump would come a bounce in the big leather chair!

Hal was quite willing to be reduced to silence; but having definitely come out as an enemy of the system of things on which Robert Arthur's banking-business was built, he could not plausibly change his views at short notice. He realized, however, that the old gentleman desired exactly that; he was resolved either to reform this deluded young man, or else to make clear to him that he could not expect to be received into the Arthur family!

Before matters had got that far, however, Jessie came in; having been hovering outside the library in terror, ever since the discussion began. She put her arms around her father, and wet his white side-whiskers with her tears, ending the argument by sheer force of emotion. The old gentleman had never denied her anything in her life-time, so he could hardly let luncheon get cold while he debated Socialism with her lover. And after luncheon, she played the piano for him, and laughed and chatted all the time. Hal did his part—but it was like playing the piano and chatting on the slopes of Mount Vesuvius!

[15]

From this adventure Hal went for a call upon Adelaide Wyatt and the revolutionary parlor-maid. Already the miracle he had foretold had happened to Mary Burke. She had taken her mistress's hints, with a result that her finger-nails were clean, her treasure of auburn hair shone dazzlingly under a white lace cap; she was the very picture of a parlor-maid! Her clean, straight figure had filled out in this new home, and her cheeks had again that vivid Irish color which had struck Hal when he had first encountered her, a rose in a mining-camp, taking in the family wash!

Mary was watching things about her, he could see. Yet she was sensible about it, free from self-consciousness; she would learn what a lady might have to teach, but without losing her head, or forgetting her people at home. Her steady grey eyes met Hal's with the old frankness. Apparently it was not an impossible thing for one to be a servant with dignity.

Mary was happy; for the first time since Hal had known her the burden of care was lifted from her life. Her brother and sister were living with Mrs. David, who was good to them; Tommie's wages as a "trapper-boy" sufficed for his keep, and Mary was sending three dollars a week for Jennie. As for Old Patrick, he was living in the cabin alone, and his drinking harmed no one but himself. Mary had long ago been forced to give him up, so the pain of this did not cut too deeply into her soul.

Yes, Mary was happy! That sense of fun which is never very deep below the surface of an Irish mind bubbled over when Hal told about Little Jerry's adventures at the New Year's party, and about his own adventures with the chancellor of Peter Harrigan's College. He told about the "social study" nights, and all the interesting things that went on there. As he talked, the thought came to him, How Mary would have liked to go to one of those "social study" nights! And how he would have liked to take her! But he could not take her, he could not even tell his friends about her. He had to keep her a dark secret—more so now than when she had been in North Valley. For a college boy to be interested in a rose in a mining-camp might conceivably be a romance; but for him to be interested in a parlor-maid could not possibly be anything but a scandal!

Hal ought to have been satisfied, as Mary was, now that she had a good home; but instead he was thinking about this dirty world which drove people to cringing and cowardice, when they had nothing of the kind in their nature. Mary was as fine and straight a girl as you would meet in a life-time; but nevertheless, there would be a line drawn, and Mary would be on the other side of that line. And this canting world would go on calling itself a democracy, a land of equal opportunity!

Poor Mary! She was so happy in her wonderful new role—so charming with her Irish fun! But let her not make the mistake of thinking that she had really been taken into this world of cleanness and ease! Let her not get the idea that she was dusting Mrs. Wyatt's bric-a-brac out of love and gratitude! Mrs. Wyatt herself would be as human as she dared, but the ladies who came to her tea-parties would be quick to put a revolutionary parlor-maid in her "place"! Hal was wondering how long it would be before something happened to break "Red Mary's" bubble of happiness. Suppose it was to occur to her in this new and wonderful prosperity to ask her friend "Joe Smith" to take her to a picture-show!

Just now, however, Mary did not need picture-shows. She was revelling in a more wonderful world to which she had secured admission—her mistress's library. Hal had told her what to read, and she had been sitting up half the night to finish "Comrade Yetta". He was interested to see the effect upon her of this story of a Socialist working-girl. She told him about it with a thrill in her voice. "Joe, I never knew there were such things, so many movements, so many people helpin' to set free the workin'-class!"

She saw now what he had meant, when he urged her to get an education. She had thought she knew enough—what did a body need to know, save that the poor were being devoured, and must stand together and put an end to it? But now she was realizing how complicated was the problem; there were many evils, many remedies offered, many courses to choose among. Hal had sent her Socialist papers, and she had read every line of them, and had got her head in a whirl! "Ye read what one man says and think he's got it right, and then ye read another man, and he sounds good too—only he says the other man's all wrong! I read about strikes in South Africa and New Zealand, places I never heard of, so I have to get the geography-book. And there are so many long words—why do they have to have such long words, Joe? Even Mrs. Wyatt don't always know what they mean!"

Adelaide put in with a laugh that she was educating herself, as well as Mary. They would be a team of revolutionists when they got through!

So Hal went away with a pleasant image of Mary. What a wonderful thing it was to see a mind unfolding, reaching out for new opportunities, discovering the world of ideas! Inevitably this brought another thought—why could not his sweetheart take hold of things like that? She really knew little more than Mary; she had the smattering of history and "polite letters" that young ladies obtain in boarding-schools, but of the great vital ideas of the day she was utterly ignorant. And when he offered her knowledge, why did she not take it?

He did not want to compare Jessie with Mary. It seemed disloyal, and he put the two in separate compartments of his mind, and strove to keep them there. But somehow they seemed always to come together! It was not their fault; for Mary never mentioned Jessie's name, and as for Jessie, she was not supposed to know about the revolutionary parlor-maid. But Hal had chanced to mention Adelaide Wyatt to Jessie, and he had noted a sudden silence, and a shadow on Jessie's fair brow. Could it be that she had heard the rumor? And if so, what was she thinking? Hal was not naturally alert to the subtleties of women's minds, but Adelaide had warned him here, and so, as he walked along, he had new quarrels with the world of caste and gossip!

[16]

Hal's next call was upon the Minettis. He learned that Big Jerry had come home, and got his head well, and gone off to resume his dangerous work. He had grown a mustache and beard for a disguise; and, how funny he looked! cried Little Jerry. They had just had a letter from him, smuggled out from the Western-American, where he was at work again. "I told you he fool them bosses!" proclaimed the youngster.

Mr. Wilmerding had called, said Rosa. He seemed to be a good man, in spite of the fact that he preached in Peter Harrigan's church. He had begged to have Little Jerry in his Sunday-school, so that the child might know the name of Jesus as something else than a "cuss-word". Rosa was not sure what Big Jerry would say, but she had let the child go, and he had been given a nice picture-book; also a lady had called on them, bringing good things to eat, and a spinning top, and a rattle for the baby. Rosa, who was a Socialist like her husband, and a shrewd little body for all her child's face, remarked that perhaps Mr. Wilmerding wanted to make his conscience feel better, by getting some poor people into his fine church!

During this visit to Western City Hal went also to see Jim Moylan, the secretary of the district organization of the miners, who had just got back from a trip to the "field". There had been a sudden flare-up of revolt at Harvey's Run, and it had spread to Pine Creek and Bonito, in spite of Moylan's best efforts. He had called Harmon, his chief, to his aid, but a number of the men were still refusing to go back to work. They were cursing the union officials up and down the line, calling them cowards and weaklings, some going so far as to call them traitors. They, the men, were on strike, and others wanted to join them;

who was paying the leaders to hold them back? Such was the situation in the coal-country, at the time that everybody in Hal's world believed that the labor leaders were trying to stir up trouble.

Hal met John Harmon, the executive of the United Mine Workers in charge of the district; a man who had gone to work in the coal-pits at the age of eight, educated himself at night, and risen to this position of leadership. Hal found it thrilling to talk with him; as if a private who had been fighting in the trenches were suddenly taken to staff-headquarters and allowed to see a map of the battle. Such stories as he had to tell, not merely of open fighting, but of secret sapping and mining! There was a powerful strike-breaking concern, the Schultz Detective Agency, which had fought the miners' union in many a field; they were sending in scores of "spotters", and bribing and buying right and left. Only a short time before this, one of the prominent men in Harmon's office—he would prefer not to say who—had been invited to a conference with a well-known politician, and found himself in an automobile with no less a person than the great Schultz himself. The detective had something to propose; he was diplomatic and cautious—but properly led on, he disclosed the fact that he desired to pay the union official a hundred and fifty dollars a month for selling out his organization. And when the official showed reluctance to close with the offer, the other intimated that if necessary the sum might be increased.

[17]

In other places than union headquarters there were signs that Peter Harrigan and his associates were preparing for trouble. There was, for example, their effort to cripple the Western City "Gazette", which, with its thirty thousand working-class readers, made it impossible for coal-camp rebellions to be suppressed in secret. The other papers of the city were already combined against the "Gazette" to the extent of black-listing dealers who sold it; the paper had to be delivered by carrier, or sold by enterprising boys. And now suddenly came a mysterious eruption of rowdies, who attacked these carriers and boys, beating them and scattering their papers in the mud. For some reason the police were blind to this eruption, and powerless to find the rowdies when called upon. Likewise the other newspapers maintained dead silence.

This occurred just before Hal's commencement. Happening to be in the city he went to see his friend Billy Keating, and heard the details. "For heaven's sake," he cried, "why don't you get some rowdies of your own?"

Billy answered, with a laugh, "We did, but the police saw them!"

"And what are you doing now?"

"We're printing the news. When you're printing the news, the people will get your paper, even if they have to walk to the office for it."

To one who had been down in the coal-country, and realized how vital to the miners' cause was the little publicity the "Gazette" could give, this situation was intolerable. Hal got hold of Lipinsky and two other members of the "Social Study Club", who constituted themselves a committee to get legal evidence for the "Gazette". Needless to say, the "Gazette" did not fail to "play up" this support of its cause by "prominent young collegians". And so Hal got himself into the hottest tub of hot water yet!

In this far western country men were still close to the frontier days; when they fought, they fought fiercely, and were not squeamish about the weapons they used. In Western City the "interests" maintained an underground scandal-sheet for the purpose of intimidating those who might threaten them; "Simple Simon" was the name of the mysterious organ, and all "society" somehow got hold of it, and rejoiced when it enabled them to believe the worst about those who attacked their privileges. In the issue of this paper following the raids upon the "Gazette", there appeared a paragraph to the effect that a certain youth of too much wealth had become active as a labor agitator; his family, who were distressed about his behavior, might possibly find the source of his extreme ideas if they would inquire as to the visits he paid to a mining-camp damsel in the home of a sprightly member of the "smart set", soon to be a grass-widow, if reports were to be believed.

Hal went wild, and set out forthwith on a hunt for the editor of "Simple Simon". But the editor was not an easy person to find—he had many people hunting him, week by week! Having failed to discover anyone more to the point than an office-boy and a janitor, Hal went to interview his cousin, Appleton Harding, a rising young lawyer of Western City. "Appie" pointed out the obvious fact that all Hal could accomplish by proceedings for criminal libel was to give the story wider circulation. But Hal was not to be restrained, he would not quit till he had that editor either in jail or in hospital! He went off, threatening to consult another lawyer; and so "Appie" made haste to warn Brother Edward, who had to threaten to take the matter to their father.

And even this lesson was not enough; the youth of too much wealth would not give up being a fanatic, and trying to overthrow the foundations of society! There was another conference, this time between Edward Warner and Garret Arthur, the "bond-worm", and a plan was worked out for the salvation of the young fanatic's future. But it was cautiously agreed by the conspirators that they would not let the young fanatic know the origin of the plan. Garret Arthur could realize that it was no job for a "bond-worm", to handle a youth gone mad on socialistic moonshine. If they hoped to save him, they must be wise as serpents and harmless as—well, as a girl with star-dust in her hair, and eyes wide-open, questioning, full of wonder!

The day after the conference Hal received a note on beautiful stationery, in one of those tall hands whereby young ladies of fashion demonstrate their indifference to the cost of ink and paper, asking him if he could not come to see her at once about a matter of importance. When he came, he did not find Jessie wrought up about hideous insinuations in scandal-sheets; no, for if young ladies of fashion know of the existence of scandal-sheets, or even of scandals, they do not mention it to their brothers and sweethearts. Jessie was lovely as ever, and radiant with happiness, because she had just heard the most heavenly news—her mother was going to take her abroad in a couple of weeks, to stay for a long, long time; and, wonder of wonders, she invited Hal to accompany them!

Hal could easily perceive what was back of this project. But he saw that Jessie's heart was set upon it, so he contented himself with replying that he would have to take time to think about so vital a change in his plans. What were his plans, Jessie asked; and when he answered, he saw tears of distress come into her eyes. Hal had been meaning to spend the summer investigating conditions in the mining properties of the Warner Company!

[18]

When Hal thought the matter over he found himself inclining to give way to his relatives. There were movements in Europe which a young revolutionist might well afford to know about. There were men who had been grappling for generations with problems to which America was just awakening; and to meet these men face to face would be a wonderful experience. Also, Hal told himself, he would be giving Jessie a chance—he could devote more time to her, he could take her about and try to make her understand what he was doing. On the other hand it was clear that if he declined the invitation there would be real trouble with the Arthur family. Mount Vesuvius had been in eruption again, and Jessie lived in an atmosphere of sulphur and brimstone. Possibly that was why her eyes were full of tears, Hal suggested. But it was no joking matter; her father had actually proposed that she should drop Hal from her life. On that occasion the old gentleman had made the alarming discovery that his favorite daughter was subject to attacks of hysterics; so now he was as much afraid of his daughter as of Socialism!

In their concern, the family even called in Hal's father, whom it was strictly forbidden to trouble with cares. Sitting on the porch of their home one sunshiny afternoon, Edward Warner Senior brought up the subject of his son's unfortunate attitude to life. In his pitiful childish voice he pointed out what a serious thing it was to their business to have the hostility of Peter Harrigan. There were certain courtesies one owed in the world of affairs; and while it was certainly true that working-people ought to be kindly treated, one must remember that they were foreigners, ignorant and excitable, and that it was very wrong to stir them up to disobedience.

Hal was careful and gentle in his reply. He gave a few details about the evils he was opposing. Yes, the old gentleman admitted that American business-men worked their employes too hard; they worked themselves too hard, they lost the enjoyment of life. Also, there was no doubt that Peter Harrigan was a harsh man; a good man in his own peculiar way, generous if you came to him right, and useful in the church—but thinking a great deal of his worldly power, and driving his business machine at a cruel pace.

Hal knew that in the days before his father's illness he would not have had such an easy time in a discussion. Edward Warner Senior had himself been a business-man, and had driven his machine at the usual American pace. But in one dreadful night he had been turned into a feeble old man; his hair was white, and the very gentleness of his smile wrung your heart. He was getting stout, because he sat about all day, or was driven in his car. His ideas had a tendency to wander, and he craved to be entertained.

More than anything else in the world he craved the companionship of his favorite son. His delight in life was to have Hal and Jessie play tennis with Laurence Arthur and his fiancée; the old gentleman would come to the court and sit, watching every stroke, keeping up a constant run of comment, applauding gleefully the good shots—especially if they were Hal's. There was one business-man in the family, Edward Junior, and the old gentleman's idea was that Hal should be its ornament, its holiday part. Hal was so designed by Nature, with the grace and the charm; but instead of filling his proper role, he went off and bound himself in a treadmill of killing toil! He cast away and trampled upon the heritage his father had won for him, he put himself before the world as a living indictment of his father's life-work; and to keep him from such madness there was no way save to ship him off to Europe, five thousand miles away from a lonely old invalid!

Hal announced that he would go; and great was the joy of Jessie, and the satisfaction of those two master-diplomats, Edward Junior and Garret Arthur, who met in secret at the club and drank high-balls in tribute to their own astuteness. They had got the young madman safe, for a few months at least; for of course neither of these capable and hard-headed young business-men had any idea of that "Europe" to which they were sending their patient. To them the name meant a place for boarding-school-girl tours, personally conducted, for honeymoons, and such play affairs; the home of a venerable thing called "culture"—cathedrals, guide-books, endlessly multiplied Madonnas. There would be Jessie, to administer to the patient the daily medicine of love, and Mrs. Arthur to watch his symptoms and report; surely a promising course of treatment for a youth gone mad on socialistic moon-shine!

[19]

There now began a whirl of excitement for Jessie. She would buy things in New York, of course, and still more things in Europe; but it was necessary to buy some things at once, in order to develop the holiday atmosphere. As for Hal, he had to see his radical friends and explain his decision, making them understand that he was not running away.

In particular he had to explain to Mary Burke. Why should he have been so much embarrassed to tell Mary that he was going to spend a summer in Europe, studying the Socialist and Syndicalist movements? Was it because he was to be with Jessie Arthur? Or was it because he could not keep out of his mind the preposterous thought that Mary too would have enjoyed spending a summer in Europe, studying the Socialist and Syndicalist movements?

Here is one of the inconveniences incidental to the adopting of revolutionary ideas by a member of the leisure-class—that the member can no longer take the most obvious things of his life for granted. Of course a man cannot change the system all at once; but he will change as much as he can—and feel uncomfortable because he cannot change more! If he is befriending a revolutionary parlor-maid, and trying to be entirely democratic, he will find himself asking why the daughter of a banker should go to Europe, while the daughter of a miner remains at home. Under real democracy, obviously the daughters of all men would have equal access to the opportunity of going to Europe; if the purpose of the going be a study of the Socialist and Syndicalist movements, then the question of who should go would be determined by some kind of competitive examination. But here there had been no examination; Jessie was going, because she had the money, and Mary was staying, because she hadn't.

But Hal found that Mary had not thought of anything like that; the parlor-maid was not that revolutionary! Her only thought was of him—that he was being lured away from his work. His family was getting him under the spell of Jessie, with her softness and her clingingness, her beautiful clothes and her expensive charms. They would have their way with him, they would tone him down!

She did not say any of this; she would not mention Jessie to him. But there was pain in her face as she bade him good-bye. "Ye'll not be forgettin' the miners, Joe?"

"No, Mary," he answered, with conviction. "You don't need to worry about that."

"They'll be needin' ye so bad, Joe! Perhaps they'll call ye back."

"I made them a promise, Mary; just as you did. We're going on working and studying, so we can give them the kind of help they need."

"Joe," she cried, with passion, "sometimes I think it's more than I can do to stay here, where things are beautiful and clean, and I have all the good food I want." She turned to her mistress, who was in the room. "Don't think I'm not grateful, Mrs. Wyatt; but 'tis bound to be so when ye've lived among people that never have enough of anything."

"I understand," said Adelaide, gently.

"Ye can't get them out of your thoughts! The men ye know that go down into the pits, and may come out on a plank! Ye think of them, this time and that—now they'll be going down in the cage, now they'll be eatin' their dinners, now they'll be comin' out. Ye tremble when ye pick up a newspaper—ye think maybe there'll be a piece, somewhere off in a corner, about another mine-disaster, and some woman ye know will be lonely all the rest of her days, and her young ones will be hungry and cold! Think of them, Joe, the men that went back to work at North Valley, after the strike! They're waitin' for ye to keep the promise ye made! They'll not forget ye, what ye did for them; they'll think: 'What's he doin' now? When's he comin' back to us?' "

Hal did not fail to think of them; Europe would make no difference, he assured Mary. He pledged his faith anew—and then he went off to Jessie, to try to share her holiday mood, to admire her travelling trinkets! Jessie wanted him for herself, and Mary wanted him for the miners; the two of them pulled and tugged at his thoughts. A trying thing for a young man to be so very much in demand!

[20]

The day before the party set out for New York, something occurred to bring the miners and their struggle even more vividly into Hal's thoughts. The mail brought a note from one of his workingmen friends, whom he had not heard of for a long while—Tom Olson, the organizer who had come to North Valley and given Hal his first impulse towards unionism. Olson was in Western City for a few days' rest, and Hal went away early from a fare-well dinner-party at the Arthurs' and spent half the night talking with the organizer.

For three years Olson had been at work in the coal-camps, and not once had he been caught. But many times he had come close to it! It gave one a thrill, just to sit here in his home, for when he came, he had to steal in by night, and shave off his beard and change his clothes and his accent. His wife was a school-teacher, and told her friends that her husband was a "travelling-man". Only one or two intimates knew the truth about him.

Camp after camp the young fellow told of, the strange experiences he had had, the personalities he had encountered. Sometimes he was afraid to trust anyone, but would slip union literature into men's dinner pails or their coat-pockets, and not stop to ask results. Sometimes advances would be made to him, and then he would have to make up his mind whether he was dealing with a *bona fide* workingman or a company spy. No less than three times he found himself sparring for an opening with another organizer, a man he had never hear of, and who had never heard of him!

Olson was a fellow with a sense of humor, and knew how to tell his adventures. He would reproduce the manner and dialect of different personalities, so that you saw them before you. His pretty young wife would sit and listen, not trying to hide her pride in him. She was a miner's daughter, and this fight was hers.

Throughout the district, he said, there were now hundreds of men and women working to spread the union message. Few of them knew each other; even in the same camps there were separate groups, having no contact. Old Johann Harman, the secretary in Sheridan, was the only man who had a complete list—and he had it in his head! But some day the moment would come; these scattered rebels would spring up and discover one another, these frail strands would be woven into one cord!

"When will it be?" Hal asked; and the other answered that it could not be put off much longer.

"This summer, you think?"

"I hope not. We've just had a big strike in West Virginia, and you know how it was smashed. Cost us a million dollars, and nothing to show for it."

"Well," said Hal, "I'm going abroad, but I want to tell you, when the strike comes in this district, cable me, and I'll take the next steamer. I'm telling my brother that, and my father, and everybody. I don't mean to miss it!"

"We'll let you know," said Olson; and the two of them shook hands on the bargain. The scene was stamped upon Hal's mind indelibly—embodying to him the mystery of the thing which men call fate. Men walk blind-folded into the future; they grope in darkness, amid lurking destructions;

and is there anywhere a power that foresees, that could perchance be persuaded to warn? Here, on this soft June evening, Hal parted with the organizer and his happy young wife; and the next time he heard the name of Tom Olson was in a cablegram from Billy Keating, three months later, telling him that his friend had been murdered in cold blood upon the main street of Pedro, shot through the body by a coal-company detective, Hal's old enemy, Peter Hanun, the "breaker of teeth"!

[21]

Mrs. Arthur and her charges set out for Europe by the Mediterranean route. They landed at Naples, and went up the slopes of Vesuvius, through the ruins of Pompeii, and out in a launch to Capri. Hal escorted them upon these jaunts; and then, while they were resting, he made a discovery. The peasants of this poverty-ridden region, victims of exploitation since the dawn of history, had come at last upon the road to power. They had combined into agricultural laborers' unions, against the absentee landlords who spent the fruit of their toil in the cafés of Paris and the gambling-palaces of the Riviera. Having declared a strike, and been shot down by their own brothers and sons in army uniform, the laborers had put the funds of their union into the purchase of an estate, which they worked co-operatively. When the next harvest season came, and labor must be had at all hazards, there was plenty of labor for this co-operative crop—and a strike against the adjoining landlords! After a few seasons of such double-edged warfare, the landlords were glad to sell out for anything they could get; and thus step by step, the degraded peasantry of Southern Italy were raising themselves to the status of citizens in industry. And a prosperous industry, with modern machinery and unimpeachable credit—an industry founded upon the rock of labor solidarity!

This was the most interesting thing in all Italy to Hal. In a little office in the business part of Naples were men from whom, with the help of an interpreter, he could get the full story of this revolution, with picturesque details of the strife. Why was not this enormously important story known in America? What power had been able to keep it from the knowledge of a press that boasted of enterprise, from magazine-editors who ransacked the world for "live" material?

They went to Milan, and there Hal found that the glass-blowers had made the same discovery as the farm-hands, and were conquering the industry with this new double-edged weapon. Could anything be more fascinating, to a young man who, only a year ago, had been watching Italian workingmen ground up in Peter Harrigan's profit-mill? Here was Jerry Minetti's own home city; and here were artisans, proud, erect, masters of their own product and their own destiny, with the light of human brotherhood shining in their eyes!

Hal's revolutionary thinking had so far been conducted under Socialist auspices; and when you explained the Socialist scheme of things to a business-man, you always met one objection: to turn over industry to the control of politicians! To be sure, the Socialist argued that he meant to put into office a new kind of politician; but the American system of graft was so firmly established, it was hard for the man in the street to believe that people elected to political office could ever run any business but retail liquor.

And now, here was a new method of bringing about the change; and not a matter of theory, but of fact—going on whether your theories allowed it or not! The person to run the industry was not the politician, who had never been inside a shop; it was the man who had worked in the shop all his life, who had done the real job of building it up! Co-operative ownership and democratic control in business!

Local self-government in the factory, with representation in an assembly of the trade, and a congress of delegates from all the industries of the country to regulate terms of exchange among them!

Such was the solution of the Syndicalist. Why, he asked, should the management of industry be committed to a legislature, elected upon a basis of geographical location? In a great city you might live all your life without knowing the people in the next apartment. But you never failed to know the men who worked in the shop with you! You knew who was honest, who was competent, who was really on his job! So there, in the democracy of toil, was the unit upon which to build the industrial republic, just as the political republic had been built upon the town-meeting.

[22]

Mrs. Arthur and her party moved on to Paris, and here again Hal found that the word of the hour was Syndicalism. The French workers had tried electing politicians; they had had soul-stirring revolutionary speeches made to them, but when it came to a big strike, the revolutionary politicians ordered out the military, just as the bourgeois politicians had done. To be sure, the Socialists repudiated those who had committed this crime; they put up new candidates, who made fresh promises, and were sure they would never sell out. But the workers were embittered, and Paris, the center of the world's new thought, was swarming with the advocates of "direct action".

Hal tried his best to be interested in art-galleries and monuments; he went patiently and looked at the multiplied Madonnas, and walked through cathedrals with his head tilted backwards and a guide-book in his hand. But how could a man think about art, with the whining in his ears of pitiful beggars on the steps of these temples of submission? How could a man shut his eyes to the pitiful faces of starving children in the slums through which he drove? An American could not encounter such things without having his soul one cry of determination to save his country from these old paths of misery and oppression. The more Hal tried to contemplate the past, the more he found himself impelled to seek out the pale and undernourished men with restless, burning eyes, who had offices in obscure quarters and homes in slum-garrets, and who were building the cathedrals of the future, the mighty arches of labor solidarity, the towering spires of proletarian dream!

Hal had as much French as is imparted in American colleges—enough to order a meal in a restaurant where the waiters speak English. Now, wishing to take part in revolutionary conversations, he hired a student-contributor to working-men's papers, and the two of them went about jabbering. Jessie liked this young man, because he had a face full of melancholy, with a dear artistic little beard. Her boarding-school French did not admit her to the conversation; and this was just as well, for the young Frenchman's revolutionism was of a kind that startled even Hal.

Pitiful were Jessie's efforts to follow her lover, to overcome her instinctive shrinking from the sordid and shabby, to understand his hostility to the elegant and refined! The Arthurs had brought letters of introduction to rich Americans, and were taken up by the "colony", and even began to break into the diplomatic set. Jessie was invited to a thrilling reception at the embassy—and made the discovery that Hal had arranged to attend a congress of railwaymen!

Now and then she went with him to these terrible places. In the Salle Wagram they listened to an address by Gustave Hervé, who had just begun to recede from his position of anti-patriotism. His speech sounded wild to Jessie—but apparently it was too tame for this audience, for there was a constant clamor of protest, and at the culmination a man leaped upon the platform and began to exhort the crowd. A terrifying figure, with long black hair and a face of ashen grey, the pallor of prison; Malatesta, the anarchist, expressing his opinion of renegades and compromisers! His supporters danced about and howled; the supporters of Hervé rushed to the platform, there were scuffles, blows, chairs uplifted and hurled about. In the midst of the tumult the lights went out, there were screams, and then half a dozen shots. Hal fought his way out, with his fiancée fainting in his arms; and that was the end of revolutionary activity for the daughter of Robert Arthur!

[23]

Jessie kept this episode from her mother. But little by little the painful truth was becoming clear to both ladies that the trip abroad was doing Hal no good at all. They crossed the channel, but only to meet worse trouble. England was in the midst of an historic labor convulsion, the strike of the coal-miners. Here was Syndicalism in action—gripping the world's second-greatest industrial nation by the throat! How could anyone go about doing tourist-stunts, while such an event as this was shaking civilization?

So Jessie and her mother wandered alone among the tombs in Westminster Abbey, while Hal was in the Westminster Hotel across the street, where the delegates of the miners were meeting. He had brought letters from John Harmon, and his story won the trust of these men. There was one, especially, who made him a friend—a twenty-two-year-old miner from South Wales, whom Hal picked out for the future president of the industrial republic of Great Britain! Such a mind as this youth had, and such a will! Frank Bollett was his name, and he had been trained at the expense of the union at a labor college in London. Two years ago he had written a pamphlet, "The Miners' Next Step", which had been published anonymously and circulated among the workers of his part of the country. Now he had come to London to see to the carrying out of his program—"the Mines for the Miners"!

Young Bollett seemed to have one hatred in all the world, and that for a politician. He did not hate the "masters", any more than he feared them; the banded workers would settle with the masters very quickly—if only the politicians would keep their hands away! Whether they were capitalist politicians, or called themselves representatives of labor—only let them keep their hands away, and the workers would decide their own destiny!

Hal went to a gathering one Sunday afternoon, in the home of a well-to-do sympathizer with the miners, and heard young Bollett pitted in impromtu debate against a member of parliament, the editor of a leading liberal weekly. How fascinating to see this great man backed up against the wall and speared through by the logic of a mine-boy!

To the boy, who had faced the realities of industry, the only person who counted was the producer; he saw the problem from the producer's point of view, he planned a society in the producer's interest. But the editor was concerned about the consumer, he cried out for the consumer's right. It was quite impossible to get him to see that in a just society there would be no consumer who was not also a producer; so that if you made certain that every producer got his full product, you could dismiss the consumer from mind altogether—and with him that elaborate machinery of bureaucracy whereby the liberal statesman dreamed to chain and bind the tiger of exploitation!

The editor could not see it. But Hal noticed that he went off and wrote as his leading editorial of the next week a temperate and even-handed exposition of the ideas which the young miner had explained to him. So Hal learned something about that British devotion to fair-play, that genius for statesmanship, which is responsible for what freedom now exists in the world. He tried to imagine a senator at Washington, the editor of an organ of culture in New York, coming down from his high seat of authority and taking lessons from a mine-boy! No, the thing was not thinkable! In America they would have barred the young man's paper from the mails for "obscenity" or "sedition"; they would have sent a cursing police-official to suppress the young man's meetings, and likely as not to bash in the young man's head. And the editor of the organ of culture would have stayed in his sanctum, not mentioning, hardly even knowing of these proceedings, but writing bewildered editorials on the spread of the dynamite plague in the American labor movement!

[24]

In the middle of these events Hal's thoughts were turned suddenly towards home. There came a letter from Jim Moylan, telling him that there no longer seemed hope of staving off a conflict in Pedro County. The discontent of the workers was mounting, and the operators remained as stubborn as ever. A formal set of demands had been drawn up and sent to them, but the communication remained without answer. Now a call was out for a convention of delegates from the various camps, and there could be little doubt that this convention would declare a strike.

And the same mail brought a letter from Jerry Minetti, giving the news from the field. Jerry had been "fired" from another camp, but still he was able to do work for the union, for the men had taken to meeting secretly in the canyons. There had been a walk-out at San Rafael last week, and one was expected any time at Greenough. So many men were being turned off that a big strike had become a matter of life and death to the union. If Hal had any idea of helping, he had better be starting.

Jerry stated the facts without comment; but Hal could read between the lines and realize what tension must be in the coal-towns. And the day after the receipt of the letters came a cablegram from Billy Keating, telling that Pete Hanun had shot and killed Tom Olson. Hal decided that the time for holidays was past. He took the letters and cablegram to Jessie and her mother, and informed them that he would have to leave for home.

There was a terrible scene, of course. Jessie wept hysterically, and Mrs. Arthur was indignant. It startled Hal to see this placid lady, whom he thought of as amiability incarnate, suddenly transformed under provocation. It was not merely a personal discourtesy, because Hal was an escort, and their trip would be spoiled if he deserted them; it was an offense to Mrs. Arthur's class prejudices—for Hal was going to mix himself up in a strike, and it was stupid folly, because he could not do the least good to poor ignorant people who were being deluded by selfish agitators.

Before the discussion came to an end, Hal had been made to understand that in taking his departure in this way he might be giving up all hope of becoming a member of the Arthur family. And then Jessie flung herself into his arms, weeping like one possessed. "You are killing me for your old miners!" she cried; and when he tried to argue with her, she started back from him, her eyes flashing, her hands clenched. "Oh, how I *hate* your miners! How I *hate* them!"

Mrs. Arthur sent a cablegram at once, and next morning there was a message for Hal from his father, imploring him not to take this mad step. Hal saw the hand of his brother in this, and appreciated the tact which had led Edward to dictate a plea instead of a command. But he cabled back that he had already engaged passage.

Hal was humble and apologetic, and Jessie was haughty—that is, until the day before he sailed, when she broke down, imploring him almost on her knees to abandon his project. She was pleading not only for their love, but for his life. She had been impressed to this extent by the awful tales he told of the coal-camps—that she wanted her lover to keep out of them. Rough and untutored men might face such dangers, that was their part in life; but she and Hal belonged to a world in which such things had no place. What madness to throw away one's birthright of safety and ease!

All the way home, on the steamer and the train, Hal carried these images of Jessie with him; the feeling of her tears upon his hands, the sound of her sobbing in his ears. This was love, no doubt; but it seemed to him a cruel thing, not the brave thing he had dreamed. Much as it hurt him, he had to turn his back upon that love.

In the ten days' journey he had plenty of time to think about the course he was taking, to examine stone by stone the foundations of his beliefs. He was at the parting of the roads, he could see; casting away the life to which he had been born, his family, his friends, his whole "world" of privilege. To live in a beautiful house and eat at a well-appointed table; to wear elegant clothing and carry a full purse; to have a host of cheerful friends and be popular with them; to marry a charming girl and raise a loving family—these things were not to be lightly discarded by any man. But there was a price a man had to pay for them, the price of his conscience. He must know that these pleasant things came to him through the enslavement and degradation of thousands and tens of thousands of other people. He must know that the food he ate was the flesh of other people, the wine he drank was other people's blood!

There was no escaping this. Could any man of sense persuade himself that the privileged classes were doing or ever would do anything to compensate the masses for the misery and despair in which they lived? No, in this world of economic anarchy it was no matter of justice, it was a matter of power; some had the things by which others had to live, and they sold these things for a price, and enforced the price by the clubs of policemen.

So Hal was going back to the coal-country, to pay for his privileges—his health, his freedom, his culture; he was going to pay the only price that would satisfy the God within him. That health and freedom and culture had been made from the labor of other people; they belonged to other people, and there was only one honest use a man could make of them—devoting them to putting an end to the system of parasitism, and leading the world's wage-slaves into the future of brotherhood and co-operation.

BOOK TWO

GOVERNMENT BY GUNMEN

[1]

The office of the Western City "Gazette" was the place of places for a European voyager, reaching home on a week-day morning, and desiring to catch up with the stirring events of a labor war. The big, disorderly "city-room" was a-clatter with typewriters and telegraph instruments. There was no office boy to bar the door, and Hal strode in, and made his way to a corner where Billy Keating sat pounding his old machine. He looked up; his good-natured, full-moon face lighted with pleasure, and he sprang to his feet, and grabbed Hal by the two hands and shook them till they ached. What a pile of news he had! Here, take this chair—he swept a heap of dusty newspapers to the floor. Just a minute, while he hammered out the tail-end of a story about the gas fight for his first edition; then he would be free for an hour. Or perhaps in the meantime Hal would go in and chat with Pringle.

Inside a glass partition sat Larry Pringle, managing editor, an emperor in shirt-sleeves, upon a throne heaped round with newspapers, pamphlets, clippings, letters, proofs. Lean and white-faced, with much gold in his teeth and a cigarette always between them, Larry's restless eyes searched you as he talked, as if suspecting you had a story concealed in your vest-pocket. He made Hal think of a hawk—though some would have called him a buzzard; it was his business to search out uncleanness in the business-life of the great metropolis of the mining country. A strange kind of heroism, requiring a strange kind of mentality: a man who would live in poverty, refusing fortunes in bribes, for the satisfaction of telling how other men took them; who would imperil his own job, and the existence of his paper, fighting a battle for workingmen who had not sense enough to fight for themselves!

Yes, said Pringle, there was going to be a strike in the coal-country. The operators were going to force it, if necessary, in order to wipe out the labor movement in their three hundred camps. They had raised a fund of half a million for a start; Pringle gave the amounts which the "Big Three" had assessed themselves and the "little fellow". He had had Keating on this job for a couple of weeks—in spite of a fight for cheaper gas which the paper was making.

Billy came in, and two fountains of information poured forth. The reporter had the "real dope"; it was apparently a simple matter for him to ascertain what went on between Peter Harrigan and his lieutenants—Schulman, his general manager, Judge Vagleman, his chief counsel, and the heads of the two other companies which ran the coal-business of the state. At first you might doubt if Billy really knew the things he told; but you would not doubt that for long, for Billy proved his knowledge by the test of exact science, the ability to predict.

Presently he revealed, in confidence, one of the sources of his information. So harsh was Old Peter's rule that the "little fellows" were being driven to revolt. Did Hal know Perry White, president of the Red Mountain Company?

Hal answered that he knew him very well. He had been superintendent of St. George's Sunday-school when Hal was there—a tall, white-bearded old gentleman, very dignified, and with an exacting sense of honor. Hal remembered having heard his father say that "Perry" would not use his pass on a railroad of which he was a director, unless he happened to be travelling on business for that railroad.

He was too good a man to be in the coal-business, said Billy; and it looked as if he would soon be out of it. At last meeting of the secret association of the coal-operators, the president of "Red Mountain" had dared to oppose Old Peter, pleading with the assemblage to meet some kind of a committee of their men and work out a settlement. Peter had flown into a rage with him, and now was seeking to buy him out, for fear he would not "stick".

"Why doesn't he kick over the traces?" asked Hal. "Meet his men and sign a contract with them, as they do in other states!"

"I asked him that," said Billy. "He couldn't do it."

"Why not?"

"He might as well quit at once. There was a fellow, Otway, of Central Fuel, who tried that plan, and it took Old Peter's crowd about six months to land him in bankruptcy. He couldn't get cars from the railroads; and when he investigated, he found interlocking directorates, railroad officials owning coal-stock and working deals."

Pringle took from his desk a clipping from the "Gazette", showing what sums were at stake in this controversy. The miners were demanding a ten per cent increase in wages, which would cost half a million a year; they were demanding a check-weighman, which would reduce the weight of a ton of coal from three thousand pounds to two, and cost one or two million more. So it would pay the operators to spend almost any amount, if they could break the union movement once for all. That was the economics of the situation, said the editor, and Hal might as well know it before he went in.

"You don't think the men stand a chance?"

"You know what happened ten years ago," said Pringle. "Peter Harrigan has the same power to-day. What's to prevent his using it?"

"I don't know," said Hal—"unless it's public sentiment. If people who care for justice and fair play could be roused, they might manage to cast some odium on him—more than a million dollars worth, possibly—"

The other smiled. "You don't know how low a price the old fellow puts on odium!"

And Keating put in, "You should have heard Perry White tell about him at that meeting, with his cigar in one corner of his mouth, and his ugly lower lip sticking out. 'I have a coal-business, Perry, and I intend to run it. Any time you want to pay my price, you can have it; but so long as it's my business, it's mine, and no damned union agitators are coming in giving me orders!' "

[2]

Hal had seen the members of his family, and heard their views. His brother Edward had met him at six o'clock in the morning, having taken a four-hour train journey and spent the night in a junction-hotel, in order to be the first to get the young fanatic's ear. While the train was speeding over the flat prairies, covered with beet-fields and bordered with towering red-brown mountains, Hal was hearing his brother's impassioned arguments. He was surprised by the amount of energy Edward was willing to expend. Could it be that Edward had inside news about what was coming?

In the course of the stormy scene, the truth came out. Edward had heard from Old Peter! Just how, he would not say; but the understanding had been caused to reach Edward Warner that the Coal King had heard of Hal's return from Europe, and the reason for it, and wished to make clear that the young man's meddling in the coming strike would be taken as a declaration of war between the two families. There would be no excuses accepted, no plea that Hal could not be controlled. It was Edward's business to find a way to control him!

And Edward had thought of a way; in his desperation he revealed it soon. The father would be called in, and Hal's income stopped. At present he was accustomed to signing his father's name to checks; but now that pleasant custom would cease.

Hal considered this proposition, and answered, gravely, "I think, Edward, that's the thing for you to do."

"*What*?"

"Cut me off, and let Old Peter know that you've cut me off. That's the way to avert his anger from you."

"Damn it!" cried Edward. "I'm not afraid of Old Peter!"

"You aren't? But I thought—"

"Naturally, I don't want to go hunting trouble—"

"Well, that's what I mean—and it's not fair that you should have to. It's not fair for me to use Dad's money to make trouble for his business. Cut me off, let everybody know that I'm acting on my own. That will settle it."

Poor Edward! He was a man of busines, to whom money is the blood of life, and he was facing the perplexity which confronts such a man when dealing with a matter of conscience, whose life-blood is drawn from another source.

"What would you do?" he demanded.

"I'm going to the coal-fields," said Hal. "I think I can get a job."

"When they know who you are?"

"I don't mean a job with Old Peter. I'm a member of the union, and I think I can earn my salary as an organizer."

When Edward answered, the energy was gone from his voice. "And be killed, like that fellow Olson?"

"Not necessarily," Hal replied. "I should say that might rest with you."

"With *me*!"

"If you should see fit to send a return message to Old Peter, letting him understand that your brother is to have his rights as an American citizen—"

"Oh, my God!" broke in Edward.

"I know," said Hal, "it sounds mad to you, but perhaps it won't as time goes on. I'm going to the coal-country to do my duty, and before I get through I may have to call on you for help. If I do, you'll give it, I'm sure."

"I'll see you in hell first!" cried the other.

Hal answered, "There's no use trying to get mad, Brother. I'm right, and in your heart you know it. You've been to that coal-country, and you know that the men who're running it aren't fit to run a pirate-ship. And just remember this, whenever you are worrying about my safety, whenever Dad is worrying about it—any time you get ready to go to Old Peter and talk to him straight, you can make me fairly safe in the coal-country!"

The train was gliding through a tangle of tracks and yards and coal-sheds, and came to rest in the big depot. Hal alighted and strode down the platform—and there was his father, toddling towards him with stretched out arms. How unutterably pitiful he was, in his mingled happiness and anxiety! He caught the boy in his arms and kissed him; and while they sat in the automobile, he would run his trembling hand over Hal's sleeve, or catch the boy's fingers in his feeble grip, looking pleadingly at him. "You're not going back to the mines, Hal!"

Hal answered, "Let's not talk about that now, Dad. I'm so glad to see you!"

But they could not avoid talking about it. The old man could look at Edward's face, frowning and tense, and read the story there. Tears came into his eyes, and when Hal looked at him, the tears came into Hal's eyes also. It was no joke, this being a revolutionist!

[3]

In the office of the United Mine Workers John Harmon sat at his desk—a man Hal had come to know well, and for whom he had a deep admiration. Harmon was a miner born, his Scotch parents and grand-parents having been miners as far back as he knew. At home Hal had been taught to think of a labor leader as a noisy and pushing person, thriving upon trouble; but Harmon was exactly the opposite of that—gentle of manner, slow-spoken, patient, with a quiet humor which you might miss at first. He was a man of big stature, with features so regular that they might have served as a model for a sculptor. He was not a man of imagination; he did not appreciate his own role, he could not tell his own story—but you knew that he was a solid man, who weighed the consequences of an action before he took it, and having once set forward, seldom needed to change his course.

The miners had chosen him for their best; but he was not good enough for the operators of this district, it appeared. If he had been a bandit-chief, they could not have spurned him more haughtily. In vain did he devise methods of adjustment, in vain did he write letters to the operators, individually or collectively, calling their attention to the discontent of the men, the violations of law and even of common-sense in the camps. The letters remained unanswered, and Peter Harrigan and his associates remained unaware of the existence of such a person as the executive of the miners' international.

They objected to the character of the members of the union, so the newspapers said; but Harmon pointed out that it was the operators who determined this. There could be no qualification for membership in a miners' union, save that the man was a worker in the mines. If the union was not representative of all the workers, whose fault was it—considering the methods used with anyone who sought to increase the membership?

Harmon spoke of the murdered organizer. Hal knew Tom Olson, there was no need to say that he had not been a man of violence, that he had not been in Pedro for any purpose of violence. And as for Pete Hanun and Gus Dirkett, the coal-company detectives who had shot him down in cold blood—they were out on bail, roaming the streets and terrorizing the miners with the very same guns which had done the murder! There would be a trial, some day, but everyone knew the farce it would be. "Alf" Raymond, the sheriff, would be the man who selected the jury; they would put on the stand a couple of Mexicans, who had perhaps never been in the state before, but who would swear they had seen Tom Olson draw a weapon; and on that testimony the jury would acquit the gunmen. They had been doing such things for thirty years in that "Empire of Raymond", as Pedro County was called.

No, said Harmon—and his voice trembled with feeling—there was no sincerity in the contentions of the operators. The reason they would not recognize the union was because they could make better terms with the individual man, could exploit his labor more effectively. They were doing it so effectively that the task of the union leaders was to stave off revolt; and it really seemed as if the other side must know this, and be bent on forcing the issue. Only that morning there had come a telegram from Jim Moylan, who was in the field, telling how thirty-seven men, with their families, had been thrown out of Castleton camp for having attended a union meeting in the canyon.

There was nothing to do but get ready for the struggle. The union had just lost a strike in West Virginia, and the same detective agencies which had crushed it, the same strike-breakers, even the same machine-guns, were being shipped to the West. The union had countered by shipping the tents in which the West Virginia strikers had been housed; but in this effort they had struck a snag. Gunmen and machine-guns had come through on time, but tents had been mislaid. They had been shifted from one railroad to another, from one siding to another, and no one seemed to know just where they were. Of course, said Harmon, with his quiet humor, everybody knew that freight sometimes got delayed, and that shippers sometimes lost their tempers; it would not do to make charges that one could not prove—but it was well known that Peter Harrigan was a director in several railroads, and so were other coal-company owners and officials. Hal found himself suddenly recalling Otway, of "Central Fuel", and his experience with interlocking directorates!

[4]

Hal went for a call on Adelaide Wyatt, and told her about his parting with Jessie and Mrs. Arthur. Adelaide told him about the latest rumblings of Mount Vesuvius, which had been audible to many people in Western City. And then the revolutionary parlor-maid came in—and what a time they had, exchanging news! Mary had just had a letter from Mrs. Jack David, describing the reign of terror in North Valley. Jeff Cotton, the camp-marshal, had been drunk all summer, it seemed, and his treatment of the men was atrocious. The company was getting in Japs and Koreans, and the decent men were going out in a stream. "Joe," said Mary, suddenly, "how do Koreans talk?"

"I suppose there's a Korean language," said he. "I learned a few words from Cho, the 'rope-rider'." And then he laughed. "Are you wanting to study it?"

"I was just wonderin' what we'd do, if they filled up the mines with people like that."

"We'd find a way to reach them, never fear, Mary. There's no kind of people in the world that don't want freedom, and that don't find out sooner or later about standing together."

Mary had been reading a history of the trade-union movement, and also a pamphlet about industrial unionism, the wonderful new idea of "one big union" of all the workers in an industry. So she and Hal had many things to talk about; Mrs. Wyatt said she wished that every well-to-do person in Western City might be provided with a revolutionary parlor-maid, and have such interesting discussions in her home! Mary laughed–she could realize the strangeness of this situation as keenly as any well-to-do person. But then her brow clouded; it was so hard for her to stay here and be comfortable, when she read what was going on in the coal-country! She wanted to know what Hal thought about her going back.

"There's nothing you can do just now," he answered. "The money you send home is more important."

"I might help to wake up the people, Joe!"

"They don't need that–there's enough bitterness and blind discontent. What they'll need are ideas; and if you stay here a while, and study and think, you'll be of more use later on."

"I know," said the girl. "Mrs. Wyatt tells me that. But 'tis so hard, when ye hear about all the sufferin'! It seems like ye could hardly bear to sit down to a table with plenty to eat on it!" She sat with her hands clenched, and there was a quiver in her voice, that went to the souls of both her auditors. Hal knew these qualms–they had brought him home from Europe and his sweetheart. As for Adelaide–she lived the life of her class, she did not want people to say that her interest in new ideas had made her into a "crank"; but when she got this thrill of Mary's, she must have had moments of doubt about her costly clothes and her gracious home!

Hal asked about the Minettis, and learned that they had returned to Pedro, where Jerry was now working. Rosa had written a post-card; she had a new baby, which kept her busy. Another person who had written was John Edstrom. "The old gentleman's been sick again," said Mary. "If it hadn't been for what ye sent him, Joe, he'd 'a starved!"

"I suppose they'd have fed him at least," said Hal. "Or don't they take miners into the poor-house?"

"I never heard," said Mary. "When a miner gets too old to work, he generally drifts away to some other job. Mr. Edstrom says that cold weather's coming, and he's hoping to earn a bit tendin' furnaces. He's sure paid for the help he gave us at North Valley!"

Hal went away with the thought—how many thousands of other men there were all over the country, obscure, unheeded men, paying the same desperate price for loyalty to their class! And all the comfortable, kindly people Hal knew, who went about their affairs of pleasure and profit, leaving these obscure, unheeded ones to be rolled down by Peter Harrigan's machine of greed! Comfortable, kindly people, who had no revolutionary parlor-maids, but who had formulas, whereby they justified themselves in leaving the world as it was. Religious formulas—they were having the poor always with them, they were rendering unto Caesar the things that were Caesar's! And economic formulas—they were maintaining the beneficent system of freedom of contract, *laissez faire* and the "open shop"—while eleven thousand men, with thrice as many women and children dependent upon them, were bracing themselves in anguish and despair for a struggle against annihilation!

[5]

Hal did what one man could. He sought out the comfortable, kindly people he knew, arguing, pleading, adding to the reputation he had won as a fanatic. He went out to Harrigan; but the professor of economics to whom he appealed was reading the proofs of a book on the theory of value. He listened politely while his former student told him that he knew nothing at first hand about industry; but for some reason he did not feel inclined to drop his book and complete his education in Peter Harrigan's coal-camps.

Hal went to St. George's. Will Wilmerding, the assistant, was away; and Dr. Penniman, the rector, had no patience whatever with "agitators". White-haired and dignified, polished and urbane Dr. Penniman was on the surface, but when you dug below you discovered a zealot out of the seventeenth century. There was a ritual and a system of salvation, and these were the things that mattered to erring and mortal man; if the ritual and the system were right, it made little difference what wages a man got, or what kind of house he lived in. It was obvious, of course, that Dr. Penniman did not apply that doctrine to the clergy; his check came regularly the first of each month, and his house was warm and sanitary. But if you should venture, in the most tactful way imaginable, to point out that aspect of the matter, you would stir what seemed an unchristian set of emotions in the bosom of the white-haired and dignified rector.

Everywhere Hal went, these same unchristian emotions seemed to rise to meet him; at his club, at his father's office, on the street. Arguments would be started, and people would show exasperation at the connecting of their ideas with their pocket-books. "Appie" Harding, Hal's cousin, for example—a rising young lawyer who disliked labor leaders, and took coal-company cases when they came along! And if "Appie's" angry dignity annoyed Hal, he might get his comfort from the cynical good-humor of "Bob" Creston, who grinned cheerfully when Hal suggested that his indifference to conditions in coal-camps might be influenced by his engagement to Betty Gunnison, Percy Harrigan's pretty cousin.

Hal happened to run into Miss Betty, coming out of a confectioner's; and what a snapping of black eyes there was! She could not quite refuse to speak to him, and he thought it proper to make friendly inquiry after Percy; he was very anxious for a chat with Percy!

"Percy's where you can't get hold of him!" was Miss Betty's response.

"Where's that?" queried Hal.

"You find out!" the young lady replied, as she stepped into her electric. He helped her in, as he was duty-bound to do, but she did not thank him—she started up the smooth-running, aristocratic machine, and glided haughtily away. And Hal made inquiry and learned that Percy was indeed quite safe. He was traveling with his mother and sisters—just now taking his ease where palm-branches rustle and ukuleles charm the air!

On one of the busiest corners of the business district of Western City stood a tall brown office building, and if you went in and studied its directory, you discovered that the eleventh and twelfth floors were occupied by the General Fuel Company. The "G.F.C." had no need of flaring signs to advertise its presence—it was the master, and if you wanted it, you found out where it was. So Hal came; outwardly calm, but inwardly trembling, he entered the elevator and ascended to the twelfth floor, and walked along the corridor to a door with the sign: "Office of the President." He turned the knob, and entered the Coal King's ante-chamber.

A page took his card, and pretty soon a smooth and decorous young chamberlain appeared—a chamberlain having the modern title of secretary. Mr. Warner desired to see Mr. Harrigan personally? What was the nature of his business? Mr. Warner reminded the secretary that he was known to Mr. Harrigan; and would the secretary kindly present the card? The secretary answered that he would do so; although Mr. Warner must realize that Mr. Harrigan was extremely busy at this time.

Hal took a chair; and presently the secretary came back. He was sorry to have to report that Mr. Harrigan was too busy to see Mr. Warner. He was so very busy that he feared he would not be able to make an appointment to see Mr. Warner. Could not Mr. Warner explain his business to the secretary? The young man said this with perfect politeness, and without a quiver of an eyelash; Hal answered, with the same politeness, and the same absence of quiver, that he would not be able to explain the matter to anyone but Mr. Harrigan. He went out, and retraced his steps to the street; John Harmon's letters were still unanswered!

[6]

The convention of the miners was to meet in Sheridan on Monday morning, and Billy Keating was going down on Friday, to report the situation for the next day's paper. He had suggested that Hal go with him; it would be safer travelling in pairs. So they set out—in the smoking-car, where there was education to be got.

The car was crowded with passengers, of a type easily recognized by one who had lived in the coal-country. Rough, evil-faced fellows with revolvers and whiskey-bottles bulging their pockets, they sprawled over the seats, filling the air with the odors and sounds of the bar-room; they leered at the women passengers, making jests and singing ribald songs. There went a load of them every trip, said the conductor. There had been a fight on the last trip, and two had been thrown off the train.

Billy Keating knew more than one of these "huskies". In his capacity as reporter he had frequented their haunts, and could tell anecdotes about them. They were in the pay of the Schultz Detective Agency. The great Schultz himself had come to Western City, and made his headquarters in a basement-room of the Empire Hotel, where the "tough" citizens of the West were welcome; there were free cigars, and to a limited extent, free liquor. The ward politicians, who marshalled the gangs to stuff the ballot-boxes and slug the reformers on election-day, were now recruiting for Schultz, and no man who was handy with his gun need go thirsty. From top to bottom, the political machine was being got ready for service; even up to the Governor, a gentleman who had been given his nomination at a secret dinner conference in Old Peter's home, and who would now have a chance to pay for that costly dinner.

In the course of the day's ride, Hal got into a chat with half a dozen of these "huskies". There was no shyness about them, they were entirely willing to tell about themselves, their histories, and their intentions. One could gather wild tales of adventures in every corner of the world; there was an ex-policeman from South Africa, discharged for drunkenness; an ex-soldier, who had demonstrated the "water cure" upon Filipinos; an adventurer from Central America, who had fought wherever there was loot; a pickpocket from the "Barbary Coast" of San Francisco; a couple of gangsters from "Hell's Kitchen", in New York—men who had not been out of prison long enough to grow their hair. Only one question was asked by the Schultz Detective Agency: "Do you know how to shoot?"

Arriving in the evening at Sheridan, the travelers found the station crowded with outgoing parties. There was no reason why any man should stay and face the coming trouble, if he had the price of a ticket to some other job; so here were miners and their families, natives of a score of lands, with huge bundles on their heads or slung upon their backs; there was pushing and jostling, messages of farewell in many tongues, crying babies, shouts of hack-drivers. Not far away was a street-meeting, with an Italian orator haranguing a cheering throng.

Hal and Billy drove to the headquarters of the union, where they found another Babel; swarms of people who had been turned out of their homes and had no place to go, with distracted union officials trying to make them understand why the union had not provided a place. Old Johann Hartman, secretary of the Sheridan "local", and Tim Rafferty, his assistant, were besieged. They had not slept for the last four nights, said Tim; the telephone never stopped ringing—and they had reports to make out, letters and telegrams to answer, a hundred organizers to keep in touch with, and twice as many spies and detectives to dodge.

Hal sought out Jim Moylan, the district secretary, a long, tall, black-haired Irish boy, who had come to take charge of this chaos and bring it to order. Eager and sensitive,

Moylan was a fountain of news, poured out in a torrent. He made you see it and feel it—the enthusiasm, the pent-up energy, the thrill in the souls of these toilers, who were hoping, daring for the first time in their lives. His black eyes would blaze as he told of some fresh outrage; but then he hastened to add a word of caution—one must not believe everything, for there were no end of spies posing as miners, and they too had stories to tell. Now and then one would come in to headquarters, declaring that he had been robbed or beaten, and must have a gun to protect himself. Would not the union give him a gun? Or perhaps he had discovered a *cache* of weapons belonging to the operators, and wanted some of the miners to form a raiding-party to take possession of this treasure!

[7]

Billy Keating made notes, and then with Hal went out to wander about the streets. There were meetings on every block, it seemed—the ordinances of the town of Sheridan had been temporarily forgotten. A man stood upon the tail of a truck, addressing a little group in some strange tongue, and as Hal came near and made out the orator's face, he exclaimed: "It's Mike Sikoria!"

They stood and listened to the flood of Slovak eloquence. All Hal knew of the language was its commoner swear-words, but he had heard Old Mike discuss short weights and coal-company graft in English, so he had no trouble in imagining the speech. Presently at a pause, he hailed the orator—and then what a time there was! The old fellow clambered down from his platform, and gave Hal one of his grizzly-bear hugs, and half a dozen of his tickling, hairy kisses. "My buddy! My buddy!"

He shouted something to his fellow-countrymen; and again Hal could imagine the words—here was the rich young fellow who had come to North Valley and got a job and helped the miners! The other Slovaks grinned, and Mike patted Hal on the back, and would have had him make a speech—he was so proud of his American "buddy"! But the "buddy" lured him away by the suggestion of a lunch room. Was Mike hungry? *Pluha biedna!*

Billy Keating went to the hotel, to lock himself up in a room and get his story ready; and meantime Hal and his old instructor sat gossiping away. Mike heard with amazement that Hal had been abroad, and had come back for the strike. So that was the way these rich fellers did—running about over the world! Mike had done some travelling himself—but after the fashion of poor fellers. It was the old story, he said; he could not keep his tongue still while he was being robbed. But now was the chance of his life—he could talk about his grievances all he pleased, and there were throngs on the street to listen. "And I talk to them, you bet!"

"Have a hunk of apple-pie?" said Hal; and Old Mike grinned and nodded. His mouth was full of "sinkers" and coffee.

"I fill myself up," he said—"if you think you got money enough." He had thrown up his job at Barela because he would not miss this convention; he had started before dawn, and walked all the way—having only forty-two cents in his pocket, his balance on the pay-roll of the mine. He had worked there five months, worked like a mule, by Judas; but there was a son-of-a-gun of a pit-boss, that had charged him twenty dollars for his job, and then gone ahead and loaded him up with "dead work". And the worst of it was, the union wouldn't pay strike-benefits until the strike had been on for a week! They must have made that rule for some other part of the country, where a man could get a week's living ahead!

The old Slovak accepted a loan of two dollars from Hal, but he declined to stay at the hotel with him; he was lousy, he said—they charged you a dollar a month for wash-house privileges at Barela, but all they had was one tub to wash in, and when you saw the diseases some of them fellers had, you'd rather keep your own dirt, by Judas!

[8]

All day Saturday the human floods poured down the canyons; and in the afternoon there was a great procession, with a brass band at the head, and painted signs to proclaim men's feelings. Alone, they had been helpless, but in this throng, with the big national organization of miners behind them, they would assert their self-respect and win their rights. Men whose shoulders were bowed, whose figures were deformed by a life-time of cruel toil, marched here with their heads up, making the street to ring with their "union song":

> "We'll win the fight today, boys,
> We'll win the fight today,
> Shouting the battle-cry of union!"

All Saturday and Sunday there was oratory; and on Monday, the day of the convention, few orators went back to the camps. It would be of no use, for there were spies watching them, making lists. So they thronged into the convention-hall and listened to the proceedings, and backed up the delegates with their applause. One could feel their excitement, the pressure of feeling that burst forth in murmurs of indignation, cries of resolve.

Those who came as delegates to the convention had been chosen by secret ballot, and knew that their appearance here meant the expulsion of their families from their homes. Nevertheless they came—more than two hundred men, sober and determined, driven by a sense of intolerable wrong. They came as representatives of eleven thousand toilers, who sent to the great world outside an average of eleven million tons of coal each year.

They were uneducated men, with no gifts of oratory, no experience in affairs. To them this gathering was the event of a life-time, the moment when they were called upon the platform a mighty crisis. But this also was a duty; one by one they came forward, and in the best English they could muster told the story of their grievances. They came from two hundred different camps, their destinies were under the control of more than seventy different companies—yet the stories they told were all alike! Hal Warner sat and heard them, and found himself thinking that if he had shut his eyes, he would have been unable to tell which of them was the delegate from North Valley.

There was, first of all, the issue of poor pay, the inability of a man to earn a living for his family. Said Delegate Gorden, "There's too much rock in the mine. When I was working there, I couldn't make my day's wages in the place I was in." Said Delegate Obeza, "If a man goes to work at three o'clock in the morning, he can make three dollars a day. If he goes to work at seven o'clock he make about a dollar fifty, because all his time is taken up cleaning rock on the roads." Said Delegate Lamont, "The little children go bare-footed and are half clad."

There was the old story of "short weights", with its endless variations. Said Delegate Talerbeg, "The cars hold from forty to forty-two hundred, they give us from twenty-six to twenty-nine hundred." Said Delegate Dominiche, "We have from seven to eight hundred pounds of coal stolen every day, and we don't get paid for laying tracks." Said Delegate Miller, "When they're in need of coal for the boilers, they stop a trip near the boiler-room and unload coal off the miner's cars." Said Delegate Madona, an Italian with a grin, "Never load by the ton; load by the acre."

These men, like Hal Warner, had made test of the check-weighman law. Said Delegate Harley, "I know of four men who asked for a check-weighman and were fired." Said

Delegate Duran, "Never ask for a check-weighman, because we would be fired if we did. Weight is very bad." And Delegate Salvine revealed a new device: "Pay thirty-five cents a month to company check-weighman. The boss put him up there, and we pay him."

There were the complaints of miscellaneous grievances, stories to which Hal had listened on so many occasions from all over the district. Said Delegate Costo, "House is in bad condition, when it rains we have to get under the bed to keep from getting wet." Said Delegate Fernandez, "Conditions are very bad. We can't travel through the man-way, and have to risk our lives going through the haulage-way." Said Delegate Miller, "There's a company saloon, grocery-store, and doctor in that camp. This doctor has caused a number of people to be cripples."

Everywhere was the same treatment of men who protested, the old story of "down the canyon with you". "Men are fired as soon as it becomes known they are members of the union or have an inclination to be." "If a man appears to be a union man he is fired." "The boss found out I was a union man, and every time he gets a bad place he puts me in it, and as soon as I get it cleaned up he puts me in another bad place." And so on, man after man, camp after camp, for fifty miles up and down the line! "They told me this was a free country," said one English miner. "But I have found out that it is not a free country!"

[9]

On the evening of the first day the chairman announced that the convention would listen to an address by "Mother Mary". There broke out a storm of applause, which swelled into a tumult as a little woman came forward on the plat-form. She was wrinkled and old, dressed in black, looking like somebody's grandmother; she was, in truth, the grandmother of hundreds of thousands of miners. The masters had put her in prison, sometimes they had beaten her, on one occasion they had shut her in a stockade with smallpox patients. But nothing daunted her spirit; when "her boys" called, she answered, even though it was all the way across a continent.

Hearing her speak, you discovered the secret of her influence over these polyglot hordes. She had force, she had wit, above all she had the fire of indignation—she was the walking wrath of God. Her address was what the cultured classes would describe as a "harangue", but it suited "her boys", it swept them to ecstasies of resolution. For so many years they had endured—now they would stand together and endure no more!

Her purpose was to lift the spell of fear which lay upon their souls. "Don't be afraid, boys; fear is the greatest curse we have. You fear because you don't know your power; but it is only because of your fear that you are powerless!"

She told about the strike in West Virginia, the fierce revolt in Cabin Creek. There had been a stone-wall built there, and no organizer dared go beyond it, or he would come out on a stretcher. But two young lads had come to her, and asked her to attend a secret meeting in that canyon. "The men came over the mountain with their toes out of their shoes and their stomachs empty; fifteen hundred men of every description gathered there. Some of these men looked up at me, as much as to say, 'Ah, God, is there a grain of hope for us?' Others would look at the ground, thinking, 'All hope is dead.' When I was about to close the meeting, I said, 'Boys, let mother tell you one thing; freedom is not dead, she is only resting; she is sleeping, waiting for you to call.' The voices of those fifteen hundred men rang out, 'Ah, mother, we will try to be true! Will you organize us?' They lost all fear, they came forward as one man. And when I organized them, I said, 'Put on your mining clothes tomorrow, don't say a word about this; don't speak of it in the mines. Take your picks and continue to dig out the wealth; be good and don't make any noise about it.' But they were discharged, of course."

There was a "lady" who had been keeping a rooming-house in North Valley; and last week the superintendent had said to her, "What are you going to do when the strike comes on?' She had answered, "I'll not feed any scabs." And he had given her two days to move! "What do you think of that?" cried Mother Mary; and you heard a fierce murmur from the crowd. They would not feed any scabs—not they! Men who would steal other men's jobs in such a crisis, who would deprive their fellows of a hope of freedom and life! There was a strike of the brewery-workers in Sheridan, and Mother Mary warned them not to drink any beer that had been made in the town. "There are scabs in that beer!" said she.

[10]

On the second day, the convention listened to the report of John Harmon, with the list of demands which the policy committee had sent to the operators. The majority of the delegates were for a strike at once, but Harmon pleaded for delay. Perhaps there might yet be some concession; perhaps when the operators read reports of this convention, and saw how wide-spread and intense was the feeling of discontent, they would recede from their present position.

At least, they should have the opportunity; Harmon urged that they delay action for a week, and send one more letter to the operators. There was a hundred thousand dollars a week in wages at stake; there would be thirty thousand dollars a week in strike-benefits to be paid by the "big union". With sums such as these involved, the officials in charge were not apt to fail in caution. A strike was like a war; when the fatal word had been pronounced, it must go to the bitter end, and no one who had had experience of its cruelties could be eager to speak the word.

The convention voted as Harmon asked; but Billy Keating, talking with Hal after the convention had adjourned, declared that the action was a mistake. It was only giving the enemy more time; there would be no concessions, the strike was inevitable. In his character of reporter, Billy met both sides; he had been that day in the sheriff's office, and seen the hard citizens who had come down in the train being made over into guardians of law and order. Many of these men, the day before, had cringed and slunk away at sight of a policeman; but now they had mumbled a magic formula, they had the powers of government on their side—and three-fifty a day and board in the bargain! There was a state law providing that no man should be a deputy-sheriff until he had been a citizen of the state for a year, and had lived sixty days in the county; but Billy had watched the sheriff-emperor violating that law in batches—twenty times a minute!

They were sitting in the lobby of the hotel, and saw a man go past—a tall, lean person, black-browed like a villain in a stage melodrama, dressed like a gentleman from the far South as he is imagined in romantic fiction. He peered about him through eye-glasses with thick lenses, and stared hard as he passed the two young men.

"Know who that is?" said Keating.

"No," replied Hal.

"Can you guess?"

"It looks like somebody who thinks he's Sherlock Holmes."

"That's Schultz."

And Billy went on to picture the great strike-breaker in the sheriff's office, marshalling the deputies, giving them orders over the sheriff's head. Schultz was a powerful and sinister figure in the labor world; in order to find anything like him you would have to go back to the days of the Italian *condottieri*. Someone had told Billy about these hireling armies, and he had looked them up in the encyclopedia, and it was the very same thing; this Schultz might be a reincarnation of Francesco Sforza, of the fifteenth century! Schultz commanded a private army of five thousand men, horse, foot, and artillery; the whole of it could be shifted three thousand miles in a week, and wherever it went it took over all the powers of government. He paid his enlisted men three dollars and a half a day, and he charged the corporations whose work he did at the rate of five dollars for each man. So, said Billy, the profits of the business depended directly upon the amount of trouble there was; and from this grew the most sinister fact about this system of "Government by Gunmen". Without exception these big strike-breaking concerns maintained a secret department, and while they were putting down violence with one hand, they were fomenting it with the other. In these coal-camps they had had spies among the miners for years; so that men who could present the best of credentials as strikers were occupied in inciting their fellows, putting murder into their hearts and revolvers and dynamite into their hands.

Which added peril to a situation already perilous enough. Union officials might plead as eloquently as they chose against disorder; but with eleven thousand strikers to handle, speaking more than a score of tongues, and scattered over hundreds of square miles of mountain country, it was not possible to censor every orator, nor to make sure of the honesty of every leader who might come to the fore. When you had, as you had in this district, knavery and oppression enthroned for a generation, with the deliberate bedevilment of every good instinct of humanity—then you had created a volcano of passion whose possibilities you could hardly imagine. Hal was soon to find what it meant to be a labor leader, held responsible by the public for everything that might be belched forth from such a volcano!

[11]

The convention adjourned, and Hal went up to Pedro in company with Jim Moylan. The union had opened a headquarters in Pedro, and a steady stream of discharged union men and strikers-to-be poured into this place. In the week before the strike practically all the delegates to the convention were turned off. The General Fuel Company had a clause in the lease to its houses, giving it the right to evict the tenant at three days notice; but in most cases they did not even give this time, but dumped the family and its belongings out on the street. The company was making a census of its men, asking them if they intended to strike; those who answered yes were turned off immediately. Under this cheerful system, it was hardly surprising that the company was able to publish in the newspapers a statement to the effect that ninety-five per cent of their men were opposed to a strike.

Hal went first to call upon his friend John Edstrom. The old Swede had never really recovered from his beating by the mine-guards; his kindly old face was more nearly the color of dough than ever, and his dark, sunken eyes made you think of a friendly skull. He was trying bravely to become self-sustaining, but he was a marked man in this district. Old and feeble as he was, the "G.F.C." seemed to consider it necessary to have a detective keeping track of him everywhere he went.

The garret-room in which he lived had boards laid on the floor, and a curtain to divide off a portion, where the Minetti family cooked, ate and slept. Big Jerry had become a well-known organizer, doing work among the Italians in the town. Rosa had her new baby, another boy—the cutest little round-headed Dago doll that ever you laid eyes on, with sharp black eyes, and the softest silky black hair. The girl-mother was divided between rapture over this treasure, and terror for the fate of its wonderful father. Since the killing of Tom Olson, many organizers went armed; they were pledged to a defensive attitude, but it was not always easy to define what such an attitude should be. There were never less than half a dozen of the gunmen hanging about the door of the newly-opened headquarters, and when an organizer entered, these men would bristle like angry dogs. One could stand the bristling; but suppose one of them reached to get his handkerchief at that moment? They carried "handkerchiefs" of a large size, making a conspicuous bulge in the side-pockets of their coats.

Big Jerry would come home and tell about these things, and Little Jerry would listen, thrilled beyond utterance. He played all day at stalking gunmen with the little boy of the landlady; but they were always disputing, because neither wanted to be gunman! The Dago mine-urchin was blood-thirsty in his intentions, and it was comical to listen to his disputes with John Edstrom, who, for lack of a grown-up audience, would expound his pacifist philosophy to Little Jerry. The youngster was naturally in awe of this white-harired old man, whose stories of labor-strife he had heard so often; but he had a hard time comprehending the program of turning the other cheek to Peter Harrigan's gunmen. "Suppose one of 'em beat up my mother?" he would cry. And Hal would smile, hearing the echoes of an age-old controversy!

[12]

As Billy Keating had predicted, the operators paid no attention to the latest communication from the union, and so everyone knew that the strike was inevitable. The union leaders were making frantic efforts to get ready, especially to provide shelter for the tens of thousands they would have on their hands. The tents were still missing—and in September it is cold at night in these mountain regions! Hal, who had taken a great "shine" to the long tall Irish boy, Jim Moylan, and was trying to help him in every way possible, spent part of each day interviewing freight agents, and telephoning and telegraphing railroad officials. When his efforts broke against the sýstem of interlocking directorates, he took to telephoning to Western City, and started a miniature war-boom among the awning manufacturers of that metropolis.

The strike was scheduled for Tuesday, and as early as Saturday and Sunday the great exodus from the camps began. All the way down the line for fifty miles the canyons poured out their human flood. They poured into the towns, thousands upon thousands of them; as one observer phrased it, it was the migration of a race.

It seemed as if the elements had entered into a conspiracy with the forces of capitalism, for the flood-gates of the heavens were opened on that day. Up in the canyons it was snow, in the valleys it was a deluge of rain mixed with sleet, which drenched everyone to the bone, and soaked all their pitiful belongings. The stream of wagons came into Pedro—farm-wagons, express-wagons, broken down carriages and hacks, even hand-carts—their wheels solid with thick, greasy mud, the piles of furniture and bedding dripping like trees in the forest. Here and there you saw faces peering out, faces of hollow-eyed and shivering children, of babies who seemed to have been suckled upon fear.

Jim Moylan and his college-boy lieutenant were in charge of headquarters that Saturday, and it was an experience never to be forgotten by any man. It seemed that everybody arrived late in the afternoon. The union had hired a hall, and to this the throngs came; there must have been three thousand people packed into the place—it was hard to see for the steam of so many water-soaked bodies. If you had anything to do with the union, you were besieged by fifty men and women at once, trying to ask you questions in what appeared to be fifty different languages. The Tower of Babel was the only thing one could compare with it; but there could have been no such desperate haste in the Tower of Babel, for if a stone was not got into place until the next day, it would not matter so much—it did not snow or sleet in that neighborhood, and there were no shivering women and children to be got under shelter.

Yet, incredible as it might seem, it was not for food and shelter that these people clamored first; it was for their souls they wanted food—their souls which had been starving for years in those lonely mountain fastnesses. Wet, shivering and hungry as they were, they were aflame with excitement and enthusiasm, and their cry was for "talk". Jim Moylan had to "make talk", in the kind of pidgin-English which passes in the coal-camps; and then he had to find men to "make talk" in various other languages. Because he did not always know these talkers and could not trust them, he had to find interpreters and act as censor to these floods of foreign oratory. Hal had had this same experience in the little strike at North Valley, and knew all about it. In one or two cases he even knew the orators—the Italian orator, the Bohemian orator, the little Greek orator who tore all his passions to tatters, fastening his fingers in his long black hair and tugging desperately, demonstrating the ancient lesson of solidarity. "Pull one hair, he come out; pull all hairs, no come out!"

They worked out a system for handling the throngs, and when the oratory was over, Moylan gathered the interpreters around him. He asked one question, and each interpreter in turn shouted it aloud in his language. Everybody who had to have food must come to a certain part of the room; and when that had been supplied, there was a cry for everybody who had no place to sleep. The first tents from Western City were due to arrive in the morning, and the people from certain camps must be sent to certain places. So on through hours and hours, while the crowds thinned out; but meantime new crowds had been coming in, and they must hear new speech-making. All the while the rains continued without let-up, and water and mud and strikers poured out of the canyons in one turbid flood.

[13]

The little town of Horton had been selected as the site of one of the biggest of the tent-colonies. A tract of land had been rented, not far from the depot, and it had been arranged that the people from the Northeastern, from Barela, Greenough and North Valley should make their homes here. The lumber for platforms was already on hand, the tents were on the way, and one evening Hal and Jim went over to help get things started.

When they got off the train they were met in the dark, ill-lighted depot by a Polish miner, Klowowski, who had been a member of Hal's check-weighman group. He was a frail little man, or seemed so just then, hollow-eyed and haggard of aspect. His clothing was soaked through with rain, and his teeth chattered as he talked, which made his broken English still more difficult to understand. But his face lighted with pathetic delight when he saw the rich young fellow who had come to help the miners. There was a "big lot strikers" wandering about the streets, he explained; they were "very scared", but now the union had come to them, they were saved!

"No eat two days," declared Klowowski. "No got place for sleep, sleep out two nights in rocks." But he had not let the other strikers know this, for fear it might discourage them!

Here again, however, it was not food and shelter that men wanted. Truly has it been spoken that man does not live by bread alone! The little Polack wanted to be told that he was right in defying the tremendous power which had been God and king as well as master to him. He and his friends wanted to be told about the "big union" that was back of them, and would stand by them. If only they were made to feel the reality of this "big union", they would endure such things as rain and cold and hunger.

So Moylan must "make talk" once more. Let the men gather in the vacant lot by the blacksmith's shop, and he would come in an hour. Hal gave the little Pole a quarter and warned him to get some food if he did not want to be ill; but Klowowski had too much on his mind to think of food. While Hal and Moylan sat in an eating-room, a place crowded with strikers, the door was pushed open and the rain-drenched form of the little Pole appeared in the entry. He did not see the two, but lifted his voice and shouted: "Meeting by blacksmith shop! Big men from union come make talk! Ever'body come! Tell ever'body!" And away rushed this Paul Revere of the coal-camps!

It was impossible to tell how many people attended the meeting. It was pitch-dark, and they had only two pit-lamps, turned upon the face of the speaker. Beyond beat an ocean of human sound, murmurs of indignation, cries of resolution, long salvos of applause. And this in a drizzling rain, with mud and slush underfoot, utter darkness everywhere, and half a dozen gunmen, armed with "thirty-thirty" rifles, prowling about on the outskirts of the crowd!

After the meeting Hal went into a saloon, where he found a group of his old North Valley friends, trying to dry out around a stove. There was Wresmak, the Bohemian miner, and "Big Jack" David, the Welshman, with his wife and their two young children; also Tommie and Jennie Burke. Hal had been inquiring for the children, having promised Mary that he would look after them. He had written to North Valley, and now he learned that his letter had not been received. Could it be that the power of the Schultz Detective Agency affected even the United States mail?

A crowd had been turned out of North Valley that morning, because there had been a union meeting in Jack David's home, and a spy had got in. Mrs. David had refused to leave without her belongings, and had given the camp-marshal a tongue-lashing; there had been a mix-up, and "Big Jack" had nearly got a broken jaw. It hurt him to talk; but the condition of his black-eyed and hot-tempered little wife was such that it would have hurt her not to talk. There were more than seventy men in North Valley, she declared, who wanted to quit work and were being held inside the stockade. The bosses were turning out the active and intelligent men, from whom they had things to fear; they were holding the ignorant and defenseless, on the pretext of debt to the company. Most of them had been in debt ever since they had come to the camp—and what chance would they have to pay it off while there was a strike? Being non-English speaking men, for the most part, they believed what the bosses told them—that they would be sent to prison for long terms if they tried to get away without paying what they owed the company.

"What chance have we if such things are allowed?" cried the little Welshwoman. "If they can work their mines with slaves, the whole strike will go to pieces."

"The men will get out somehow," said Hal. "They'll not be able to hold them long."

"But they're doing it! They're doing it everywhere!" insisted the other. "Up in Barela they won't let anybody out, only those they think have been agitating! There's a Greek fellow here who was fired yesterday—hear what he says!"

So Hal had his first meeting with "Louie the Greek", a man of whom he was to see much. Louie had been a teacher in Western City, but seeing in this situation a chance to uplift his fellow-countrymen, he had come down and gone to work in a coal-mine. He was a man about thirty years of age, quiet and rather shy, but having under his scholarly appearance unflinching resolution, which showed itself in times of trial.

Louie substantiated all Mrs. David had said. He had seen women and children thrown bodily from their homes, and all their belongings flung out into the mud and slush. He had seen families crouching out in the open, not knowing where to turn, because the doors of their homes had been locked behind them, while at the same time access had been refused to wagons to take them away. The object of these proceedings was to intimidate those who were still wavering. They were notorious at Barela for cruelty to their workers; it was in this mine that more than five hundred men had been killed during the past three years. And now they had a machine-gun mounted at the gate, and a fellow named Stangholz would handle this gun, turning it about and aiming it at the strikers, jeering at them and telling them how with this "baby" of his he would wipe them out.

Among Louie's countrymen Hal encountered the breaker-boy, Androkulos, who had been one of the first of his acquaintances in North Valley. "Andy" was an ardent striker, and had been "shipped" the week before, together with his old mother and his two young sisters. They had been all this time without shelter, without blankets or even a change of clothing. And when at last the young fellow had got a wagon, and had made the long trip up the canyon to get his property, Jeff Cotton had refused him admittance to the camp.

The boy had just got back, almost beside himself with rage. "I pay eight dollars for that wagon!" he cried. "And I got to pay it just the same for nothing! What right they got to keep my things, hey? What right they got?"

Hal tried to comfort him. "I'll help you find some place for your people," he said. "We'll get you dry clothes–"

"But I wany *my* clothes!" protested the boy. "By Christ, I kill them fellers inside that fence! If I had gun, I kill Jeff Cotton yesterday."

"No, Andy, don't kill him, that won't do you any good. They'd only hang you."

"They might as good hang me as treat me like they do! My father work in that mine ten year, he get blowed to pieces in that mine, and what I got for it? Only some old things in that house, and I can't get them! I tell you I get a gun and go back there, if I don't get my things I send somebody to hell!"

Louie broke in, speaking his own language. There was a voluble argument between the two, from the gestures and tones of which Hal gathered that it was another such controversy as he had heard between Edstrom and Little Jerry. The older man was gentle and persuasive—and gradually Hal saw the storm die down. Without having understood a word of Louie's utterance, Hal was brought to realize that here was a remarkable man, a personality to count upon in time of trouble.

Presently the Greek turned to him, and shook his head sadly. "It will be dif-fi-*cult*," he said—(he pronounced the word very carefully, even if not correctly). "It will be dif-fi-*cult* to man-age our people." Hal noticed that Louie used many book-words, and did not always get them straight; the reason being that he had taught himself most of the English he knew. He was a man of reading, had studied medicine in the University of Athens.

[14]

Hal got Billy Keating on the telephone, to talk the matter over with him. Billy urged him to come to Sheridan and try to get some satisfaction from the sheriff of the county; Jim Moylan, who stood by, backed up the suggestion. Hal could perhaps accomplish more than one of the strike-leaders, because he was the son of a well-known millionaire.

So Hal took the morning train to Sheridan, and visited the courthouse, which was like a military camp, with men carrying rifles everywhere. They glared at him as he went in—evidently having "spotted" this son of a well-known millionaire. He sent in his card to Alf Raymond, the Emperor of Pedro County, and without delay was ushered into the throne-room.

Hal had heard much about this sheriff-emperor, but had never seen him before. He was a large man, with sandy hair and a red face with many freckles; pot-bellied and gross—you could see that he was in the wholesale liquor business in more than one sense of the phrase. He was a man of no education, but his eyes revealed animal cunning.

He was polite at the start. His wealth and power had been got by serving the rich, and this young man belonged to the class he was accustomed to defer to. He listened to the young man's stories, and then began to explain the difficult position he was in. How could he get the truth about anything, when both sides were so busy telling lies? But of course it was absurd on the face of it that the companies should attempt to force men to work. You might drag a man down into a coal-mine, but you couldn't make him dig. Moreover, the sheriff had made inquiries about matters at North Valley and was satisfied there was nothing to the tales. He knew Jeff Cotton, the camp-marshal, and had written him a letter.

Hal answered that he had seen the letter—a very polite communication, calling the camp-marshal's attention to the fact that he ought not to keep people from getting their belongings. The communication had got into the papers, and had made some talk.

Yes, replied the sheriff-emperor—there was a newspaper-fellow by the name of Keating, who had got hold of a copy in his office. The sheriff-emperor made it clear that newspaper-fellow would not get into his office again!

Hal had before his mind a vision of women and children without blankets and clothing; so he was not to be put off with empty words. He insisted that the sheriff should go up to the camp without delay, and see that every person who wanted to come out was allowed out, and that every person who wanted his property was allowed to have it. But the sheriff had no idea of taking any such steps, and as Hal continued to push him, he became annoyed; when finally Hal began threatening him with public opinion, he turned red in the neck and brought his fist down on his desk. No "outside agitators" were going to come in to run this strike! When Hal inquired as to whether the head of the Schultz Detective Agency was regarded as an "outside agitator"—"Young fellow," said the sheriff, "you get this through your head: you needn't think because your father's got a lot of money, you can come into these here camps and stir up trouble! Take my advice and quit this game; you may get in too deep, even for the stakes your old man can put up!"

And that was all the satisfaction Hal got in his first and last interview with the sheriff-emperor. The truth was the monarch had abdicated; it was Schultz who was running the empire now, and all the sheriff had to do was to swear in deputies. He had sworn in some four hundred so far, and there were four hundred more waiting for commissions. Later on, when these events became the subject of government investigation, "Alf" Raymond admitted that he had known nothing about any of these men, and had made no effort to find out anything about them.

Billy Keating was waiting outside to hear the results of the interview. He had telephoned to Pringle, and been told to get a camera and a rig, and go up to North Valley and see for himself how matters stood. Of course he wanted Hal to go with him, and to figure in the story. So far, at Hal's request, the "Gazette" had printed nothing about the presence in the coal-country of the son of a well-known millionaire; the other newspapers had also kept silence, for a different reason—they were reporting things from the point of view of the operators, and if they mentioned Hal's presence, they would have to give more of the strikers' side. The situation was a trying one for reporters and editors; for if there is anything that American newspaper readers wish to hear about, it is the doings and sayings of the sons of well-known millionaires.

Hal was willing to accompany Billy, but predicted that they would not get much of a story, because they would not be allowed near the North Valley camp. And sure enough, about half way up the canyon, just above the fork to the Northeastern, they came upon three men in a shelter-tent by the roadside. The men had rifles in their hands, and one of them stepped out and held up his hand. "Got a pass?" he inquired.

"What sort of a pass?"

"Company pass."

"No."

"Then you can't go by."

"Why not?"

"It's orders."

"Whose orders?"

"Company orders."

Keating started to argue. "This is a public road! It's a United States mail-route; it leads to a United States post-office."

"You can't go by," said the man.

"I have business in North Valley, and I have a right to travel on this road. I'm a reporter from the Western City 'Gazette'."

"I don't care if you're God Almighty, you can't pass this place without a permit from the 'G.F.C.'."

"And suppose I insist on going?"

The man swung forward his rifle, and laid it in the crook of his arm. The two in the buggy could see the round black eye of the muzzle, looking at them unwinkingly. There was no other reply, and no need of any.

[15]

Hal and Billy consulted, and were about to turn and make their retreat, when there came a diversion from up the canyon. Some figures appeared round a curve in the road—a procession of men and women walking; with the familiar bundles of bedding and other belongings a-top. At the end of the procession came two men on horse-back, wearing ponchos, and carrying rifles. Hal and Billy backed the buggy out of the way; the sentry watching them meantime, never once taking his eyes off them.

It was still pouring rain, and the figures and the bundles were soaked through and dripping. The scene was one for a painter—a painter who, like Hal, had been up to the head of the canyon and knew what was to be found there, and could read the story of each individual face and figure.

There came Charlie Ferris, the big miner whose steam-siren voice had supported Hal on the night of the North Valley uprising. His face was now set and desperate; he looked, and saw Hal, and seemed about to speak, but thought better of it and marched on. There was a bandage about Charlie's forehead, with a splotch of blood showing through. Behind him plodded his wife, a stoutish woman with three children at her mud-draggled skirts; the children slipping and staggering in the mud, which in places was over their knees.

Then came—of all people—"Blinky", a young fellow whom Hal had come to know as the vaudeville specialist of North Valley; a lad from the East Side of New York, with sandy hair and merry eyes, and a mouth organ always in his trouser-pocket. Nothing ever seemed able to repress his high spirits; he could execute amazing fancy steps, and everybody and everything in sight was the butt of his wit. He was horribly obscene, of course, but Hal had been able to forgive this, for the wonder of seeing laughter upon the faces of the underworld men. The children would gather round "Blinky" like flies about a honey-comb, and even the bosses were not proof against his spell. They would curse, and bid him be off to his work; but they would grin while they said it, and he dared to answer them back, for all the world like a court-jester to a king.

And now, what had happened, that even "Blinky" had fallen from grace? Had he himself revolted, or was he sharing the fate of some other member of his family? There followed him an elderly woman, a bedraggled young girl, and a little stunted man whose back seemed breaking with his load of goods. "Blinky" himself was irrepressible, in spite of his burden; seeing Hal, a grin came on his face, and he stuck his thumb back at the two riders, and proceeded to walk the lock-step.

There came a Slav miner whom Hal did not know, followed by a woman with a baby in her arms, and another baby conspicuous in her body; then a half grown girl, dragging another child, which had evidently fallen flat in the mud, and was whimpering in a pitiful way. There followed two Italians, with their fists clenched, and hate in every feature; one of them Rovetta, a young fellow who had belonged to Hal's check-weighman group.

There came others whom Hal did not know; one bent-backed old hag, a Mexican, with silver grey hair streaming down her back, and a raw cut in her cheek, from which the blood had run down to the mud on her dress. She looked at the guards by the road-side, and Hal thought he had never seen such fury in a human countenance.

And last came the two riders: one of them slight, with a weasel face and sharp black eyes—Jake Predovich, the Galician Jew who acted as store-clerk and general interpreter to the camp, also as member of the local school-board, and "spotter" for the bosses. The other man was tall and lean, with a face that had once been handsome, but now was rotten; he sat his horse straight, his restless eyes fixed on the people ahead, watching every move. With his alert look and rather long neck stretched out, he made one think of a partridge sitting on a bough. Hal waited till he was passing, and then spoke: "Jeff Cotton!"

The camp-marshal started, and stared; then, recognizing Hal: "Well, I'll be damned!"

He checked his horse, and several of the people in the procession looked round, and hesitated. The marshal spoke sharply: "Get along there! Keep them moving, Jake." So the group moved on, leaving him with the two men in the buggy.

"Well, Cotton!" said Hal. "This is a fine job for an able-bodied American!"

The marshal flushed, but when he spoke, his voice was quiet. "Mr. Warner, we'll not have one of our arguments here. This is business."

"Well," replied Hal, "I'm here on business—with you."

"What is it?"

"There are some men down at Horton who have been turned out without their belongings, and would like to get them."

"That's what you're here for?"

"Yes."

"Well, you might as well turn around and go back. There'll no trouble-makers get into North Valley while I'm in charge."

"It's not a question of trouble-makers, Cotton; it's a question of men with teams to move furniture—"

"I told you I wasn't going to argue the matter with you. We let one team come in to move furniture, and before we knew it the driver was making a union speech. A little later, when we can get teams that we know about, maybe we'll get the stuff out ourselves."

"And meantime the people are without blankets and clothing!"

"That's their look-out, not mine. If they'd behaved themselves, they'd have had their homes still." And then, as Hal started to speak again, the marshal's voice became sharper. "That's all I have to say, Mr. Warner—except to warn you. Don't count on your social position, your being known to young Harrigan, or anything like that. We've got our hands full just now, and we'll stand no nonsense. Keep away from North Valley, and from the company's property—at least any part of it I have to do with."

And Cotton turned to the guard. "You understand, these people don't get by here!"

"Sure thing!" said the man, with a kind of salute. And the marshal started his horse and went down the road with a spattering of mud and resumed his place beside Predovich.

Hal and Billy followed; there was nothing else they could do. And all the way down the canyon they had before their eyes the two poncho-clad figures, swaying to the motion of the horses. Hal's thoughts were the thoughts inspired by that sight; he was suddenly quite willing now to turn Billy loose, to write what he pleased. If it would assist in getting this story before the people of the state, to have it happen to the son of a well-known millionaire—very well, let it happen! The millionaire, the elder son of the millionaire, the future daughter-in-law of the millionaire, the father of the future daughter-in-law—these people would all have to put up with the humiliation. Hal's mind was busy composing a telegram to the Governor of the state—the first of a series of telegrams with which he was to trouble that harrassed official.

Jake Predovich looked round several times at the buggy; but Jeff Cotton not once. "By the way," remarked Billy, suddenly, "I heard something about that fellow. They say he comes of an old Virginia family. A black sheep!"

"I gathered something of the sort," said Hal. "But he wouldn't tell me about his past."

"It seems to be no secret. He was one of a band of train-robbers—the Jesse James gang, that you've heard of. He barely escaped a life sentence, some ten years ago."

"Fine training for coal-company work!" snorted Hal. "Be sure to mention that when you write about him."

Billy laughed. "I'm not sure that it would help us. You never can tell how the American public will take things."

"You mean they'd think he was a hero?"

"I mean," said Billy, "what the eminent John L. Sullivan said about Grover Cleveland: 'A big man's a big man, it don't matter if he's a prize-fighter or president!'"

So they came to the foot of the canyon, where suddenly the two riders put their horses about. They went by the buggy without a word or a sign; and Hal and Billy drove up to the people, who seemed to be stuck in the mud. What floods of coal-camp English poured out—rage, denunciation, despair! Men and women shook their fists and cursed the retreating horsemen; while Hal and Billy got out, and loaded up the buggy with babies and children—even putting a couple of the latter on top of the horse! Walking at the horse's head themselves, they came to the Horton tent-colony, where Hal made report to Jim Moylan, and Billy shut himself up to work on his story. "A regular sizzler of a story!" he promised.

[16]

The first of the tents had come, and Hal had now several days' work cut out for him. He worked up to his ankles in mud, but it did him no harm, he was so busy and interested. There were trenches to be dug to drain off the mud, and somebody with camping experience must show people how to drive tent-pegs, and how to keep the fly of the tent from touching the roof. Also there were sanitary provisions to be enforced, among people who had no conception of the importance of such things. Streets must be laid out, and new arrivals must be tactfully dissuaded from setting up their tents in the middle of them. You could not imagine how many things there were to do until you had been there a few hours; and everything that was done had to be done over in nineteen other languages!

Rainy days could not continue forever, not even to oblige Peter Harrigan. The sun came forth, and water-soaked bedding and clothing was hung out to dry, and men, women and children emerged from shelters and holes in the ground, their souls expanding to the joy of the marvellous adventure. These men had worked all their lives under conditions where they almost never saw the sun; and now they were free to bask in it all day long, to gaze at the blue sky and the snow-capped mountains, to breathe the fresh air—even to play baseball in the fields! They made Hal think of the mules he had tended down in the bowels of "Number Two" mine, and had suddenly brought up to the open air!

Under the rule of the United Mine Workers, each striker received a benefit allowance of three dollars a week, with an additional allowance of two dollars for each woman and fifty cents for each child. This might not seem a princely sum on which to live, but the families had their rent free, which was equal to another dollar or two a week, and they had coal, which the union purchased by the carload and distributed in reasonable quantities. They had got more pay, upon paper, in the old days; but counting in the periods of unemployment, the accidents, and the sickness caused by overwork and improper ventilation of the mines, the present arrangement seemed good to them.

Especially when you took account of the spiritual factors, which made this tent-life a thrilling experience! When you took account of hope, which before had been despair; of liberty, where before had been bondage. If you do not know how much difference this makes, you have indeed missed a great lesson of life.

In the first days everything had to be done in a rush, there was no time for discussion. But all had confidence in the union officials who gave the orders; and one of the first things they did was to enroll every worker, giving him a card and a vote in the union's affairs. Then committees were chosen to run things; and so began the discipline of self-government. How proud were men and women, even children, when they were given some responsibility for the welfare of their new tent-city! There were so many things to be done—and all by people who did them for love of a cause!

In the days of their slavery each family had stood alone. They had little time for friendship or social life, their religious life had shrunk to extinction, their civic life was dead and buried and forgotten. But now suddenly a storm swept among them, they were lifted up and borne along by a tremendous force. They learned the meaning of a magical word—solidarity. They had a new interest in one another, a new meaning to one another. They had something to think about, something to talk about!

There was a general committee, and sub-committees for all the principal nationalities. There was a "headquarters tent", as it was called, and no end of coming and going at this tent. You needed skill in diplomacy to be useful there; you needed also the gift of tongues, and some acquaintance with international law.

There would come Mrs. Towakski, explaining in difficult English that she was entitled to an extra fifty cents a week from now on, because she had had a new baby yesterday. And then would come Mrs. Zamboni, declaring that Stefan Skiline was not entitled to benefits, because he had found work outside for three days in the week, and ought to be ashamed to take the union's money. After the committee had held a pow-wow and decided that men who had more than two days a week employment outside were not entitled to strike-benefits, in would rush Stefan Skiline with an interpreter, pointing out that at the Harvey's Run colony the committee allowed benefits to men who had as much as four days a week of work. He would not keep the work if he was going to lose his benefits!

And then came Frank Wagunik, claiming benefits because he had been discharged for union activity three months before the strike; and when you had decided his claim was just, there came Pete Zammakis, who had been discharged a year ago, and had been half starving ever since. And then came Mrs. Milinarich, the lady who had run a lodging-house for miners, and been sent out because she would not take care of "scabs". Surely Mrs. Milinarich was entitled to rank as a striker!

[17]

All Hal's North Valley friends were gathered at Horton. John Edstrom took the job of bookkeeper in the headquarters tent; Mike Sikoria was appointed day-policeman, and proved an excellent hand at "jollying" the intractable—the principal part of a democratic policeman's duties. Jerry Minetti came, to take control of the forty-odd Italians; he and Rosa and the three children had a tent of their own, and Little Jerry did valuable work organizing the Dago mine-urchins, teaching them to sing the "Internationale" in Italian, and to give three cheers for "il Sciopero"!

The main committee consisted of Big Jerry, Moylan, Hal Warner, and Louie the Greek. As Moylan had to be away three quarters of the time, most of the difficult decisions fell upon Hal, and this kept him absorbed. He was willing to give all the time he had—to forget the outside world entirely, in the fascinations of this new democracy. But on the third day after his trip up the North Valley canyon, a thunderbolt fell upon him; while he was directing the unpacking of a load of blankets, Jim Moylan rushed up, crying, "Have you heard the news?"

"What?"

"Jeff Cotton's been shot!"

Hal stared at him, dumb.

"Shot and killed!"

"Who did it?"

"A couple of Greeks."

"Strikers?"

"So they say. One was named Androkulos."

"Androkulos!" echoed Hal. "My God!"

"You know him?" cried the other. He had forgotten the name—having so many strange names to remember. When Hal told him it was a youth from their colony, Jim Moylan was speechless.

"Have they caught him?" demanded Hal.

"No, both of them got away. They'll be hot after them, though."

"Where did it happen?"

"Up near North Valley. They went after their things, and when Cotton wouldn't let them in, they pulled guns."

"Poor Andy!" exclaimed Hal. The face of the lad rose before him, olive-skinned, rather girlish, with big mournful black eyes. He remembered what the boy had once said to him. "Don't want to be a miner. Don't want to get kil-lid!"

Hal reproached himself—he ought to have taken Androkulos in hand. But he had so many things on his mind, so many people to keep in hand; in the confusion he had not realized the seriousness of the young Greek's threats. And now it was too late! Andy was a murderer, there was a thousand dollars reward offered to anyone who would betray him, and if they caught him, they would hang him!

His fellow-Greeks would try to hide him, of course. Hal had to think but a short while to realize that he would help to hide him, if ever the chance came. And this was decidedly a startling discovery to a young college-graduate, hitherto law-abiding. If he were to hide the boy, or even to connive at his hiding, he would become a criminal himself, an accessory to murder, liable to a long term in prison.

For some days, Hal's imagination was busy with this situation. He pictured the boy stealing into his tent some night, appealing for aid; and the two of them caught, and delivered over to Sheriff Raymond! What a sensation there would be then, for a fact! Peter Harrigan would get busy, and Edward Warner and Garret Arthur would have something really to worry them!

The imagined melodrama was never staged. Hal did not see Androkulos again; the boy disappeared, no one knew where, or to what fate. But Hal never forgot the olive-skinned, girlish face and the mournful black eyes; nor did he forget his first experience with the psychology of the law-breaker. A truly appalling aspect of the American system of "invisible government"—that men were made into criminals by the automatic operation of their best human instincts! Of the hundreds of thousands who were undergoing the tortures of a barbarous prison-system, how many had been brought to their fate by such automatic operation of fundamental human feelings?

[18]

Also Hal found himself haunted by the thought of Jeff Cotton: by the image of him sitting on his horse, alert and watchful, resembling a partridge; and again, by the image of him in his North Valley office, smiling cruelly at Hal. A merry duel of wits they had had; and at the climax of it, the mine-explosion had knocked them both over, and they had crouched on the floor, gazing at each other out of dazed and horrified eyes! In spite of himself, Hal had liked this black sheep of an old Virginia family; he had been a handsome devil, and a bold one. He had boasted himself "top-dog", and had expected to remain it. Now, Hal thought, what kind of dog was he?

The killing of a company-marshal had sent a thrill of horror through all the coal-country; it was an eruption from the rumbling volcano of anarchy, on which all men knew they were treading. From that time, every mine-guard saw a potential murderer in every striker. When Hal went down the street and passed some of the deputies, he saw scowls and heard muttered curses. They looked around to make sure he was not following them; and he in turn felt impelled to make sure they were not following *him!*

Such a tension inevitably led to clashes, especially when there was continual provocation in the form of "scabs", or rumors of "scabs". The Governor of the state announced his policy at the beginning of the strike; if *bona fide* workingmen wanted to go into the coal-fields to find jobs, they had a right to go, and he would protect them in that right, but he would not permit the wholesale importation of strike-breakers, and he would strictly enforce the law of the state, that men who were brought in to work where there was a strike, must be informed of the strike when they were hired.

This policy, announced after consultation with the union officers, was satisfactory to them. The trouble was, the agreement was not kept; strike-breakers were brought in from the beginning—precisely as if there had been no Governor, and no elaborate announcement of a "policy".

It was hard to be sure about this at the outset. The strikers could not get into the camps, and the strike-breakers could not get out. But there were rumors from every side; there came long-distance telephone-calls, telling of strike-breakers on the way; there came telegrams from places as far away as St. Louis and Chicago, telling of the wholesale hiring of "scabs". And such reports caused intense excitement. If the Governor will not enforce the law, let us enforce it! cried the hot-heads. And so the leaders would have to leave the pleasant work of building tent-platforms, and take to arguing and pleading in strange dialects.

Hal had come with an idea of what happens in strikes, derived from the reading of newspapers and the conversations of his leisure-class friends: the idea being that as soon as a strike is declared, the strikers with one accord turn out to defend their jobs with clubs and brick-bats and revolvers. But Hal saw that in this strike, at any rate, the course of events was entirely different. The leaders of the strike were convinced that they could win with the weapon of solidarity; they wanted no other weapon, and regarded a man who suggested any other as an enemy. There were rough characters among these miners, of course; but the vast majority of them were men accustomed to earning their bread by severe and patient toil. They had confidence in their leaders, and were eager to do what their leaders ordered. Their attitude to "scabs" was not always one of blind hatred; many of them could understand that the "scab" was a man out of a job like themselves, and that he might be ignorant concerning the wrongs suffered by the workers in this district. If they could explain to him, if they could plead with him, he might refuse to take the bread out of their children's mouths, and continue this system of oppression!

There came word to the Horton tent-colony that a couple of automobiles, each with two armed guards, were waiting at the railroad-station. That meant that strike-breakers were expected, and a dozen or more of the colonists set out for the village. "Go and see those fellows behave themselves," said Jim Moylan to Hal; and so Hal ran after the party. When the train came in, he stood watching, with his friend Mike Madvik, the Croatian mule-driver, at his side.

Several men who looked like workingmen got off the train. "Fellow I know there!" exclaimed Madvik, and stepped forward to speak to the man; when one of the guards seized him by the shoulder and threw him back. There were cries from the other strikers, and a rush forward; whereupon the guard drew his revolver and sprang into the midst of them, striking right and left with the butt of the weapon. He struck one man in the jaw, and with another blow he laid open the Croatian's scalp. The other guards joined in, and having driven the angry strikers back, they pushed the strike-breakers into the automobiles, leaped onto the running-boards, and sped away.

Hal took the job of binding up the head of his mule-driver friend; all during the operation the poor fellow was crying—not with pain, but with rage. Hal strove to tell him, in words of one syllable, of the wonderful vision of labor-solidarity, the one big union that was coming out of agonies such as these. But all the time Hal had an inner struggle to be convinced by his own eloquence. He really wanted to get a gun!

A day or two later a group of strikers flagged a railroad-train, to make a search for strike-breakers. They did not find any, as it happened, but they made a terrible excitement. Reading about it in the Western City papers, one got the impression that the coal-camps were in a state near to insurrection; and how could a young man of decent rearing and education give his support to such rioters and assassins? Hal could see the impression his relatives and friends must be getting, yet he was powerless to convey any other impression to them.

The newspapers were so very plausible in their accounts, that sometimes even Hal would be seized with doubts. Might it not be that some of these things were happening as described? He knew that the strikers were men with a sense of bitter wrongs, and that some of them were inflaming this sense with liquor. Might it not be that some of them had done the wild deeds which the newspapers told about? Some of them in remoter parts of the field, too far for Hal to know them!

There came one flagrant incident, a night attack upon the Harvey's Run mine, by what the papers described as a mob of armed strikers. There was a tent-colony in this neighborhood, and Jim Moylan got it on the telephone, and received most solemn assurance that there had been no fighting whatever—at least none that the strikers knew of. So Hal called up Billy Keating, who was in Pedro, and suggested that he make an investigation of this particular riot. Billy set out, and next day the "Gazette" had a report of what he had found. Over a thousand shots were fired, according to the story; but Billy had challenged the marshal of the camp, who had been able to show only three bullet-holes in the buildings—and these in an unoccupied Japanese boarding-house! It was claimed that the attacking mob had been hidden on the hillside above the camp; but Keating had stuck a lead-pencil through the bullet-holes, and found all three of them horizontal! It was the most obvious kind of "frame-up"; yet the newspapers were making it the basis of a demand for the calling out of the militia!

[19]

The reason for such proceedings was plain enough to Hal. The coal-operators were in a state of dismay, because of the completeness of the strike. Their reports had led them to expect a quick collapse; but here practically all their men were out; and many of the best were leaving the district, over five hundred having purchased railroad tickets on a single day! The efforts to keep them by force had broken down; in several cases the camp had risen *en masse*, stormed the defenses, and marched out. In other camps the union had sent in spies among the strike-breakers, and there were secret meetings, and the strike-breakers were striking!

The cost of paying and feeding the guards and deputies amounted to something like ten thousand dollars a day; and this expense the operators wished to put off on the state. They wanted the Governor to order out the militia; and to support the demand, they wanted violence.

Day by day the rifle-carrying "guards" crowded closer upon the tent-colonies, their ways becoming more insolent, their language more vile. They would stop and turn back parties of men from the post-office; solitary men they would beat, women they would grossly insult. They attacked strikers who came into the open camps to demand their back pay; they arrested men for attempting to speak to strike-breakers, even in public places. And when all this failed, they took to hiding in the hills, or in the coal-camps which happened to be near the tent-colonies, and practicing long-range firing with human targets.

There were a dozen tent-colonies, scattered along the railroad for fifty miles; and this last-mentioned experience befell them all. Again and again the miners' officials went to Sheriff Raymond, asking protection. They demanded that he take away the deputy's commissions which he had illegally issued to non-residents. They demanded that he furnish men to protect the tents—offering to lodge and feed some of the deputies, in the hope of keeping off the bullets of the others! But the sheriff-emperor said No; and when Harmon and Moylan requested that he issue deputy's commissions to the strikers, so that they might protect their own homes, he answered, "I never arm both factions!"

So naturally the miners took to arming themselves. Before this, it had been the hot-heads who bought weapons; but after the interview with Alf Raymond, the union officials themselves ordered guns and ammunition. This fact, of course, became known, and was diligently used by the strikers' enemies, in order to make the public believe that the strikers were law-breakers and desperadoes.

The public understood that the coal-companies had to employ guards to protect their properties, which otherwise would be burned or dynamited; also to protect the men who wanted to go on working, who would otherwise be beaten or shot. When you asked the public to believe that guards were being secretly used to beat and kill strike-breakers, and even to burn and dynamite properties, in order that the public might be led to think that the strikers had committed these crimes—then you went out of the realm of reality, you set yourself down for a romancer of the "penny-dreadful" order. Unless by chance you were something worse—a secret abettor and fomentor of crime!

But to the strikers these matters appeared quite differently. Not merely did they know that the "penny-dreadful" tales were true, that these strike-breaking agencies were "framing things up" on them; they knew that they did this systematically, as their regular business routine. Union labor had seen this technique of strike-breaking being developed for twenty years: from the great Chicago railway strike, where a commission appointed by the President of the United States had reported that the burning of cars and other violence had been committed by "thugs, thieves and ex-convicts" in the employ of the Railway Managers' Association; down to the recent great strike of the wage-slaves of the woolen trust, where the president of the mighty corporation was almost sent to prison—having made the mistake of employing the dynamiters himself, instead of letting a detective agency do it.

Here in this coal-country, the detective agencies were so sure of their control of government and news-agencies that they scarcely made any pretense of concealing their doings. They could not have done it, had they wished to; because their employes, swollen with suddenly acquired authority, were not to be kept from boasting. For example, Pete Hanun, the breaker of teeth, and Gus Dirkett, his pal! These two, the murderers of Olson, were under bonds of ten thousand dollars, but they still had their deputy's commissions, and carried rifles, in addition to the revolvers with which they had shot the organizer. By day they rode about the district, terrorizing all they met, and at night they got drunk in the saloons of Pedro, and told of the coming of the "death special", and the use they intended to make of this new toy.

This "death special" was the talk of all the camps: an armored automobile which was being constructed in one of the mills of the General Fuel Company. It was provided with three-eights inch steel plates, high up on the sides and about the body and wheels; it was shaped in front and back like a battle-ship, and mounted two machine-guns, one in front and one in back, each firing a hundred and forty-seven steel bullets per minute.

There was a meeting of the miners at Horton one morning, two weeks after the beginning of the strike. John Harmon spoke, and Mother Mary denounced the Schultz detectives in her coal-camp English. Some of the detectives were present, and reported her uncomplimentary remarks; and that same afternoon came the avenging "death special". In it were Schultz, Pete Hanun, Gus Dirkett, and a fourth man, the chief clerk of the General Fuel Company. They approached the colony from over the hills, took post where a ridge of bushes hid them from view, and without any sort of warning opened fire on the tents.

[20]

Hal was in the headquarters tent, discussing with Harmon and Moylan the problem of sanitary inspection, when a wild clamor broke, and the three men rushed out into the street. Women and children were running this way and that, screaming.

"What is it?" Hal cried; and Rosa Minetti answered that it was bullets—somebody was shooting at them!

At first Hal could not believe her. "Listen!" she exclaimed; and he heard a swift whirring, like the sound of a flying-machine; also a whining, buzzing sound, that might have been the hiving of bees. "Bullets!" cried Rosa. "They shoot us all!"

Now Hal had never heard the sound of a machine-gun; he had never thought of such a weapon as a possibility in his life of culture and ease. "It's an automobile!" he declared.

But Rosa persisted. "Look! Look!" And she pointed down the street, where spurts of dirt were leaping high up. "We got coffee-pot on stove! They come hole in it! Coffee run all over!"

A man rushed past Hal. "Kill them! Kill them!" he was shouting. He had a double-barreled shot-gun in his hand, and Harmon called, "Stop that fellow!" So Hal leaped in pursuit, and grabbed the man, an Italian. But it was impossible to stop him; he was like a mad creature. When Hal tried to take the gun, he leaped back and pointed it. "They killa my brud, I killa them! You stoppa me, I killa you!"

So Hal had to let him go, and follow him out into the

fields beyond the tents. Other men came running, with rifles and revolvers, even axes and picks. They looked about wildly, but could not tell where the shooting came from, even though the bullets kicked the dirt into their eyes and mouths. The little Italian, beside himself with fury, took aim at the mountains and fired both barrels at once.

Now Hal had come to this coal-country with his mind made up to a role of non-resistance. Whatever others might do, his job was to make an appeal to people's moral sense. He had foreseen that it would not be easy, but he had taken his resolve—he would fight with the weapons of the mind, and in the end the conscience of the community must rally to his support.

But here was a trying problem. What is the proper course of action for a non-resistant, when bullets are actually kicking dirt into his eyes? Shall he dance about and dodge, like a tenderfoot in a frontier bar-room! Shall he crawl under the bed with the women, or behind the stove with the children? Or shall he go about his affairs unmoved, according to the tradition of the British army officer? The latter course might satisfy a man's dignity, but it hardly satisfied his common-sense.

Fortunately this test was not a long one. Having fired about a thousand shots without hitting anyone, the guards considered that they had taught the colony its lesson, and the "death special" moved on out of sight beyond the hills. They left the inhabitants of the tents behaving like a nest of ants that has been suddenly dug out of the ground.

This episode constituted the "first battle of Horton". It had apparently been planned to cause terror, to put a stop to "incendiary speeches" in the colony; if so, it failed, for it caused only furious indignation. There was a conference of union leaders in the headquarters tent that night, and at this meeting the peace men could hardly get themselves heard. It seemed that there were only two of them left—John Edstrom and Louie the Greek—the latter being a Tolstoyan, who carried his peace ideas so far that he refused to eat meat!

The outcome of the conference was that the union leaders drew up a letter to the Governor, giving notice that from this time on the strikers would protect themselves; they would establish a guard for the tent-colonies, and be prepared to meet their assailants. The Governor did not answer this directly, but Sheriff Raymond made answer of a sort—ordering his deputies to arrest as many strikers as they could find carrying weapons. In Pedro they charged Tim Rafferty with carrying a revolver, and sent him to the filthy city-jail for thirty days.

[21]

The reports of this "battle of Horton", published in the Western City morning papers, were the cause of Hal's getting a long-distance telephone-call—a very long long-distance call, which must have cost the caller many dollars. It was Edward, demanding that his brother should get out of that strike-district. He was terribly excited, and more profane than Hal had ever known him before. "If you don't come, by God, I'll come and fetch you!"

Hal answered, "I've been wishing that you might see things here with your own eyes!"

"If I come," declared the far-off voice, "it won't be to see anything—it will be to have you locked up in a lunatic asylum."

"If you'd only come, Edward, you might stop some of the shooting! They wouldn't take chances of killing a business-man!"

"I tell you to take the night-train!" persisted the tones of distant indignation. "I mean for you to do what I say! If you don't, I'll have Dad cut you off! I'll denounce you in the newspapers! I'll make it clear that I have nothing to do with this tom-foolery!"

"There's no use wasting telephone charges on that sort of talk, Edward. I'm not going to desert these people in their trouble. If you want to save my life, the thing for you to do is to go to Governor Barstow and make him protect these tent-colonies."

There was an interruption in the telephone service. After some delay Hal heard the faint voice, seeming pathetic in its helplessness: "Why don't you come up and see the Governor yourself?"

That seemed a real idea. "Wait a minute," said Hal, and he turned and asked John Harmon about it. "By all means!" Harmon said, and Hal answered, "I'll take the train at once. But I want you to understand that I'm coming back—straight back!"

"Take the train!" insisted Edward. "Promise me you'll take the train!" If he could only get his hands on this madman!

"All right," was the reply—"but I'm coming back!"

Hal made a dash and caught the morning train, and got into Western City in the evening. His brother met him in an automobile—and such a row as they had! At home there was Hal's father, who had heard of the fighting and was in terror for his boy. There was nothing for Hal to do but tell his side; and this, while it overwhelmed the old gentleman, did not lessen his distress in the least.

Hal found that his brother hoped to back him down from his idea of seeing the Governor. Peter Harrigan would be so furious! But though Edward argued until midnight, Hal was not to be moved, and early next morning he climbed the hill and entered the white marble State House. The Governor's secretary took his card, and after reading the name, said politely that he would endeavor to arrange an interview. Soon afterwards he ushered Hal into an inner office where the chief executive stood behind a flat-topped desk.

He was an extraordinary chief executive for the people of a great state to have chosen. No one could credit an account of him, without first coming to understand the political and social system of which he was a product. This mountain state possessed enormous wealth in coal and minerals, and for fifty years, ever since the Indians had been driven out, its politics had served as a weapon in the struggle for the control of this wealth. At the present time the question had been pretty well settled: the mines, the railroads, the franchises of every sort were in the hands of a few great corporations, which managed both political parties, subsidizing their leaders and providing them with money to bewilder and corrupt the public. The corporations would support one party for a few years; then, when the actions of that party had made it odious, they would shift to the other party, starting a fresh campaign on the issues of "economy", "public honesty", "law and order".

Being in possession, all that the corporations now asked of the state was to be let alone; and in order to be sure of being let alone, it was their custom to choose public officials who had too little intelligence to interfere, even if they wanted to. If you let a clever crook get into office, you could not tell what turn his crookedness might take; but if you chose an honest and well-meaning imbecile, you were safe. Thus it happened that the "invisible government" had contributed to the portrait gallery in the white marble State House a long row of studies in human futility.

The present study went by the name of Elon Barstow, and had been an obscure politician, a ranchman member of the state legislature. He had debts, and a mortgage or two, according to rumor; it was everywhere known that his nomination had been decided upon at a dinner-party in the home of Peter Harrigan. In appearance he was a small man, with a curious sort of face—childish, but wizened, making you think of a prematurely old infant. In manner he was nervous and uneasy, peering at you as he talked, as if groping to get hold of you. Hal had been told that he was half blind, and could not make out a person's features at all.

[22]

Hal explained his business; he had just come from the strike-field, where he had seen things the Governor ought to know about. But at once the Governor interrupted; he had received Mr. Warner's telegram, about the forceable detaining of men in the mines, and he would not believe such tales. He considered it very wrong for a young man of good family to take such a line as Hal had taken—giving encouragement to dangerous and unruly foreigners, who were destroying property and making resistance to the state, because of the violent talk of outside agitators, especially that incendiary old woman "Mother Mary"—

Hal had been told about Governor Barstow by Billy Keating, who had interviewed him more than once. He took it as a personal grievance that you should oppose his policies; his voice would rise high and angry as he complained about the troubles you were making for his administration. If you argued with him, his excitement would increase, until he would be pounding his desk. The thing you must do was to get even more excited; to shout louder than he shouted, to pound harder on the desk. So you might make him quail—make him admit there was something in your argument, and that something must be done about the matter.

Hal had never learned to pound desks; but he had practiced arguing on Edward, and having once got started, he kept on, so fast that the little Governor could not get in a word. He told of incident after incident which he had seen with his own eyes—the violation of law after law, even of fundamental constitutional rights. After he had gone on for half an hour, arguing, exhorting, he saw tears of distress in his victim's eyes. "Mr. Warner, what can I do? I tell you frankly—I don't understand about coal-mines. I'm a ranchman—that is the only business I know. They keep telling me that conditions are peculiar in this state—there are reasons, for instance, why there should be so many accidents. What can I do?"

As Hal argued, it came to him what the little Governor ought to do—to go down to the strike-field himself. Let him see with his own eyes what was happening! Let him hear the stories of the people! Even if he did nothing positive, his presence would have a moral effect. It might remind the deputies that there was such a thing as law!

Hal made the suggestion. Yes, said the other, he had thought of that—several people had urged his going. He had engagements at the moment, but he might go next week.

"But if you wait, you may be too late!" argued Hal. He went on, remembering another remark Billy Keating had made—that you might labor for hours and convince Governor Barstow, but he would remain of the conviction only so long as it took somebody else to get into his office, and to shout at him still louder, and pound on his desk still harder. The loudest shouter and hardest desk-pounder of all was Peter Harrigan, and so it was that he ruled. Not long after this one of the Coal King's letters was published, in which he told about this himself—referring to the chief executive of his state as "our little cowboy Governor", and declaring that he treated him as one treated a child, first "spanking" him, and then "giving him candy" to soothe his feelings!

So Hal tried to persuade the Governor to act. There might be another "battle of Horton" at any moment! But the other would not promise; he had grave matters pressing for his attention; but meantime he would telephone to Sheriff Raymond, and give him strict orders to enforce the law impartially. Hal might rest assured there would be no more attacks upon tent-colonies.

[23]

Not many of Hal's friends read the "Gazette", but it seemed that all of them had somehow got hold of the issue containing Billy's account of his trip up the North Valley canyon. Now as he walked down the street, he had to face the witticisms of youth, and the admonitions of maturity. After an hour or two of telling things over and over, he recalled the existence of the art of printing, and dropped into the office of the United Mine Workers, to suggest that they ought to get out a pamphlet, so that a man could have something to hand to his friends.

In the office were Harmon's secretary, and a couple of officials of the union. They agreed that Hal's idea was a good one. Would he write the pamphlet? Obviously, he was the man for the job—he was so full of the subject. Hal agreed to do the writing that very night. Next day was Sunday, but the secretary knew a union printing-place that would stay open and do the work, and on Monday morning Hal might walk down Broadway, handing out pamphlets right and left. Or he might give a list of names, and the secretary would mail copies.

Next Hal went to see Adelaide Wyatt and her parlor-maid. He remembered how, after the North Valley disaster, he had got a new sense of the horror of the thing from seeing Mary's grief; and now again, watching the girl, he realized afresh the passions of this coal-camp war. Mary sat with her hands clenched till the knuckles shone white, and her lips were quivering with pain. He tried to spare her some of the details, but she dragged them out of him. She wanted to know everything, to suffer every pang.

"Joe," she exclaimed, "I got to go back! I can't stay here and live easy while this is goin' on!"

Now it sometimes happens that a man who is stern and rigid in dealing with himself, is soft and cowardly about making others suffer. Hal recalled the hum of machine-gun bullets in the "battle of Horton"; he did not want Mary within the range of those bullets. "There's nothing you can do—" he began.

"I got no right to stay away from it, Joe!"

"Mary," he argued, "be sensible. Nobody ought to be there that can be anywhere else."

"And my little brother and sister—"

"You ought to get them up here!"

"That's what I tell her," put in Adelaide.

"What would I do with them?"

"You can find them a home," said Hal. "Go and see the labor men—they'll know some working-class family where they can stay. Jennie can go to school, and we'll find a job for Tommie."

He thought for a moment. There was the "Emporium", which seemed a store-house of jobs. "Mrs. Pattie" would not be thinking kindly of Hal, since he had withdrawn his foot from her embrace; but Adelaide could go to her, and tell the pitiful story of two mining-camp children in terror of machine-gun bullets. "Mrs. Pattie" would have a thrill of the philanthropic nerve, and would see that some other boy was discharged to make a place for Tommie; and this would give her such a sense of goodness and usefulness that she would feel justified in a thousand new acts of vanity and extravagance. Or so Hal put it—having become bitter under the influence of machine-gun fire.

Adelaide agreed to the program, and it was arranged that Hal should send the children as soon as he got back to Horton. Mary began to ask questions again, and he told about his interviews with the sheriff and the Governor, and about his further plans. Adelaide was deeply moved by his narrative, and declared that she would say something about gunman outrages at the next meeting of the Tuesday Afternoon Club. If you knew anything about Western City society, you could picture what a flying of fur there would be, when the coal-strike was discussed before the wives and daughters and sisters and cousins and aunts of the operators!

There was another person upon whom Hal had designs; he would sow seeds of dissension in the home of his brother. Lucy May, Edward's pretty young wife, was a real human being; moreover she came from Philadelphia, where people have top-lofty notions about themselves, and are not to be awed by Western coal-magnates. Hal would tell her about things that were happening to children; Lucy May had three, whom she loved to the point of frenzy, so it ought to be easy to bring tears into her eyes.

The nefarious project was under way when Edward came home at six o'clock in the evening. It was just all he could do to respect the decencies of hospitality; all through the perfect dinner which his wife had seen to providing him, he nagged at his brother, and three or four times Lucy May put in some remark which indicated that the seeds of lunacy had begun to sprout in her mind also. Could it be that the strikers were *all* ruffians? And had Edward heard what some of the mine-guards had admitted to Hal about themselves? The effect of all this was to confirm Edward in one of his strongest convictions—that the secret purpose of all Socialists is to break up the home.

Then it transpired that Hal intended to spend most of the night writing an incendiary pamphlet; and that Lucy May had invited him to stay and do his writing in the guest-chamber of Edward's home! "I suppose next thing you'll be signing your name to the pamphlet!" growled Edward; and Hal looked interested. That might not be a bad idea! What did Lucy May think about it?

The pamphlet was written between the hours of eight o'clock and two next morning; and at eight on Sunday morning a messenger boy took it off to Harmon's secretary, and Hal washed the weariness and wrinkles from his face, and came downstairs to a breakfast of kidney-stew and waffles, which represented the supreme effort of Lucy

May's Virginia cook to restore cheerfulness to the distracted household. In the course of the repast Hal made the announcement that he was planning to accompany the family to St. George's.

He saw a frown on his brother's face.

"What's the matter, Edward? Old Peter won't order me out, will he?"

"How can I tell what you'll be up to?"

"Well," said Hal, "if you think it'll make trouble for the old man to see me in your pew—"

"Nonsense!" broke in Lucy May. "Of course you shall sit in our pew!" And Edward gave a glance at his pretty little lady, and swallowed the rest of his kidney-stew and waffles in silence.

[24]

There was a family reunion in St. George's that Sunday morning: Edward Warner Senior, and his sister, Aunt Harriet, who took care of him, and Edward Junior, and Lucy May, and little Lucy May, their eldest child, looking like a tiny white fairy in a white silk hat and white silk dress and white silk stockings and shoes. Hal stood in the aisle and let all this family in, and then took the outside seat; he noticed with an inward grin that his brother was nervous about this. Was Edward afraid that Hal might interrupt the service—telling the congregation what the Coal King was doing to his serfs?

They were ultra "high church" at St. George's; catering to a class of people who were used to elegance in their homes, and assumed that their God would expect it in his. They had candles and processions, a costly choir, and most gorgeous altar-cloths and communion plate; and two or three times on Sundays, and often several times on week days, they gathered together in honorific clothing, and went through hallowed formalities in honor of an ancient Hebrew divinity.

And they would listen to texts out of ancient Hebrew literature! The tirades, the clamor of ragged old shepherds and vagabonds, solemnly intoned before this refined and aristocratic congregation! You might hear Dr. Penniman, white-haired and dignified, polished and urbane, reading Isaiah on the subject of those ladies of fashion who go with stretched forth necks and wanton eyes and a mincing gate; you might hear him repeat without a blush or quiver the old prophet's obscene and terrifying threats against these vile creatures—and with never a suspicion that the remarks might have application to members of the congregation who had been dancing the tango on the previous evening! You might hear Mr. Wilmerding, quoting the incendiary utterances of the carpenter's son: "Woe unto you, you lawyers!"—and in the presence of no less than three members of the legal department of the General Fuel Company! Or the needle and the camel's eye—in the presence of Peter Harrigan himself!

It was Wilmerding who preached that morning: dear "Uncle Will", with his rugged, knobby face, and his brown beard which he wore upon principle, a symbol that he was not quite so "high church" as his rector. Everything was a symbol to the clergyman, Hal knew—every shade of color in his robes, every gesture of his hands, every intonation of his rolling voice. He preached upon faith, hope and charity, and the greatest of these; and Hal sat and listened, marveling to realize how the world had changed. He loved "Uncle Will" as much as ever; yet it was impossible to keep his unruly mind from formulating hateful images of his boyhood's friend. He was one of the Coal King's lackeys; he cleaned up, with this thing he called charity, the human wreckage of the Coal King's profit-machine. He was, quite literally, the Coal King's old clothes man; twice a year he collected the garments which the congregation found no longer suited to its social position, and distributed them to the grateful poor, who had no social position to consider. Or he was a clerk of the Coal King's conscience-department; a soul-physician, administering spiritual soporifics, ecclesiastical anodynes. Strange that a young man could sit through a service in St. George's, and think thoughts such as these, and not have them manifested in the form of sulphurous fumes, or a glare of flame, or other infernal phenomena!

Time came for the collection, and half a dozen gentlemen in dignified frock-coats arose and marched up to the altar-rail, while the organist played soft music to which they did not keep step. Dr. Penniman put into their hands the collection-plates, mahogany with red plush centres, and they turned and came down the aisles. One came to the Warner pew to begin his round: a thick-set, elderly man, partly bald, with a smooth-shaven face and cheeks that seemed to sag at the bottom. His lower lip thrust out, and the heavy lines about the sides of his mouth, gave him a grim expression. He walked with a firm tread, staring before him, wholly bent upon the service of the Lord. But when he came to the Warner pew, his eyes met suddenly with those of the young man in the outside seat; his grim face flushed, and his lips became set in a way that was suggestive rather of week-days, than of the service of the Lord, and gratitude for the round silver dollar which Hal was donating to that service!

There were members of the congregation watching, and noting the little drama. The faintest possible thrill went about, and Edward Warner's face grew red with embarrassment. Why had he not had the wit to put his mad brother in an inside seat, so that he could not stare directly into Peter Harrigan's face?

The offertory was sung, and the benediction given, and the singing choir marched out; then, while people gathered to greet one another, and to look rather than whisper their little thrill, Hal went into the vestry-room, where Will

Wilmerding was taking off his robes, and putting on the detachable white cuffs which he wore—possibly a symbol of the fact that he was a servant of the Most High, and not of the fashions of this world. He took Hal in his arms, and gave him one of his tickling hairy kisses, exclaiming, "What's this you have been doing, boy?"

Hal answered, "I have been about my father's business." And this when he saw Judge Vagleman, chief counsel of the General Fuel Company, standing within hearing! "I want to come to lunch with you," he went on. "I've something important to talk to you about."

It was embarrassing for the clergyman, but Hal had no mercy; he meant to "get" Will Wilmerding, he told himself. The clergyman came of an old English family, and under his professional humility was buried real human pride. He would not remain Peter Harrigan's old clothes-man, if once he could be made to see himself in that role! There would be an old clothes-man's revolt, possibly even an old clothes-man's union!

They had their first bout that night, after the evening services, in the study of the little home where Wilmerding lived with his old mother. Hal told his story, and they argued back and forth for hours. Hal did not succeed in his project of persuading his friend to come at once to the strike-country; but he left him pacing the floor of his study, with his hands tightly shut and his face strained and haggard. Hal went home in the certainty that he had planted the seeds of a mighty spiritual crisis in the soul of Peter Harrigan's assistant rector!

[25]

Next morning Hal set out as a distributor of pamphlets; one of those wistful, pathetic persons, with the pockets of their overcoats bulging, who encounter you timidly, and try to accomplish their purpose without letting you know that they are out especially to save your soul. At least, that was the picture of the situation which Bob Creston took into the Merchants' Club, to the vast hilarity of the younger crowd—excepting Appie Harding, who happened to be the wistful propagandist's cousin, and Garret Arthur, who happened to be his prospective brother-in-law.

But in the course of the day Hal lost his wistfulness and timidity. He took a copy of his pamphlet to Larry Pringle, and found the "Gazette" office in a state of excitement, owing to the arrival of a telephone message from Keating. A stenographer had taken it down, and Hal sat by this man's typewriter and read the words as they popped into view on the machine.

There had been another "battle of Horton" that morning, and Keating had been present, and gave his description as an eye-witness. A mile or so down the railroad-track from the colony was a steel bridge where the county road crossed, and under this bridge three gunmen had concealed themselves and opened fire on the tents. There were at this time about twelve hundred people in the colony, and probably two hundred of the men had weapons of some sort, most of them shot-guns and cheap revolvers. As before, they rushed out and began firing blindly—the little Italian had fired his shot-gun into a pile of steel-rails and nearly blinded himself. Pretty soon one of the men, who had gone in the direction of the firing, came back with one eye knocked to a pulp by a spent bullet; and then came a second man, with his hand to his cheek, and blood pouring in a stream through his fingers. The account gave the name of this last man—Klowowski; it was the Paul Revere of the coal-camps, the little Polack who had met Hal and Jim Moylan on the first night of their coming to Horton! Hal saw in a swift vision his pitiful, eager face; above the click of the typewriter keys he heard the shrill voice: "Big men from union come make talk! Ever'body come! Tell ever'body!"

The shooting had continued for a couple of hours; and in the course of it a ranchman who lived near-by had ridden out to try to bring his horses to a place of shelter. He had a wife and three children, who besought him to stay at home, but he insisted that no one would hurt him. He rode toward the steel bridge, not knowing the gunmen were there, and one bullet pierced him through the forehead, and two others through the body. Billy explained that this man had earned the enmity of the guards by leasing to the union the property on which the tent-colony stood.

Later on the reporter had gone over to the railroad-station, and heard one of the mine-guards denouncing the three sharp-shooters under the bridge. They were hardly hitting the strikers at all, he said; they were filling the water-tank of the railroad full of holes!

With the receipt of this news, Hal set out on the jump for the white marble State House. This time the secretary declared that the Governor was very busy; finally he took Hal's card, but the answer was that the Governor could not possibly see anyone that day. Hal, however, was not to be put off; he laid siege to the office, and late in the day caught his victim trying to escape by a side-door. He literally held him up on the grounds of the State House, where they had another vehement argument.

Hal knew his man by this time—he had got to the desk-pounding stage! He told the Governor that civil war was coming in the strike-district; there would be wholesale murder if he did not go at once. He kept at the distracted official until he went back to his office and ordered his secretary to get ready, and telephoned his wife to get his grip packed. He would take the train that night.

Whether it was the secretary that leaked, or the wife, Hal never found out; but it was a fact that Peter Harrigan knew about that decision within fifteen minutes after it was taken. The first thing he did was to get Hal's brother on the telephone and give him a tongue-lashing; the second thing was to get hold of his general manager in Pedro and order him to meet the Governor, and never let him out of his sight while he was in the field.

Hal went down in the same train with the Governor and secretary, and his heart was so full of hopes and his head so full of plans that he could not sleep. But in the morning all these hopes and plans were dashed. When the party alighted from the train, there was Schulman, the general manager, with a "G.F.C." automobile; and in this automobile there was room for the Governor and the secretary, but no room for Hal! The young man saw the trap in an instant, and made protest in no timid language. Had the Governor come down here to make his investigation under coal-company supervision? If so, his visit would be a farce—his time worse than wasted.

It was very awkward; for there was a crowd of people on the station-platform, and the populace is not supposed to be present at the settling of grave affairs of state. The Governor hesitated; but Schulman, a manager who was used to managing, was at his side, with matters of very great importance to lay before the chief executive at once. And so the Governor stepped into the car, and Schulman and the secretary followed, and they rode off with a rush—leaving Hal standing there like a fool.

And so it was that the visit, for which Hal had worked so hard, came to nothing. The Governor rode about and interviewed company-marshals and guards; he never met one of the strikers, nor heard a word of their story. When Mother Mary got word of his being in Sheridan, she called on him, and when he refused to see her, she organized a procession of women and children, and marched them to the hotel where he sat at dinner. The poor little man had to leave his meal, and make his escape by a back door!

[26]

Meantime Hal had returned to Horton, where he found the spirit of things much changed. One no longer heard from pacifists; one heard a cry for men who knew how to shoot. Even Billy Keating had become military. He was going on sending in stories, but merely from loyalty to his chief, and without much faith in the work. All one could accomplish by writing stories was to stir up a few soft-hearted women—while meantime the hard-headed men went on with their bloody work! Billy was a new kind of "Ironside"—with a reporter's note-book in one hand and a gun in the other. There were fellows in this tent-colony, Greeks and Bulgarians, who had been fighting each other at home; get them some real rifles, and they would teach these mine-guards a lesson!

There was no rest for anyone in the colony these nights. At the Barela mine, a couple of miles away, a searchlight was kept playing all night, the light streaming into the tents and making them bright as day. This light made it possible for any vengeful or drunken man to fire into the tents from the darkness; also it kept the women terrified, for there was a machine-gun mounted beside the searchlight, and they knew that their enemy, Stangholz, was in charge of that gun; they could see, in their nightmares, his devilish face leering at them, as he shouted what he was going to do with his "baby". Some of the strikers took to digging holes in the ground for the women and children to hide in; others, under Billy Keating's advice, were studying the terrain, planning what students of military science know as the "offensive defense". Why not occupy the heights which commanded Barela? Why not send a raiding party some dark night, and shoot that searchlight to pieces?

Hal had not forgotten his promise to Mary; but when he inquired he learned that Tommie Burke was one of the unnamed casualties of the second "battle". Running to a near-by arroyo to hide, the trapper-boy had been shot through the foot; the bones were badly shattered, the doctor said, and he might be more or less lame all his life. At present he had a fever, and it was out of the question to move him.

So Hal called up Mary, and told her the bitter news. The girl declared instantly that she would come to Horton; she would take the train that night. Hal said that he would meet her, if he could. One never knew, in setting out for the railroad-station, whether one would get there or not.

There were two guards on hand when the train came in, but they only glowered, and stood by to watch what Hal did. When they saw a fresh, handsome girl with a treasure of auburn hair and a neat grey travelling suit step from the train, and talk to him with great intensity, and then begin to cry, and have to be led away by the arm, they understood suddenly the solution of a mystery which had troubled them—why a rich young fellow should have come to this tent-colony to live with Dagos and Hunkies and "wops". One of them uttered his thoughts with a guffaw; and Hal turned upon him, exclaiming, "You go to hell!"

—Which, one must admit, was not exactly a pacifist utterance. "That's all right, young rooster," was the guard's reply. "We'll get you before we quit!"

Hal had not foreseen this consequence of Mary's coming, but there was no help for it, and they did not talk about it. He tried to console her by saying that John Harmon had gone that day to have an interview with the Governor in Sheridan, to try to have orders issued to the sheriff to provide guards for the tent-colonies. At this time it was still impossible for Hal to believe that the forces of law could be continuously used in the service of anarchy.

But when the sheriff-emperor came to Horton, it was not to guard it. Two or three days after Mary's arrival, there was an uproar down at the railroad-track; a box-car with seventeen Mexicans in it, the sheriff-emperor in charge, and a crowd of strikers jeering and shouting. At first Hal supposed the Mexicans were strike-breakers; but soon the situation was explained to him—they were men brought in from the state of New Mexico, in order to be turned into deputy-sheriffs! The strikers had stopped the train, and the sheriff-emperor fled to one of the passenger cars.

There were Mexicans among the strikers, who talked to the would-be deputies, and learned that they had not even been told there was a strike! They were told it now, with the result that they declared their wish to join their brothers, and descended from the box-car, and marched to the tent-colony amid hilarious rejoicing. The sheriff-emperor retired, vowing vengeance upon the enemies of "law and order". In order to appreciate the humor of his indignation, one would need to be reminded of the law of the state providing that no man should serve as deputy-sheriff until he had been a citizen of the state for a year, and lived for sixty days in the county!

[27]

A couple of days later occurred another incident. There were rumors that carloads of strike-breakers were being taken into the Northeastern, and a party of strikers was told to do picket-duty. The union leaders were most strenuous that no man should be armed while on the picket-line, and on this occasion every member of the party had to submit to a search in the tent-colony before he set out for the mine. News of the disarming must have got to the gunmen; for while the men were parading up and down the road in front of the stockade, there appeared Schultz and his "death special" on one side, and a party of a dozen men on horseback on the other. They held up the pickets and marched them away—forty-eight workingmen driven eight or ten miles along a public highway, with armed horsemen in front and on each side, and an armored automobile with machine-guns in the rear! The procession came into the town of Pedro, which was swarming with strikers; but in spite of all the uproar, the prisoners were landed in jail; after which, under the personal escort of the sheriff, two carloads of strike-breakers were run into the mine.

The "Gazette" wanted some photographs of these scenes, and Billy Keating employed a photographer, who made so bold as to snap the "death special", with Schultz at the machine-gun; whereupon Schultz leaped from the car and took away the man's camera. The photographer came back to Pedro, demanding from judges and police-officials the arrest of his assailant. Having failed in all his attempts, he went out on the street, and encountering Schultz, Hanun and Dirkett, made a demand for his property. He thought that he was safe, there on Main Street, with scores of witnesses present; but at Schultz's command the two gunmen held him up and searched him, and after they had made sure he had no weapon, Schultz struck him over the head with his cane, and beat him in the face with his fists.

Next day came another "battle", this time at Harvey's Run. The "death special" was turned loose for a quarter of an hour, and a couple of thousand shots were fired into the tents. One man was shot through the head, and a boy running down the street was shot through the foot; he fell, and attempted to crawl away, but the machine-gun followed him, until he had got five bullets through one foot and four through the other. Billy Keating, who hurried over to Harvey's Run in an automobile, counted a hundred and forty-five bullet-holes in one tent.

Not content with this achievement, next night the "death special" came again, with reinforcements; three other machine-guns, posted at commanding points, and more than a hundred deputies. They surrounded the colony, ordered out the strikers, and with kicks and blows and curses drove them down the road, herding them like sheep into a distant arroyo. Meantime the under-sheriff, with Schultz and a crowd of deputies, were searching the tents for arms. They tore up the platforms, they chopped open trunks with axes, and scattered the possessions of families in the mud. They confiscated about a dozen rifles and revolvers, and incidentally helped themselves to pocketbooks and jewelry and whatever else they fancied.

It was to be the turn of the Horton colony next, so all rumors declared. There was especially bitterness against Horton, as it commanded the roads to half a dozen important mines, and was a strategic point for "picketing". There was never an hour of the day that guards were not to be seen prowling about; at night you heard men shouting from the neighboring hills, cursing the strikers and threatening them with annihilation. It was impossible to find these men, but the men could always find the tent-colony, because of the searchlight; until a raiding-party of strikers, adopting Billy Keating's suggestion, crawled up and shot this searchlight to pieces.

The miners had few guns, and it was difficult to get more, because the stores in the coal-towns had been cleaned out by the companies. Some of the Horton men went down to Pedro, telling the towns-people of their plight, and begging from house to house for arms. They came back with rusty shot-guns, old pistols, little "twenty-two" rifles such as boys use to shoot sparrows. Mike Sikoria carried a muzzle-loading musket, which antedated the Civil War; he had got powder and buckshot for the weapon, but could not fire it because he had no caps! Rovetta and his Italian friends had not been able to get guns of any sort, but had armed themselves with pitch-forks and axes, and meant to charge upon machine-guns with these implements of industry.

[28]

These developments increased the difficulty of Hal's position in the colony. He could not deny the right of these people to defend their homes; yet he wished to stand by his resolve to make his appeal to moral forces. And how could he do this, how could he make his personal position clear? He had become a leader here; irresistibly, in spite of himself, he had been pushed to the fore, and now the world held him responsible for the behavior of the strikers. Nor did he wish to repudiate them, to say, I have no part in their acts of self-defense. And on the other hand, how painfully easy it was for the strikers to misunderstand him. Oh yes, they would say, he's a rich fellow; he's only playing at striking! He's willing to help us, when it comes to appealing to his rich friends, or getting his name in the papers; but when it comes to getting a bullet through him, or laying himself liable to a long term in jail, or maybe to being hanged—then he thinks better of it, naturally!

In his own soul Hal suffered the qualms of the "conscientious objector", whose way seems so easy—too easy! Yet it is not really easy; these qualms of the spirit are real things—as real as bullets or jails. To stand forth as weak-minded, a futile person, crying peace where there is no peace; appealing to moral forces, when there are no moral forces anywhere apparent! And when it would be so simple and so satisfactory to become a normal human being, to seize a weapon and go after those you hate! Yes, it takes courage and daring, even to believe in the existence of moral forces, in a community which seems to believe in nothing but bullets and jails, which presents itself to the eye of the soul as a mighty fortress of falsehood, pouring out upon its assailant a deluge of slander and abuse!

Each day made it harder; each cry of distress that Hal heard, each face of pain that he saw! He must identify himself more completely with each victim, he must be sure that he suffered in soul as much as they suffered in body! Also he must explain himself to them, he must keep their respect!

Mary Burke's respect, for example! Mary was Irish, and the blood of ten thousand ancestors called out in her for a share in this shindy. When she thought about Hal's attitude—yes, he was a gentleman, and this was "dirty work", from which a gentleman shrank instinctively, whether he realized it or not. So Hal would have to argue with her; it was not really that—at least, he thought it was not that, he hoped it wasn't! They would have long searchings of soul, trying to settle deep problems of philosophy under the shadow of Stangholz's machine-gun!

What was there, really, in this busines of pacifism? Mary wanted to follow the best light she had; she wanted to learn from Hal, to be as good as he was, in the moral sense. Should she, too, place her trust in moral forces? Or were moral forces, perchance, a luxury of the rich and respected? There was really a difference, Hal must admit. If anything untoward were to happen to the son of Edward S. Warner, it would make a fuss in the newspapers; whereas, who knew or cared anything about Tommie Burke's lame foot, or the four teeth which had been shot out of Ike Klowowski's jaw? Really now, did it not seem a farce to preach non-resistance to common working-people? It was like telling a man to submit himself to savages on some lonely cannibal island. The man died and was eaten, and that was the end of it; nobody knew about him, and nothing came of his action—save that some cannibals grew fat and hearty!

Unless you believed in God, of course! Did Hal believe that there was a God, who watched what you were doing, and would do something to help you? And if so, what would He do? Hal saw that it was really a serious matter to preach brotherhood to the poor and lowly! Almost as serious to preach it in this far Western mining-country, as it had been to preach it in Judea, way back in the days of imperial Rome!

What Hal had to do was to nerve his soul to new moral efforts. To write more fervent appeals to the newspapers, and when the newspapers refused to print them, to have them printed at his own expense–or rather at the expense of his father! To sit up till all hours of the morning, mailing these appeals to his friends, and adding personal messages, more excited than good taste might seem to permit! To hurry off to Sheridan and throw himself, a sort of moral dynamite bomb, at the head of the little cowboy Governor!

The Governor was still riding about in the coal-company automobile with Schulman, well guarded from dynamite bombs both moral and material. "You cannot see the Governor!" declared Old Peter's general manager; and the general manager was, as has been said before, a person accustomed to managing–one of those "forceful" men, who look you straight in the eye as they speak, and whose sentences hit you like blows between the eyes. "No sir, the Governor is busy."

"How do you know he's busy?"

"I'm keeping him busy, if you must have it!"

The two of them stood with clenched fists–there in the corridor of the American Hotel, outside the Governor's rooms. When Hal would not go about his business, the Governor's secretary came, and repeated the general manager's words, as obediently as if he had been an office-boy. And finally came the Governor himself, hurrying past, with Schulman on one side and Atchison, chief clerk of the "G.F.C." on the other, and one detective in front and two in the rear. And here was Hal, following along and arguing, while the Governor shook his head and almost ran, and one of the detectives rammed his elbow into Hal's ribs, almost pinning him against the wall. A most undignified scene–and a most unsatisfactory moral effort!

[29]

The little cowboy Governor had issued a statement in the newspaper that morning, saying that everything was now all right in the coal-country. But that afternoon he was so unwise as to ride out in the automobile without either Schulman or Atchison—with only his secretary and two detectives. He was going up to the Pine Creek mine, to make sure that there were no strike-breakers held in "peonage"; and half way up the canyon he ran into a sentry, who brought up his rifle and commanded, "Halt!" When the question was asked, "Why?" the answer was, "Company orders!" No one was to pass by that road. When it was explained that this was Governor Barstow, the answer was, "Nothing doing." How could the guard know it was really the Governor? And anyhow, he had no business on the property of the "G.F.C."!

So the chief executive went back to Sheridan, from which place he issued another statement, to the effect that all the strikers now had to do was to obey the law. He took his departure for Western City; and that same day a party of gunmen, going through a crowded street in the town of Sheridan, turned upon a group of strikers and opened fire on them with rifles, killing three people and mortally wounding two others. The strikers had hooted at them, calling them "scab-herders", they declared; later on they added the claim that someone had thrown a stone. But none of the strikers had been armed, and no one ever asserted that they fired a shot in the entire affair.

That was the climax; there was little revolution in Sheridan that night. Mobs of strikers swarmed the streets, and the gunmen fled, some fifty taking refuge in the courthouse, with the sheriff-emperor in command, and machine-guns mounted in the windows. Young Vagleman, son of the judge, and an attorney of the operators, got together a crowd of deputies in Pedro, with three machine-guns and a ton of ammunition, and stopping a United States mail-train, piled this miniature army into the sleeping-cars. When a brakeman protested, the deputies threw him off the train and knocked him over the head. But when they got to Sheridan, they found the situation in charge of the citizens of the town. As a general thing in strikes the tradespeople sympathize with the employers; but in this case, the employers had gone too far, and about four hundred armed citizens held the streets of Sheridan for six days, while the sheriff-emperor and his courtiers stayed within the walls of their castle.

At Horton the leaders were holding anxious conferences. It was a desperate crisis which confronted them. Should they purchase arms in earnest, and get ready for what might turn into a revolution? Or should they join with the operators and their newspapers, in calling for the militia to keep order? The strike-leaders distrusted the militia, for they knew how it had broken the strike ten years ago. Would the operators want it now, unless they knew what it would do?

Hal listened to tales about the militia, but could not believe them. His cousin, Appie Harding, was an officer in this body, and a number of his friends belonged to it—Bob Creston and Dicky Everson among them. These young fellows were self-indulgent, and entirely without idea of the meaning of the strike; but they were not depraved, and they had a sense of fair-play—surely they were more to be trusted than brutes like Hanun and Dirkett and Stangholz! Surely anything would be better than this present arrangement of "Government by Gunmen"!

[30]

They were threshing out these questions on Friday evening at a meeting in the headquarters tent. Half a dozen times the discussion was interrupted by strikers rushing in to say that parties of mounted men were gathering. They were getting ready for their threatened raid! They were going to wipe out the colony! All that night excitement continued—alarms and excursions, telephone-calls from all sides, the digging of shelter-holes and trenches.

And sure enough, soon after daylight a party of twenty armed horsemen came down the canyon. With Mary Burke and Jerry Minetti at his side, Hal stood at the edge of the village, and saw one of the horsemen raise his rifle and fire into the tents. The provocation was deliberate; and it at once became evident that every preparation for attack had been made. Parties of the guards had been posted on all sides, and opened fire.

The strikers rushed into the open, to draw the fire from the tents with the women and children; but the enemy meant business this time—their volleys were deadly. Several men fell, and the rest retreated, firing as they went, to the steel railroad-bridge. From this shelter they continued shooting all day, until darkness fell and the gunmen retired up the canyon.

But the strikers were now thoroughly aroused, and gave ear to Billy Keating, with his program of offensive defense. They followed the guards, and sent out a call for help all over the district. Men walked all night—twenty or thirty miles in some cases; they posted themselves behind rocks, and laid regular siege to Barela and North Valley and the Northeastern.

The guards too, called for help—and needless to say they did not understate their peril. Rescue parties with machine-guns were loaded upon trains both at Sheridan and Pedro; but behold, a new development in strike warfare—the crews of the trains refused to move them!

So the strikers had things in their own hands next day. One may judge how little pleasure they took in fighting, by the use they made of their advantage. It was Sunday, and there was a festival scheduled in the evening, and they came home to dance! To celebrate their victory, to tell their wives and sweethearts how they had made the enemy run! There was music and feasting in the big school-tent, until the small hours of the morning—when word came that the "death special" was on the way, and an army of gunmen leaving Sheridan in a train of steel cars, which they were running themselves!

So the strikers proceeded to organize and put themselves under military discipline. They mounted pickets, and when the steel train made its appearance in the grey dawn, it was received with such a hail of bullets that it was forced to back away. And all that day bodies of strikers continued to arrive, and new companies were formed and new leaders appointed. They offered a command to Hal Warner, and more than one man looked at him with wonder and distrust when he declined. One of those who accepted with alacrity was the new "Ironside", with the gun and the reporter's note-book; another was Jerry Minetti, and a third Jack David. "Big Jack" got himself a horse from one of the neighboring ranches, and with it the title of "General". It was a small horse, and when the "General" sat on it, his feet all but touched the ground. But nobody smiled at the sight; the silent Welshman was become suddenly bold and determined, giving his orders like a veteran commander.

That night a blizzard descended out of the North. But even this made no difference—in the early morning a thousand men marched forth, and proceeded in business-like fashion to besiege the coal-camps. They cut the telephone and telegraph wires, so that it was no longer possible to send out news; they made ready to dynamite the railroad-tracks, so that the steel train could not make another foray. They beat down the resistance of the guards, killing several; the group under the command of "Captain" Keating was actually in possession of the North Valley stockade, when word came to Horton that the "policy committee" of the union had had a session with Governor Barstow, and had worked out an agreement for the ending of the strike.

The militia was to come to the field immediately, and both sides were to submit to its authority. The Governor gave his solemn assurance that the mine-guards would be disarmed as well as the strikers; also that the importation of strike-breakers would cease, that the laws would be enforced and the strikers protected. So messengers were sent to the various war-parties, and the wearied skirmishers came in, and piled their guns in the headquarters tent.

There was general relief; but it died with the day-break, for no militia appeared, and more and more mine-guards kept coming. Toward the middle of that day, seven automobile loads were turned loose in the hills around Horton; the strikers rushed to the headquarters tent, declaring that they had been betrayed, that it was all a trap. They clamored for their arms, and marched out to fight again.

But now the guards were well armed and well led, and the strikers were repulsed, and the tent-colony was again under siege. All night the terror continued, and in the morning the strikers resolved to send their women and children to Pedro. The militia was all a myth, the colony was doomed to destruction! So, while General Jack and his men were digging trenches about the tents, General Jack's wife was marshaling women and children, in a weary and pathetic procession through the snow. Hal put Tommie Burke on his back and carried him over to the village of Horton, where a room was found in which he could stay with Jennie and Mary. As for the strikers, they settled down to a siege. Their supply of fuel had failed, their food was failing, but they would die rather than surrender to the guards.

[31]

That day, however, the militia arrived. They detrained and camped a few miles below Horton, and the commander rode up to the colony under a flag of truce.

Adjutant-General Wrightman was an old man, with heavy white mustaches and beetling eyebrows; in civil life he was an oculist, but he looked imposing in his big military overcoat, with a sword and revolver in his belt, and he did the soldier business sufficiently well to impress the untutored strikers. He laid his hat and gloves on the table of the headquarters tent and made a little speech, in which he informed them that he had come to carry out the Governor's orders, disarming both factions in this wicked civil strife. He assured them that they would have fair treatment, that no further attacks upon them would be permitted; he ended by solemnly adjuring them, in the name of the law, to surrender their weapons and abandon their disorderly activities.

John Harmon replied to him, declaring that the strikers were peaceable people, who asked nothing but to be protected in their rights as citizens. He, for one, was willing to accept the General's assurance, and he believed the other leaders would do the same. They would do their best to influence their followers; the militia might come in without fear of trouble, and all weapons in the tent-colony would be given up as soon as they could be collected.

That evening Hal went down to the militia encampment. It was Company C which was coming to Horton—the company under the command of Captain Appleton Harding. Appie, needless to say, did not relish having to give up his law-practice and come to the coal-country in the midst of a blizzard; but he came, because it was his duty. He was a slender young man, wearing gold-rimmed eye-glasses, and looking more like a college professor than a soldier. But he was a man of rigid honor, and Hal felt sure that no outrages would be perpetrated under his authority.

Also there were Bob Creston and Dicky Everson, and half a dozen other of Hal's friends; for it happened that Company C was the "society" body of the State Guard. These young fellows had joined the militia because they liked to wear dress uniforms, and to invite their young lady friends to "hops" in the armory; but here they were trapped and ordered away on disagreeable duty! They vented their discontent upon Hal. What the hell was this stunt he was doing? Talk to them about brotherly love for "red-necks" and "wops"! Was he getting ready to go into politics? But when they heard a bit of his story, their attitude changed. Yes, it was rotten, they must admit; Hal might assure his "wops" that nothing of the sort would happen while Company C was on the ground!

Other circumstances contributed to restore the strikers' hopes. The Governor seemed so very much in earnest; he repeated his pledges publicly, and in such convincing fashion, that the doubters were silenced. The women and children came back from Pedro, and the men who had been hiding in the hills came in and surrendered their guns. Parties went down to visit the militia-men at their camp, and there was joking and good humor, and music and feasting in the tents.

It was decided that the coming of the troops to the colony should be made the occasion of a demonstration. The strikers resented the false accounts of the fighting which had been published in the newspapers, and they wanted to show the country that they were decent, law-abiding men, with proper respect for the flag and the government. There was a day of hurried preparations; over a thousand children were gathered from the near-by tent-colonies, and the women set to work to dress them for a holiday. They bought all the white goods they could find in the little store at Horton, and brought from Pedro as much as several men could carry; there was cutting and sewing and fitting all night, and next day came two brass bands, and there was a pageant so pathetic that Hal could hardly keep the tears from his eyes. Little children of the coal-camps, marching out all in white, in spite of cold and snow on the ground; little children of twenty nations, waving American flags! Cheap flags, stiff and shiny, some of them no bigger than a child's hand—but meaning the same as if they had been big and made of costly silks. They meant that America had come to the coal-camps! The dream of liberty and self-government, which these children, or their parents, had come from the far ends of the earth to seek—this dream was at last become reality in their lives!

Little Jerry had on those wonderful clothes which Hal had bought him for the New Year's party; also he had two wreaths of pink paper roses which his mother had made him. He gave one of these wreaths to Hal, and made him put it on; he clung to Hal's hand, and clamored for him to fall in with the procession. Also the two David children and the swarm of little Rafferties crowded about him and clamored; so he went, with children dancing about him, and children's voices singing in his ears, "My country, 'tis of thee".

And here came the soldiers, tramping through the snow—all in dress uniform, with glorious big flags of silk flying in the wintry gale, and a fifer and drummer making shivery music. The children divided, and the soldiers marched through, under a shower of paper flowers and confetti: General Wrightman, his horse prancing, the rider sitting stiff and solemn, staring ahead, his white mustaches looking like snow on his rosy face; Captain Porter, a real-estate man, with Troop A of the cavalry; Captain Smithers, a leading hardware merchant, with Battery One, two rumbling cannon and a mounted machine-gun; Captain Harding, with his "society" company of infantry, the men marching by fours, some of them looking stern and self-conscious, others grinning, and dodging the paper roses.

The children fell in behind and followed the procession, through densely-packed lines of strikers. Such cheering and yelling—you could hardly hear the two brass bands, blowing for dear life, each a different tune! Hal looked at the faces in this crowd as he passed: Old Mike Sikoria, with a cap of purple tissue paper on his head, capering like a trained billy-goat; "General" Jack David, having dismounted and resigned his commission; John Edstrom, white-haired and benevolent looking, smiling feebly; Louis the Greek, black-eyed and olive-skinned, his mouth open, singing "America" in Greek; Charlie Ferris, with his head still bandaged where Jeff Cotton had hit him; poor pathetic little Klowowski, with his face almost hidden in bandages, where the bullet had taken his four teeth!

There was many a bandage in this throng, not only on men, but on women and children; but now the wounds of war were forgotten, the hatreds of war were swept away. The American flag had come in, there would be no more Government by Gunmen! That night there was a big dance and a feast, and everywhere you went through the tent-colony, you heard men singing their song of hope:

"We'll win the fight today, boys,
We'll win the fight today,
Shouting the battle-cry of union!"

BOOK THREE

LAW AND ORDER

[1]

The next two weeks were among the most interesting of Hal Warner's life. This wonderful experiment in democracy which he was watching came as it were from under an evil spell, and found freedom to express itself. Collectively, the strikers had the sense of power, of certainty for the future; while individually they were happier than they had ever been in their lives before. They had comfortable homes, they had enough to eat, and they made their life one long celebration.

All sorts of musical talent discovered itself in the tent-colony; there was an impromptu concert going on somewhere all the time, and on two evenings a week the big school-tent was cleared out, and there was a dance with a regular orchestra, made up entirely of strike talent. To these festivities all the world was welcome—especially the militia, whom the colonists looked upon as their deliverers. The militiamen were glad enough to come to the dances; even the young society men, who found the military routine wearisome, and were surprised to find how gracefully the daughters of these foreign races could dance.

Then in the day-time, the soldiers and strikers would go a-hunting. The soldier would take his gun, and the miner his pick and spade, which was quite as necessary to hunting in that country. A man could not get much meat upon the strike allowance of three dollars a week, and the militiamen soon grew sick of their rations of "bully beef" and "salt horse", canned corn, beans and "hard tack". So, when the hunters came in with a load of jack-rabbits in a bag, there would be a grand feast, and afterwards there would be music and dancing in the square in the middle of the tent-colony, and the Greeks and Bulgarians and "Montynegroes", who were war-veterans, would take the militiamen's guns and show how they drilled "in old country".

Also the militiamen and strikers played baseball together; and it is impossible to despise anyone with whom you play baseball. When it came to mingling the races, a baseball bat was better than any wizard's wand. Jerry Minetti, who in his organizing work had had trouble with race prejudice, sat watching a game among these Balkan peoples, who at home had been at one another's throats. "Somebody ought to make Balkan Baseball League," said he. "Then they don't fight over there!"

The colonists had leisure also for study and discussion. Hal now saw in the case of many, what he had seen in the case of Mary Burke—the discovery of the intellectual life. Someone started a class in English, and in a few days a tent was crowded with pupils. They had put up a school-tent for the children, but it had been impossible to get a teacher, on account of the continual fighting. But now Mrs. Olson, the young widow of the murdered organizer, gave up her position in Western City and came to help the colony. Also there developed local talent; white-haired and benevolent old John Edstrom liked to get the youngsters about him and tell stories, and Mrs. Olson insisted that that was a fine thing—children ought to hear stories all the time. She got books for him—poetry, and all sorts of fairy tales and legends, and three times a week he would read to a whole tent-full of youngsters. Also he would tell them stories about the lives of workingmen, about strikes that he had seen; giving them little Sunday-school lessons in the new religion of solidarity.

It was a great occasion for all sorts of people who had things to teach. The Socialists gave out pamphlets, and Socialist papers in all languages were read until they were frayed to pieces. The miners were publishing a daily paper in Pedro, giving their side of the strike; you might see one copy of this paper being read aloud to half a dozen men. There were the Syndicalists, with their vision of one big union of all the workers; there were all sorts of men "making talk"—even employes of the Schultz Detective Agency! One of these gentry, posing as a striker, made a speech in which he urged the miners to burn all the coal-camps; and he had a photograph taken of himself making this speech, which photograph he sent to the agency as evidence of his services. It happened that this letter and photograph came into the hands of a union sympathizer in Schultz's office, who forwarded them secretly to John Harmon. If that speech happened to be reported in the capitalist newspapers, what a to-do there would have been about violence in the tent-colonies! And what a time Hal would have had persuading any of his friends to consider the idea that the speaker might have been in the hire of the coal-operators!

Upon first coming to North Valley, Hal Warner had been struck by the rigid race-lines, the little social sets into which the population was divided; but now before his eyes he saw these lines being wiped out, he saw twenty nations being melted into one. Old Mrs. Rafferty, who in times past had been for Irishmen only, was now heard to remark, "I never knew them Greeks was such gentlemen! Sure now, they're a fine lot, ain't they?"

Then again, among the wonders of camp-life was the musical power revealed by the once-despised "Dagos". They had most marvelous choruses, with complicated voice parts; they had revolutionary songs, which Jerry Minetti and other Socialists taught them. The words were Italian, but the spirit could not be missed by any striker;

Irish and Welsh and American miners would listen and marvel, saying, "Would you ever have thought that Dagos could make music like that?"

There were leaflets of union songs to be had in half a dozen languages; and those who knew how to sing gathered groups of the others and taught them. These songs were full of the vision and the resolution of labor; they taught the new morality of brotherhood, of mutual helpfulness, of useful work as opposed to parasitism. Perhaps they would not have passed a severe test as music or as literature, but they were adapted to their purpose of impressing untutored minds. Hearing men chanting these lessons over and over, Hal thought of "The Island of Dr. Moreau", where the scientist, working by strange and terrible feats of surgery to turn animals into human beings, wishes to cure the monsters he has created of their habits of lust and murder, and to this end composes maxims in rhythmical form, which in lonely places they chant in chorus all through the jungle-night.

This singing served another purpose—a substitute for whiskey. Up in the coal-camps men had got drunk, because there was nothing else for them to do, because overwork and misery predisposed them to it, and social feeling and business interest pushed them on. But here was an enemy to be fought, and drinking men do not make good soldiers. The managing committees could not forbid a man's going off and getting drunk in the towns, but they could forbid him to come to the colony when he was drunk, and they could bar liquor from the tents. What excuse had any striker for getting drunk on alcohol, when he might get drunk upon faith and hope and solidarity?

Hal remembered how, at the beginning, he had shrunk from these people, his Anglo-Saxon prejudices offended. But now that he had come to know them intimately, how cheap such feelings seemed! Yes, they were dirty—but how glad they were to be clean, how quickly they responded to suggestions! They were ignorant—but so eager to learn, if only you came to them with sympathy! Hal was sanitary officer, among his many occupations, and he saw how glad people were to take his advice. They had leisure now, they were not exhausted when they came home; so they could take pride in the place where they lived. Above all things in the world they desired to be like Americans. Did not every man of them, with his first spare wages, make haste to get a suit of store-clothes, and a celluloid collar, and possibly even a fancy ready-made tie? Only give them a chance, give their children a chance, and in a little while you would think their ancestors had come over in the Mayflower!

[2]

Mary Burke came back from Horton, bringing her little brother and sister. Mrs. Wyatt offered her money, so that she could take care of the children outside the tent-colony; but Mary did not wish to be under obligations. There were wounded and sick people to be taken care of, and she could work in the hospital-tent and earn her living, at the same time that she was helping her brother. Mary had some clothes now, and could keep herself neat, as a nurse ought to be.

Hal saw in her no trace of that melancholy which had so impressed him at their first acquaintance; none of those moods of bitterness and despair. Mary had now her job to do; she and Mrs. David and Mrs. Olson were the three most intelligent among the strikers' women, and the executive committee looked to them in all matters which involved the children. So Hal saw a great deal of Mary in the headquarters tent, and saw his faith in her justified.

Here, as in North Valley, young couples went to stroll in the moonlight; but Hal and "Red Mary" were no longer among them. They realized that the mine-guards believed the worst of them, and they felt that they owed it to the union to give no cause for criticism. Besides, they were really not thinking about themselves now; they were lost in this fascinating new experiment in democracy. You may think that you know about life, who live isolated and selfish lives, seeing to the welfare of one little family; but you have no real idea of the possibilities of your being, until you have become part of some great body of people, working for some great cause—until you have merged your life in a social life—until you know what it is to tremble with the fears and thrill with the hopes of a great mass of human beings.

Such was the condition here—brought about automatically by the strike. The cleverest and most capable person in this tent-colony realized that he was nothing by himself, he was able to do nothing by himself. When he sang a song about brotherhood, he had a real feeling about a real fact; it was true that the man who stood by his side and sang was his brother, bound to him in sufferings, in fears and in triumphs. And this was a tremendous discovery, something which you could only understand by experiencing it, and which then you could never forget. That was what old John Edstrom had meant when he said, Once a striker, always a striker! After this experience, the lowest and most degraded man in the tent-colony would know himself for something different, he would have a new outlook upon life.

And this gigantic effort of thirty thousand people was succeeding! The mighty machine of oppression and greed which had made their lives a nightmare for so many years was going to be broken at last! Their masters would have to come to them and make terms; the miners would go back to work with the knowledge that they had rights, and the power to enforce those rights! "We got them! We got them!" cried Jerry Minetti; and Old Mike would caper with glee and exclaim, "We teach them fellers a lesson!"

Mary Burke, hardly able to face the thought, would clench her hands and whisper, "Do you think it's possible, Joe? Do you think it's possible?"

At that time Hal really thought it was. The operators had proclaimed to the world that what kept their employes from going back to work was fear of the violence of "agitators"; but here the militia was on the ground, and rigid order was kept—yet scarcely a man went back to work, scarcely a man! All through the district the mines were dead; no "empties" went up the branch-railroads, no loads of coal came down; and here was winter coming on, and the world was clamoring for coal. In Western City the small consumer was paying a cent a pound for it, and there were meetings of protest, demands that if the operators could not run their mines, the state should take over the job.

But among the older and wiser of the strike-leaders there were no illusions about the situation. They knew the power of their enemies, and realized that these enemies would not surrender easily. Billy Keating especially was full of forebodings; they would find that Old Peter had hardly got started to fight! Could not Hal see that the respite the strikers were enjoying was a pure gift from the Governor of the state? It depended upon his order against the bringing in of strike-breakers. So long as he enforced that order, he was using the power of the state to give victory to the strikers; while if he rescinded the order, he would be using the same power to give them defeat.

[3]

There being no more news in "the field", Billy Keating was recalled to Western City, and from here he wrote Hal about the war which was being fought in the white marble State House on the hill. On one side were the "Big Three"—

Peter Harrigan, Allen of the Western-American and Harbridge of Central Fuel—together with their henchmen and allies, the big bankers and captains of industry of the state; on the other side were the miners' leaders, with the support of the State Federation of Labor: the scene of the war being the mind of that most pitiful creature, the "little cowboy Governor".

It was at this time that Old Peter wrote the letter which fastened this name upon the chief executive of his state, telling how he "spanked" him and then "gave him candy". The awarding of such punishments and favors is usually done in the privacy of the family; but up at the State House the reporters were swarming, on the alert for news, and once a group of them got a glimpse into the family's affairs. Someone going out left open the door of the Governor's office, and the voice of Peter Harrigan was heard: "You God-damned coward, we aren't going to stand for this! You've got to do something, and do it quick, or we'll get you!"

And then came the whining tones of the pitiful chief executive: "Don't be too hard on me, gentlemen, I'm doing it as fast as I can!"

When this news came to Hal, in a letter from Billy, he did not show it to anyone in the tent-colony. There was no one he could bear to torment with such a terrifying picture. But he wrote to John Harmon, who had gone back to Western City, warning him that now was the critical time, and that he must move every influence he possessed. So there was a great mass-meeting of protest, called by the State Federation of Labor, and this meeting sent a delegation to interview Governor Barstow. The trouble, the delegation insisted, was all because the operators refused to negotiate with the union, even to meet the union leaders. If a conference could be held, a settlement might be arrived at. Was it not the duty of the Governor to insist upon such a conference, rather than to let himself be used as a tool for the crushing of the strike?

The unions had one threat which they could make—that of a general strike throughout the state. The Executive Committee of the State Federation was authorized to take such action; and when they held up this spectre before the "little cowboy Governor", he would lose his self-control and plead with them, as abjectly as a few hours before he had been pleading with Peter Harrigan. A monstrous and wicked thing for men to talk of crippling the industries of a great commonwealth!

This cry was taken up by the newspapers, which said nothing about Peter Harrigan's desk-pounding, but pictured the labor men as holding a pistol at the Governor's head. It was unheard of arrogance; and apparently Peter Harrigan and Judge Vagleman thought so too, for they saw the United States prosecuting attorney, and caused a Federal grand jury to indict forty-two of the leaders of the coal-strike for conspiracy in restraint of trade!

There was a struggle going on for the sympathy of the public; and in this contest the operators had all the advantages. They could publish whatever they pleased in the papers, not merely their arguments, but their news; they had unlimited funds to be used in propaganda, public and private. Just now they were publishing in newspapers throughout the state a series of advertisements with the sarcastic headline: "STARVATION WAGES IN COAL MINES". There would follow an elaborate set of figures showing the average daily earnings of all the miners in a certain mine on a certain day—which average was from three to four dollars. And how could the miners explain to the public the subtle knavery of these figures? It was true that they got three or four dollars a day—on the days when they were paid; but there were, on the average, only a hundred and ninety-one such days a year; and what about the days when they were *not* paid? The companies appeared to cover that point by declaring that the men were at liberty to work as many days as they pleased; from this the public would conclude that the men must have spent the other hundred and seventy-four days loafing in saloons. Who would credit that on those extra days—even the Sundays, often—the miner was down in the pit, slaving as hard as a man could slave, but without having a cent to show for it?

This was the "dead work", about which so much controversy centered; cleaning out rooms, "brushing", timbering, "grading bottoms", laying track, removing falls of rock! All this work had to be done before any coal could be cut or loaded; and while he was doing it, the miner's name did not appear on the payroll or anywhere else. The state bureau of labor made a computation, based upon thousands of payrolls, and upon the full number of working days, which showed that the average earnings of a miner were not three or four dollars a day, but a dollar and sixty-eight cents a day! And this sum had to be spent in company-stores at company-store prices—and in a country where the cost of living was going up ten per cent a year!

[4]

While Peter Harrigan and his associates were laboring with the Governor, the company officials and attorneys in the field were laboring with General Wrightman. They knew the General of old; he was their man, by all his class-prejudices and his personal weaknesses. Why need he be so strenuous in enforcing the Governor's "policy"? If strike-breakers came in, a few at a time and quietly, surely no great harm would be done! Here were five who wanted to get into North Valley; men who were willing to swear they were natives of the state, and had worked in coal-mines for years. Surely such men had a right to go to work! There was a bunch who started for Barela, disguised as railroad section-hands. The strikers at Horton saw them and gave chase, and the militia gave chase to the strikers, and there was an excited argument on the road. Surely the militia should have maintained its dignity in a situation such as that!

Also there was the question of the disarming of the mine-guards. There had been much said about this intention when the militia had first come into the field, and had wanted to get the strikers' arms without trouble. The strike-leaders had agreed to give them up, and Captain Harding had come to the Horton colony to get them. He had found thirty-two guns stacked up in the headquarters tent; and when General Wrightman was dissatisfied with this and ordered a search of the colony, the strikers had submitted good-naturedly, helping to take up their tent-platforms, and adding a child's air-rifle and a pea-shooter to the collection. The soldiers did not find many guns, and General Wrightman became angry, declaring that the leaders were not playing fair with him. That was not true, the leaders had done their best; it was not their fault if some of the men were distrustful, and had chosen to wrap up their weapons in oil-cloth and bury them secretly in the hills.

Hal found out from Mike Sikoria that some of the young fellows had done this. "Sure they hide them," said Mike. "They try them militias a little bit. Let them get guns from the mine-guards—they got lots more guns than what our fellers got." And Wresmak, the Bohemian, who had spent the last of his savings for a first-class rifle, admitted that he had concealed it. It wasn't doing anybody any harm under the ground, and it would stay there so long as the soldiers kept their promises and played fair.

And now it began to appear that Mike and Wresmak had not been so ill-advised in their suspicions. Schulman and young Vagleman and the rest of the conspirators were able to persuade the General to let some of the mine-guards keep their rifles, and even to return secretly the few he had taken away. There were rumors that the Governor meant to change his "policy"; if he did change it, there would surely be more trouble, and the militia would be glad of the help of the Schultz Detective Agency.

From that the conspirators went one step farther. What about the guns which had been taken from the strikers? Among them were some perfectly good weapons, and surely it was folly to leave them lying unused. Since there could be no idea of returning them to the "red-necks" who had owned them, was it not common-sense to let the mine-guards have them—quietly, of course.

But the trouble was that the thing could not be done quietly. One of the men who had given up his weapon was "General Jack". It was a new Springfield rifle, and had cost thirty dollars, but the "General" had surrendered it because Harmon told him to, and he had confidence in Harmon's judgement. But, as it happened, he had carved his name on the stock of the rifle; and now behold—down in the saloons of Pedro was a deputy, newly arrived from the copper-strike in Michigan, boasting that he had captured the rifle of "General Jack"! This was a great joke on the strikers, of course. The big Welshman went down to the town and sought out this guard, and the latter, not knowing him by sight, and having had two or three drinks, permitted an inspection of the rifle. Sure enough, there was the name, "Jack David"!

There were other indications of what was going on behind the scenes in General Wrightman's headquarters in the American Hotel. There was the matter of automobiles, for example. An automobile has come to be, under modern conditions, indispensible to the maintenance of dignity; but the state had unfortunately overlooked the need of dignity on the part of its military officers. So Mr. Schulman came forward; the company had plenty of cars, and they were at the disposal of the militia.

But these company cars were known to the strikers, who had many times been shot at from them; naturally it did not please them to see General Wrightman riding about in these cars. Still less did it please them to see the hated deputies sitting by his side: Jim Torrance, for example, a man who had come to the Harvey's Run colony with a white handkerchief in his hand, and given a signal for the machine guns to open fire—what business had such a man sitting up chatting with militia officers? What business had Schultz, and Gus Dirkett, and Pete Hanun, the breaker of teeth, being intimate with the Adjutant-General of the state soldiery? When these automobiles whirled round street-corners and nearly ran down the children of strikers, the strikers became enraged. An automobile is, at best, a fosterer of class-consciousness; it might be called a symbol of the capitalist system, in that a few people get the fresh air and the scenery, and the rest of the world gets the dust and the bad smell.

[5]

Hal had many arguments about these matters with Captain Harding. Appie was reluctant to talk with his cousin, who was apt at any time to go "off his head"; but as matters went from bad to worse, it was harder for the young officer to contain himself, and from his remarks Hal was able to piece together a picture of the situation. There were decent men in the State Guard, and they were doing what they could, but they were being shoved aside and rendered helpless by the lower element in the organization.

Company C had maintained itself in peace times as an exclusive institution; its members were young businessmen, its officers the *élite* of the city. But the work it had now to do was not the sort of work that gentlemen are accustomed to doing. Dicky Everson, for example, whose occupation had been "rushing the girls", dancing until two or three o'clock in the morning; or Bob Creston, who had been practicing up to win a cup for an indoor tennis-club—what interest had such men in pacing about in the snow with rifles on their shoulders, chasing Greeks and Dagos away from coal-mines? It was not the sort of work that gentlemen did, it was the sort of work they hired other people to do for them; therefore Bob and Dicky suddenly resigned. And then went Billy Harris, who was in his second year at the law-school, and who stood to lose a whole year of his career. The parents and relatives of these men had influence, and could get them off, on one pretext or another; and so, day after day, the character of the company was changed. New men had to be found, and the recruiting officers must take the class of men who wanted a job at a dollar a day.

Captain Harding was strenuous that none of the new men should be mine-guards; but in this he stood almost alone. The mine-guards were on the spot, a thousand and more of them, familiar with the locality, familiar with the weapons, familiar with the strikers they were to control. What more natural than to enlist them?

But then—see how this appeared to the strikers! The very same brutes who had been abusing and terrifying the tent-colonies—they were to be clad in the uniform of the state, armed with the weapons of the state, and turned loose to behave worse than ever!

When the report of these developments went out over the country, people refused to believe it; they did not wait to hear the evidence, they said such things were impossible. But the strikers had evidence of the sort which men never doubt, the evidence of their eyesight. When they saw "Butch" Andrews, one of those who had hid under the steel bridge and opened fire on Horton, strutting about the streets in khaki, they knew it was the same "Butch". When he accosted Rovetta's young sister on the street and called her a "ten-cent whore", she understood that he had not changed his rowdy nature with his costume.

And then the fellow Stangholz, who had taken charge of the machine-gun at Barela—Stangholz not merely taken into the militia, but made into an officer! Right off the bat, with no more training than he had got by swinging his "baby" about, and shouting that he would wipe out the God-damned sons-of- —! A few days later there was some trouble at Barela, and in the papers Hal read a statement by "Lieutenant Stangholz". The people at home would read this, and would be impressed—thinking of a West Pointer, perhaps! Who would dream that this "lieutenant" was a common bar-room rowdy, an adventurer who had been looting ranches with Mexican bandits?

And then came a new development. The state was paying its militia servants a dollar a day, but the mine-guards were receiving three-fifty a day; and not even the honor of wearing a uniform, and being under the protection of the flag, would persuade a mine-guard to have his pay cut in that fashion. On the contrary, he wanted both wages, and the generous Mr. Schulman saw to it that he got them. And this, naturally, caused discontent among those militiamen who had not had the good fortune to begin on the mine-guard pay-roll. If some militiamen were getting four-fifty a day, why should others work for a dollar? The logic was inescapable, and Mr. Schulman caused it to be understood that all militiamen might have the extra pay for the asking. Captain Harding admitted to Hal that the offer had been made to Company C. As a body, the company rejected it with indignation, but there was nothing to prevent individual members from getting the money, and many were undoubtedly doing so. They were coming and going at the office of Schulman all day—privates and officers alike—in such numbers that the townspeople came to refer to the place as "military headquarters".

And before long even this was not sufficient; there arose uncertainty as to the ability of the state to pay its paltry dollar. No money had been appropriated by the legislature, and the State Auditor declared that he was without authority to issue warrants. The militiamen were indignant, for they did not like to have to wait for their dollar. At Sheridan a crowd of them assembled at night and burned the over-scrupulous Auditor in effigy. So the coal-operators came forward and offered to advance the money. But this Governor Barstow would not permit—it would not "look right", he said. It had to be arranged that the coal-operators should advance the money to their banks, the banks should advance it to the Clearing Association, the Clearing Association should advance it to the State Auditor, and the State Auditor should advance it to the troops. The steps of this process were perfectly well known to the whole state; yet all seemed satisfied that "looks" had been preserved!

[6]

Once more there were troubles at the Horton tent-colony. The first of them was the experience of Cho, the Korean "rope-rider", who had a brother somewhere in America, and was expecting a letter from him; he had ordered his mail forwarded from North Valley to Horton, but this had not been done, and now that the militia was in charge, he thought it would be all right to go to the North Valley post-office. He got in and got his letter, but when he tried to get out again, the men on guard would not let him pass the gate. Go to work, they told him—the strike was going to pieces in a few days. Cho got someone to write a letter to Hal, telling about his plight, and Hal went to the militia encampment to see his cousin.

The men on duty at North Valley belonged to another company, said Harding; but he would see what he could do about it. He telephoned to the mine, and Hal listened to one end of a strenuous conversation. "You thought he was a strike-breaker, eh? Well, what if he was a strike-breaker? Is it your business to keep strike-breakers at work? How long have you been in the militia, anyway?"

So there it was! The man who had held Cho was a newly-enlisted mine-guard; he had put on his militiaman's uniform, and gone on doing what he had been doing before. And what was more significant still, Captain Harding had to do considerable arguing with the man's superior officer before he could have orders given for the Korean to be let out. "None of my business?" he cried, into the phone. "And when the man lives at Horton, and everybody in the tent-colony knows he's being held? You expect me to control them in the face of such conduct?"

And then, a couple of days afterwards, a man dashed into the tent-colony, out of breath and wild with excitement. He was one of a group of seven hoboes who had been trying to beat their way to California, where it was warm. The freight-train had stopped beyond Horton, and men with clubs in their hands had gone through the cars, poking the hoboes out; they were lined up by a group of militiamen and informed that they would go, either to the mines or to jail. "Nothing doing!" said the young fellow who told the story. "Nix on the scab-business for me." Whereupon they put a bayonet at his back and started to march him up to Barela. He made a dash for it, and after a chase of a mile or more he got away.

Hal took that story to his cousin, and again there were interesting developments. Captain Harding would not talk frankly; somebody else had been responsible for the happening—somebody apparently had given orders over Harding's head. There was a regular political "machine" inside the militia, and this "machine" blocked his efforts to punish the guilty men. The Captain was bitter about it, hardly able to face his cousin.

A couple of days later Jerry Minetti came to Hal, blazing with wrath. Rosa had been insulted by one of the militiamen, and was in her tent in hysterics. So Hal went to his much-troubled cousin yet a third time—and this time it was Appie's own affair, upon which he could take action without consulting anyone else. He went to the tent-colony and heard Rosa's story, and then brought her and her husband to the encampment, and had his men lined up, and the guilty man pointed out and put under arrest.

By which action he brought a hornet's nest about his ears. What sort of way was that to treat members of the State Guard? To line them up like criminals, to be inspected by a Dago woman who was probably no better than the rest of them! The militia were used to lining up strikers and picking out criminals; to turn the tables about was to humiliate them and weaken their authority. Captain Harding was summoned to headquarters, and there was a terrible row; he almost got into a fight with Major Curran, one of the General's right-hand men.

Major Curran had his own way of treating strikers! He was a saloon-keeper politician from Western City, a heavy-set man with brutal features and a coarse tongue. Soon after his arrival in the field, four mine-guards coming in an automobile from the Mohican mine were ambushed by strikers and shot; whereupon the Major raided the tent-colony, and under pretense of searching for arms, drove the strikers away with violence and foul language, and turned their possessions out into the mud. He added insult to injury by allowing the superintendent of the Mohican mine to drive through the streets of the colony in a buggy and give commands to the mine-guard militiamen.

When such things would happen, Hal would betake himself to his tent and write long letters to his friends: to his professor of economics, begging him to spend his Christmas holidays in Horton; to Adelaide Wyatt, begging her to start another disturbance in the Tuesday Afternoon Club; to poor perplexed Will Wilmerding, quoting Jesus and Isaiah on the subject of them that grind the faces of the poor. He did not write to his brother, but he wrote to Lucy May, and made sure that his letters were read aloud at the family breakfast-table. He wrote so many letters to Larry Pringle that for a while he saved the "Gazette" the cost of a correspondent in the field. Pringle would compose a red-hot editorial based upon Hal's facts, and then Hal would telegraph for two hundred copies of the paper. Mrs. Olson, who knew how to use a typewriter, would sit up half the night writing letters for him, and Mary Burke, who had been nursing the sick and wounded all day, would cut up the papers and seal the envelopes.

[7]

All this time the Governor's "policy" of keeping out strike-breakers was still supposed to be in force. But the operators were busy all over the country gathering strike-breakers, and the militia was doing everything to help ship them in.

At the Hazleton mine, which lay close beside the railroad-track, the Trans-continental Express stopped unexpectedly one night, and twenty men were rushed into the mine under the very noses of the soldiers. And next day a steel car was taken in to the Northeastern, with all the shades drawn and the doors locked; too late the strikers learned that this car had been loaded with "scabs", brought all the way from Pittsburg. A few days later one of these men ran away, and came to the tent-colony with a story revealing a brand-new device in the strike-breaking art. He had answered the advertisement of a company which had land to sell in the West; land which could be bought very cheaply, because it was coal-land, and only the surface rights were sold. The purchaser could pay for these rights by labor in the mines, so the advertisement stated; but when the would-be farmer got to the district, he found that the land he had bought had a surface of sandstone running up the mountainside at an angle of sixty degrees; also he found that he was in a stockaded fort with armed men at every exit!

A week later two more victims of this scheme made their escape, both declaring that they had been beaten when they refused to work, and that another man had been shot and buried at night. Affidavits were made by these men, and John Harmon took them to the Governor, who sent the State Commissioner of Labor to investigate. It was this Commissioner's duty to see to the enforcement of all the labor laws of the state, and he went up to the Northeastern in an automobile, and announced his identity and his business. But admission was refused to him; he was turned back, precisely as if he had been a reporter for a workingman's penny newspaper! And this same thing happened to him at camp after camp, it continued to happen to him for weeks; when he appealed to the Governor, he was admitted to a few camps, but denied permission to speak to the workers! The Commissioner of Labor was a union sympathizer, and therefore an "outside agitator", from Peter Harrigan's point of view.

When the strikers heard of events such as these, it was only natural that they should be disposed to enforce the law for themselves. There came a report of some strike-breakers coming to Barela, and the whole tent-colony streamed down to the railroad. Obeying orders from his superiors, Captain Harding had set a guard about the depot to keep the strikers away from it; but now came a swarm of women with weapons of domestic construction—baseball bats with spikes in them, butcher-knives tied on long poles, bread-rollers, scrubbing pails full of rocks. Shouting and cursing in twenty languages, they went straight through the militia cordon and lined up on the platform to wait for the scoundrels who would take the bread from their children's mouths!

Captain Harding came galloping up, and shouted to Hal and other English-speaking men. He had his orders, and he would enforce them. If there were any strike-breakers on the train, they would be turned back; but in the meantime the platform must be cleared, even if a battle was necessary.

He was very angry, and evidently meant what he said; so Hal and the others set to work, and with many impromptu speeches and no end of shouting and shrieking, the coal-camp amazons were driven back, and the train came in. Sure enough, there were seven strike-breakers on board—and those strike-breakers did not get to Barela. But they got into the Mohican mine that same night, escorted by Curran, the saloon-keeper Major of Militia. Much comfort there was in that for the amazons of Horton!

[8]

Over these matters there were vehement arguments between Captain Harding and his cousin. Appie was keeping strike-breakers out of the mines under his control, because as a military man he held his orders sacred; but he was not ordered to bother his head about what Major Curran did, and he did not purpose to do so, because he was privately opposed to the Governor's so-called "policy". Besides being a militia officer, Appie was a young lawyer who took coal-company cases, and had the usual prejudices of his class. Labor unions were, or tried their best to be, combinations in restraint of trade, and as such they ought to be suppressed, or at any rate kept from being of any use to their members. That this argument made property of human labor, and so virtually admitted wage-slavery, was something the young officer could not be got to see. What would become of industry if workingmen were not protected in the right to take a job where they found it?

Appie could not see that these strikers had any claim whatsoever upon their jobs. Hal showed him men among them who had toiled for twenty or thirty years in the mines—men who had lost their health, their parents, their children, slaving for the coal-companies; yet they had no share in the properties they had helped to build up, they had nothing to show for all these years but a few sticks of furniture and rags of clothing! Here was a boy whose father had worked thirty years in the mines, and in the end had lost his life in an accident caused by carelessness of the companies; this orphan had been arrested for picking up coal along the railroad-track—possibly the same coal his father had mined! His job as breaker-boy had been his one possession in the world. "And you say he has no claim on it!"

"He gave it up," said Harding, simply.

The young officer had a vision of the tyranny which would result if labor unions were allowed to have their way. If they could call strikes and keep strike-breakers away, they could control industry, there would be no end to blackmail. He had stories to tell about labor union domination—absurd disputes that had occurred in industries controlled by people he knew. A printer who was not allowed to set type in his own establishment, because he did not have a union card! A building contractor who had been held up for a thousand dollars at a critical moment in his operations! What would Hal do about such things?

Hal answered that he was groping his way toward a new solution. The present system did not give justice, whichever side won; so for the moment the sensible thing was to compromise, and in this case the only way to get a compromise was to keep the strike-breakers out of the mines. As it was, the operators refused all discussion; they refused even to be in the same room with the union leaders! Hal pointed to the morning's news—the State Editors' Association had called a meeting, desiring light on the strike, and had invited the union leaders and the operators to attend and set forth their sides. The union men had come, but had been invited to go away again—because Mr. Harrigan was outside, and refused to enter the room while they were there!

Just now the "little cowboy Governor" had evolved a wonderful scheme to settle the trouble. For weeks he had been pleading with the operators to consent to meet somebody. If they would not meet the wicked union officials, the "agitators", surely at least they would meet a group of their own workingmen, to hear their grievances and see if anything could be done toward remedying them!

After much urging, Harrigan and his associates consented to this, and the strikers selected three of their number. The choice not being satisfactory to the operators, three more were selected; and these likewise not being satisfactory, a third attempt was made; until at last there came three actual *bona fide* strikers, brought up from the coal-fields to the Governor's office, and set down to an all-day and all-night conference with the men who had ruled their lives since they were born.

It would have been hard to offer a better proof of the workers' need to be represented by some outside person, not dependent upon the operators for his job, than the pitiful struggles of these untrained and unlettered laborers with their masterful employers, skilled in controversy, and with every fact and every pretense at their fingers' ends. As Billy Keating described it, it was a combat between three small, soft beetles, and three very large and tough long-horned rhinoceri. Every time a beetle would start to raise a feeler, a rhinoceros would put down his foot!

How often the men tried to bring the discussion to an issue—and how often they were shunted aside, led down a side path, and bogged up in unessentials! They must not discuss the grafting of superintendents, for that was hearsay; when it turned out not to be hearsay, but first-hand knowledge, they must not discuss it because the superintendents were not present to defend themselves! "You understand, gentlemen," said Beetle Number One, "that we are just simple miners. We are a bit awkward, and we have not got the same expression, and we would like a little consideration on account of that." Then said the old Harrigan rhinoceros, "We will be glad to give it to you, but I do not think, gentlemen, that you need apologize." Gentlemen they were, for this bewildering day and night; solemnly addressed as "Mister" by the great ones of the earth! The passage above quoted occurred in the course of a long digression, after which Beetle Number One remarked, "I started in, but I forgot where I was." Poor beetle! It never occurred to him that this was the purpose for which he had been interrupted!

[9]

These farcical proceedings did not settle the strike; those who were in charge of the situation had never had any idea that they would. They had a quite different program—to cow and terrify the strikers, to weaken their spirit and convince them that the contest was a hopeless one. It was like a big dam, which they meant to undermine and bore full of holes; let the water get started through in one place, and you would see the whole structure collapse.

Each day this purpose became more clear; and Hal and Jerry and Mary Burke and the little group of leaders at Horton braced themselves to efforts of resistance. They must hold their followers, plead with them, exhort them to endure; they must seek out those who showed signs of wavering, they must argue with them and cajole them. With each new outrage, they must console the victims, help them in their destitution, bind up their wounds, inspire them to fresh resolve. Above all, they must find new ways of appealing to the public, of reaching the mass of the people, who were deliberately kept ignorant of what was going on in this remote and unlovely coal-country.

But it seemed as if efforts at publicity only served to exasperate the enemy, and excite them to fresh outrage. There came a case of an indignity committed upon a woman by a militiaman: such a flagrant case that the strikers made appeal to the civil authorities, who had the militiaman arrested. But this set General Wrightman almost beside himself with anger; he sent a squad of his soldiers and took the prisoner out of jail, and had him tried by a militia court and acquitted. This, of course, was serving notice upon the soldiery that they might do what they pleased, and they did not fail to get the meaning of the notice. A reign of terror broke out, worse than anything known before. The militiamen looted saloons, they danced in the streets with drunken prostitutes, they stopped women and young girls in broad daylight and inflicted obscenities upon them. They robbed and beat men on the streets of all the coal-towns; while merely to be seen talking to a strike-breaker was enough to earn a beating for a union man. How far the troopers went may be judged from the case of one miner who came into the district as a stranger, and asked a militiaman on the street where he should go to join the union. The militiaman "ran him in", and he was brought before the General, and ordered locked up indefinitely.

Under pretense of searching for arms, it became a regular custom for militiamen and guards to plunder the houses and tents of strikers. In one single raid upon the Italian colony of Mateo they robbed a dozen families of amounts which totalled five or six hundred dollars. The strike-leaders obtained affidavits from many people who had lost sums, varying from a dollar up to two hundred dollars, the savings of a life-time which had been stowed away in the bottom of a trunk. But when the matter was reported to General Wrightman, there was no satisfaction to be had. He promised to "investigate"; but if he ever did so, it did no good—the guilty ones were not punished, and the outrages went on. The General said it was impossible to get evidence; but meantime the thieves were spending the money right and left in the saloons and brothels of the town. One would have thought that here was an occasion for the Schultz Detective Agency to display its skill!

The truth of the matter was, the General did not care what was done to the strikers. They were opposing him and making trouble for him, and so there was nothing too evil for him to believe about them. Each time the representatives of the strikers came to him with protests, they got a less cordial reception; the time came when their mere entrance into the General's office became the occasion for a tirade of threats and abuse. "He's got our guns away from us now," said John Harmon, when he came back from one of these interviews. His strong features were working with emotion—so great had been the indignities heaped upon him.

It happened that Fred Norris, one of the organizers of the union, met a strike-breaker in a restaurant in Sheridan, and talking with him in a friendly way, learned that he was disgusted with his job. "All right," said the organizer, "come up to headquarters and join the union, and we'll take care of you in the tent-colony." The man followed the suggestion and went to the tent-colony; and when Schulman, general manager of the "G.F.C.", got word of this and informed General Wrightman, the General sent for Norris, denounced him furiously, and had him put in jail. And there he stayed. Nobody could get to him, nor could he get word to anybody outside. There was no charge against him, there was no knowing what was to happen to him. He was a "military prisoner", to use General Wrightman's phrase.

There were four Mexicans arrested, charged with assault. The warrants had been sworn out in the District Attorney's office, and in due course one of the District Attorney's assistants had the prisoners taken before the court and put under bonds. When the General heard of this he flew into a rage, ordered the Mexicans locked up again, and declared that he would have no courts interfering with his prisoners. He called up the offending assistant and cursed him, declaring that he had a good mind "to arrest the whole damned bunch." He actually sent a platoon of soldiers after Mr. Richard Parker, District Attorney of Pedro County, but that worthy was not about when they arrived. The assistant stood by his guns long enough to enter upon the record an important opinion—"Wrightman is not a military officer, he is a military ass!"

[10]

It was in the midst of events such as these that Jessie Arthur came back from Europe. She wrote Hal from New York; she would be home in a few days, and would he be able to spare the time from his strikers to come and see her? She had been terribly unhappy about the way he had treated her; for days she had stayed in her room in the hotel and cried, refusing to go anywhere. He must realize how hard things were for her, for her mother gave her no peace, and there were angry letters from her father and her brothers, telling how the strikers were rioting, and how Hal was disgracing both families by giving the rioters his support.

The very sight of the tall familiar handwriting made Hal's heart beat faster. He could see the brown eyes, filled with tears, the fair hair, suffused with light, making an aureole about the face. A wave of tenderness overcame him. Yes, he had really treated her badly; but she loved him, she was trying to understand him. He would accept her timid advance, he would go up to see her.

But when the first wave of emotion had passed, Hal found himself thinking about Jessie with less happiness. He had put her out of his thoughts for several weeks; and now somehow when she came back she made trouble there. His thoughts had grown stern, hardly normal for a youth of twenty-three. He had cast in his lot with people who were braced for a life and death struggle, and he was growing familiar with the feelings of anguish and despair, with the sight of wounds and death. Somehow, in the presence of such things, a man's standards of value change; things which in the old days had a spell for him lose the spell and leave him cold.

Then too, he was working side by side with Mary Burke; her strong, erect figure was always before him, her willing hands were always at his service. Mary understood and shared his moods; she did not have to make any effort to do it, she did not have to be argued with. And so, every day, she came to fill a larger place in his consciousness. They were comrades and fellow-workers in a great cause, with no idea of anything else—so Hal told himself, and really believed it. But wise od Mother Nature will not permit two eager and healthy young people to be together all the time, and be nothing but comrades and fellow-workers in a cause—no matter how great the cause may be. Hal would look at Mary and think what a splendid girl she was, how true she was proving under this test; and then he would wonder what she was thinking about him. Was she altogether as satisfied with their relationship as she appeared to be? Did she never have impulses toward him, as in the old days? If she did, and was not letting him know it, it must be hard on her; on the other hand, if she had really become indifferent—well, it was not exactly flattering to a handsome and eligible young man. It was a certain thrill gone out of his life, a romantic interest he could not help missing.

He looked back upon those old days at North Valley, when he had first met "Red Mary", and she had bared her heart, and he had seen the primrose path of dalliance stretching before him in the soft summer moonlight. Those old days somehow seemed happy days now—such is the power of life to throw a spell over itself! But then Jessie Arthur had come to North Valley and taken possession of Joe Smith, the miner's buddy, and Mary's class-feeling had blazed up like a flame; she had drawn back, and from that time on she had given no hint of anything but friendship for Hal. She asked nothing about Jessie; so far as Hal knew, she did not know whether Jessie was at home or abroad. Mary was living for the strike; and if she had been engaged in a subtle plot to lure Hal's interest to her, this would have been the wisest course she could have taken. A fine, straight girl, Hal would say to himself; a girl capable of forgetting herself! His admiration would be excited—and also his curiosity; he would be moved to spend hours with her, talking about the problems of the tent-colony and the cause.

Hal never forgot the bitter words with which "Red Mary" had once laid bare to him the soul of the class-war. She was a drunken miner's daughter, and he, who thought that he was really democratic, had shown that it was all play, that he was looking down at the working-people from a far-off height, across an unbridgeable chasm. This challenge came back to him every time he compared Mary with Jessie Arthur, as inevitably he was impelled to do. Was it so that he was in the deeps of him a snob—that he believed in those caste prejudices he was trying to force himself to fight? Was it true that a girl might have the soul of a Joan of Arc, and still be set one side all her life, because her hands were big and rough, because she spoke with a common brogue, and because no one had taught her the established way to hold a knife and fork? If so, then what was the use of a man's calling himself a revolutionist? In the days of the contest over chattel-slavery there was a test whereby the skeptic was wont to challenge the sincerity of those who professed affection for the downtrodden—"Would you let your sister marry a nigger?" And for Hal the situation had come to be summed up in a similar formula—"Would you let yourself fall in love with a drunken miner's daughter?"

[11]

There came a series of events which first postponed Hal's trip to Western City, and then quite suddenly made it unavoidable. These events hinged themselves about the personality of one of the mine-guards, who from the beginning had played an important part in Hal's affairs—Pete

Hanun, the "breaker of teeth". It had been Pete Hanun who had followed the miner's buddy when he had come down from North Valley, to try to save the men imprisoned by the mine-disaster. It had been Pete Hanun who, with Jeff Cotton, had chased the buddy into Percy Harrigan's car. And finally, it had been Pete Hanun who shot Tom Olson. Because of the prestige of this latter act, the "breaker of teeth" had become one of the leaders of the deputies, the right-hand man of Schultz; he had ridden about in the "death special", "shooting up" one camp after another. Hal saw him several times in Pedro, walking down the street with his pal, Dirkett; each of the pair kept his right hand in the side pocket of his overcoat, where the muzzle of a revolver was plainly outlined.

Such was the tension to which things had come in the coal-towns! And now suddenly the "breaker of teeth" met the fate he had so long been challenging. Stepping out of a cigar-store one evening, he was lighting a cigar, an operation which took his hand off his revolver, when someone stepped up and put a bullet into his head.

No one saw the firing of the shot, but there was an Italian striker named Dinardo staggering by, three quarters drunk, and him the soldiers seized. They must have realized very quickly that they had not got the right man; but they must punish someone, for the sake of the moral effect. Pete Hanun had been a sort of officer among the deputies, and if he could be killed, no one was safe. It was resolved to make the killing an occasion for suppressing the "trouble-makers" of the Horton tent-colony, of which Dinardo chanced to be a resident.

Next morning, when Hal was in the headquarters tent writing a letter to Lucy May, his friend Rovetta rushed in, pale with fright; the soldiers were after him! "I no got anything to do with it!" he cried. "All time I was here in camp! Mrs. Olson know, John Edstrom know!"

"What do they accuse you of?" Hal asked.

"They grab Jerry! They grab Kowalsky!"

"But what for?"

"Nothing! Nothing! We don't do nothing!" The young Italian was incoherent with terror.

"But what do they *say* you did?"

"They tell Jerry he help kill Pete Hanun! I was in next tent and I hear! I don't kill nobody, I was here in tent all time!"

"Why do they accuse you, Rovetta?"

"I don't know! He hit me that day in union hall when they make search! Maybe they think I got mad."

"Well," said Hal, "if you were here in the colony, it should be easy to prove it."

He went outside of the tent, and saw two militiamen running up—one of them Lieutenant Stangholz. "That Dago in there?" he demanded; and when Hal answered, they sprang inside and collared their victim.

"Lieutenant," said Hal, "this young man says he was in the tent-colony—he has witnesses to prove it—"

"When we want your testimony, young fellow," said Stangholz, "we'll ask for it." And with these words, and no more, they marched the frightened Italian down the street.

In front of the Minetti tent was Jerry, with a soldier holding him by the arm; and Rosa, his child wife, having the new baby in one arm, and with the other hand catching at the militiaman's sleeve, the tears streaming down her cheeks, her voice hysterical: "Mister, he don't kill nobody! Who tell you such thing about my Jerry?" There were two other guards searching the tent; as Hal drew near one of them hauled out a trunk, dumped its contents into the dirty snow, and began throwing things this way and that. Little Jerry flew at the man, hammering his legs with his tiny fists; the man kicked him to one side, and Mary Burke caught him and held him, screaming and trying to get away.

There came another guard with Kowalsky, a Polish miner, his wife and three children following behind clamoring. Why any man should suspect Kowalsky was beyond imagining, for he was a helpless and stupid wage-slave; but there was no use offering character-testimony, or even asking questions. The militiamen, having finished their search for papers or weapons, marched their three prisoners down the street and loaded them into automobiles and whirled them away.

[12]

Hal interviewed Mrs. Olson and Edstrom and a number of others, and made certain that Rovetta and Jerry had been in the tent-colony all the previous evening. Then he went for advice to his cousin, who knew the two Italians and admitted that they were decent men. But Appie did not see what anyone could do about it; he was strenuous in advising Hal to keep out of the mix-up, making no secret of the reason—that he was afraid of the extremes to which General Wrightman might go. "Wait a while," he argued. "They're bound to realize your men had nothing to do with it, and let them off."

But Hal was not to be held back. He first addressed a letter to the military commission, telling them of the evidence he had to offer; and when no answer came, he went up to the American Hotel, where the commission had its headquarters, and asked for an interview with the Judge-Advocate.

This was Barry Cassels, a lawyer of Western City, well known to Hal. He was legal adviser to the General, and in this capacity had just delivered an opinion, to which the newspapers attached great weight, that the Governor had no legal authority to enforce his policy of keeping strike-breakers out of the mines. Besides being Judge-Advocate of Militia, Cassels was an attorney with a salary from mine-operators—though, as he made haste to explain, it was the metal-mining industry, not the coal-mining industry, which paid him. If it was on a public platform, he would make this explanation with grave impressiveness; if it was over high-balls at the club, he would make it with a wink. The "metal men" of the state were perfectly good friends with the "coal men". They shipped their product over the same railroads, they borrowed their money from the same banks, they used the same lawyers and the same politicians to hoodwink the public, and the same militia officers to break the strikes of their workingmen.

After an argument with a burly guard, Hal finally got his card taken in, but that was as far as he got; the answer came that Major Cassels was "busy". Hal knew the Judge-Advocate for a gentleman with social ambitions, who ordinarily would not have declined intimacy with a member of the Warner family. Could it be that he was ashamed of his present job? Or was he sternly devoting himself to duty, refusing to be swayed by social favor?

He had every reason to call himself "busy", Hal had to admit. He was judge, jailer and prosecuting attorney to twenty-five "military prisoners", in addition to those he was now gathering in on the charge of having conspired to the murder of Hanun. General Wrightman had made announcement that none of these prisoners were to have a jury trial; the General would constitute himself a "military district", and the prisoners would be attended to by his "military court". In a country supposed to be a democracy, with a constitution providing that the military should at all times be subordinate to the civil authority, this naturally caused complications, and taxed the subtlety and legal learning of a Judge-Advocate of Militia.

This "military prisoner" idea had had its beginning ten years ago, when the president of the metal miners' union had been thrown into jail on a charge that he had "desecrated the American flag" by having union sentiments printed on it. The case had been carried to the Supreme Court of the state, a body put into office by the copper-trust for the purpose of putting its rivals out of business. This court had sustained the right of militia officers to set aside the civil authority; but it had happened that the Chief Justice of the court was an honest man, and he delivered a dissenting opinion which for its eloquence and dignity deserves to be written beside the Declaration of Independence, and taught to every school-child in America.

He declared that the court had not construed the Constitution, it had ignored it; that not one of the guarantees of personal liberty could any longer be enforced. "The accused may be guilty of the most heinous offenses. It may be that he deserves to linger in prison the remainder of his natural life; but he is entitled to his liberty unless someone, in proper form and before a proper tribunal, charges him with violation of the law. If one may be restrained of his liberty without charge being preferred against him, every other guarantee of the Constitution may be denied him. And when we deny to one, however wicked, a right plainly guaranteed by the Constitution, we take that same right from everyone. We cannot deny liberty today and grant it tomorrow; we cannot grant it to those heretofore above suspicion, and not grant it to those suspected of crime; for the Constitution is for all men—'for the favorite at court; for the countryman at plow'—at all times and under all circumstances."

And then, in grave and solemn words, he warned the people of his state concerning the perils involved in this precedent. "We cannot sow the dragon's teeth and harvest peace and repose; we cannot sow the whirlwind and gather the restful calm. Our fathers came here as exiles from a tyrant King. Their birthright of liberty was denied them by a horde of petty tyrants that infested the land, sent by the King to loot, to plunder and to oppress. Arbitrary arrests were made, and judges aspiring to smiles of the prince refused by pitiful evasions the writ of habeas corpus. Our people were banished; they were denied trial by jury; they were deported for trial for pretended offenses; and they finally resolved to suffer wrong no more, and pledged their lives, their property and their sacred honor to secure the blessings of liberty for themselves and for us, their children. But if the law is as this court has declared, then our vaunted priceless heritage is a sham, and our fathers 'stood between their loved homes and war's desolation' in vain."

[13]

Now these prophesies were coming to fulfilment; now men saw the harvesting of the "dragon's teeth". Ten years had passed, and General Wrightman was not waiting for the Governor to declare martial law; he was not even troubling to declare martial law himself—he was merely declaring a "military district", which apparently he regarded as coterminous with the state. And when the little cowboy Governor announced that he disapproved these proceedings and would not permit them, the General said nothing, but went right on doing what he had announced. The Governor sent down his secretary to see about it, but the General paid no more attention to the secretary than he had paid to the Governor; he appointed his court, and tried his prisoners, and locked them in jail for long terms.

Nor could you even protest against these proceedings! The "policy committee" of the union refused to recognize this "military court", and Jim Moylan was sent down to the field to inform the strikers as to their constitutional rights. There was a meeting in the union hall in Pedro, at which Moylan declared that General Wrightman's commission was without power in law, and that everyone should refuse to answer its questions or to pay attention to its threats. The result was that early next morning a lieutenant of militia called upon the young Irishman and politely informed him that the authorities wished to talk with him.

Moylan went down to the hotel and was ushered into a room before Major Cassels, who began to ask him questions; and when the labor leader followed his own advice of refusing to talk to this "military commission", General Wrightman burst into the room in a rage. "Take this man and lock him up in jail until he's ready to answer!"

Moylan made an attempt to argue the matter. "Has the Governor declared martial law here?"

But the other answered, "I'll have no academic discussions with you! You'll go to jail until you are ready to answer questions."

So they took the young fellow off to the city-jail, which was stuffed with suffocating prisoners, tier upon tier. But they did not keep him very long, for Jim Moylan was known and trusted by every labor leader in the state, and Governor Barstow suddenly had a hornet's nest about his ears. It happened to be just when the beetle-rhinoceros conference was taking place, and the beetles refused to proceed unless Moylan was released. So the Governor telephoned orders, and that afternoon the young man came out, and went to Sheridan and addressed a tremendous mass-meeting of the miners. Next morning Keating, who was picking up gossip at the Capitol, telephoned word that the beetle-rhinoceros conference had come to nothing, and that there was a company of soldiers waiting at the depot in Pedro, to arrest Moylan when he stepped off the train. The young Irishman took an automobile and got out through the back country, and up to Western City, where as yet the "civil authority" had some little standing.

[14]

Kowalsky, the Polish suspect, spent five days in jail before Major Cassels realized that he had got the wrong man. The one they really wanted was the Polish organizer, whose name was Kowalewsky! Kowalsky came back to the Horton colony and told how he had been taken before the "commission", and had been badgered and cross-questioned by Major Cassels and young Vagleman. Having no idea what they wanted of him, all the poor fellow could do was to fall down on his knees and plead that he had a wife and three children.

At last they had sent him to jail—the county-jail, where the murder-suspects were kept. The place was unheated all night—in the midst of winter, and with openings through which the snow sprayed over the prisoners. Kowalsky told what he had received for food: at eight o'clock in the morning, three small biscuits "no bigger than a dollar", an inch and a half slice of bacon, and a cup of very bad coffee without sugar or milk; again at two in the afternoon, a tablespoon full of beans, a couple of slices of bread, and a very small chunk of bad meat. Those two meals were all anybody was having. The Polack, trembling with illness and excitement, declared: "I don't go to toilet all them five days. I ask them fellers, they all don't go to toilet all them five days—such rotten food they give us!"

But the food was the least of it. The commission had apparently made up its mind to extract confessions by torture. Not one of the men had been allowed to sleep—not an hour during the five days and nights of horror! The guard had been changed every two hours, and they would go up and down the corridor and prod the men in the feet with bayonets. When they crouched in the corners of their cells to get away from this, buckets of cold water were thrown upon them. All day and all night you heard the moaning of agonized wretches, and the curses of their tormentors.

Now and then militiamen would come in and stand in front of the bars of the cage and question them; so gradually the prisoners learned the story they were supposed to confirm. Jerry Minetti had given Dinardo twenty-five dollars to shoot Pete Hanun, and had promised him a thousand dollars additional if he kept quiet about it. He had given twenty-five to Rovetta and to Kowalsky, to keep them quiet. "He don't give me nothing," declared Kowalsky—"only one day when strike begin he give me three meal-tickets. They say, 'You say he give you twenty-five dollars, you get out; you don't say it, you stay here, you don't never sleep no more!'"

A couple of days later they decided to drag in another man—Johann Hartman, the old "Dutchy" who was secretary of the union local in Pedro. They arrested him at two o'clock in the morning, and raided his office and took his papers; next day came a report that the truth was out at last—a list of the murderers had been found in Hartman's desk, and Hartman himself had confessed!

It was Mary Burke who brought this news. "It's a lie!" she cried, with flashing eyes. "He's got nothing to confess, but if he had, he'd die first!"

Hal read the newspaper account, which reported Hartman as declaring to Major Cassels that the reason he had remained a German citizen was that he despised America and its government. "I know he never said that," declared Hal. "I asked him that very question, and he told me that no union man could get naturalization papers in this country!"

Now this was Hal's first experience with the "frame up", and it put him in a difficult position. He knew Jerry Minetti, he had shared the young Italian's hopes and fears, and believed he could say with certainty that assassination had no part in Jerry's program for winning the strike. But on the other hand, was he to think that Barry Cassels, a "society man", supposed to be a gentleman—or young Bernard Vagleman, whom he had known at St. George's—that such men were engaged in a conspiracy to send innocent men to the gallows?

One thing was about as hard to believe as the other, and Hal swung between the two. Pete Hanun had been a wild beast, and one could not count it much of a crime to shoot him. Might it not be that Jerry, driven to desperation, had actually connived at the crime? If he had harbored such a desperate secret, he would have kept it from Hal, knowing that Hal would have opposed him.

But, even granting that this was true, why this torture of men in prison? If Cassels had evidence of murder, let him go ahead and try the guilty man; but let him not get evidence by the methods of the Spanish Inquisition!

[15]

Tim Rafferty came out, having served his thirty days for carrying a revolver. He had been in the county-jail with the murder-suspects, and was ill with the horror of what he had seen. He declared that Rovetta and Dinardo were out of their minds, and could be heard raving by the other prisoners. Perhaps they were giving them dope in their food; or perhaps it was just cold and exhaustion and torture.

Rosa Minetti heard this, and she came to the headquarters tent, weeping hysterically. Why did they not do something? They *must* do something! She stood twisting her hands together, she caught Hal by the coat, pleading with him frantically to save her husband. Hal must know that Jerry was innocent; and Hal had rich friends, there must be some way he could stop this horrible cruelty!

All this was hard on the young man. He could not eat or sleep for thinking about it. If he put food in his mouth, it would suggest the thought, how could he sit eating, while the men in the jail were starving! If he lay down to sleep, the thought would come, how could he rest in comfort, while his friends were being kept awake with bayonets and cold water? So Hal ate and slept very little; and as every physician knows, a man who does not eat or sleep becomes nervous and irresponsible–liable to go "off his head", as Captain Harding phrased it.

The young man could no longer go to his cousin to protest, for Appie had been supplanted in the control of Horton; another company had moved to the neighborhood, commanded by a Major Singleton, a crony of Wrightman and Cassels, and including some of the most hated of the mine-guards, among them "Lieutenant" Stangholz. Nor could Hal get satisfaction by wiring the Governor, for his messages went unanswered; nor by writing letters to the newspapers, for his letters were not published. What could he do?

The tension was made worse by the arrest of Louie the Greek, charged with complicity in the murder. It had been conceivable that Jerry had had something to do with the shooting, but concerning Louie there could simply be no question. And word came that they were torturing him like the others; the guards had declared they would kill him, and the jailer had knocked him down on the way to his cell.

Right on top of all this came the experience of Mike Sikoria. Mike had a sister somewhere in the East, and she had been ill, and he was anxiously awaiting a letter from her. He had got the idea that his mail was not being sent to the tent-colony, and he wanted to go to the post-office at the village to make sure. There had recently been a big stir in the newspapers over the matter of keeping men away from post-offices; the Federal authorities had threatened to interfere, and Wrightman and Vagleman had issued denials that anybody ever had been, or would be, or could be, denied access to a United States post-office. The old Slovak had heard of this, and was foolish enough to believe it, and set out for the Horton post-office, which was at the railroad-station, to ask about his mail.

Hal saw him returning an hour or so later, white and faint, with two men helping him along. His left arm hung limp at his side and he moaned with agony whenever it was touched. They took him to the hospital-tent, and there, while the doctor and Mary Burke attended to his injury, he told what had befallen him.

There had been a soldier sitting near the depot, but having waited until the train was gone, Mike thought it would be all right for him to approach. "I go right up to the post-office—over here," he said; he always illustrated everything with gestures—but now, alas, he had only one hand for the purpose! "There come two soldiers on the corner, like this, and there is the door like that. I took the door—go to open it. He says, 'Where you going, hobo?'—I says, 'No place—go to get my mail.'—He says, 'I give you mail.' And he bring down his gun on my arm. He hit me with the shaft—the wood piece—and I dropped right there. I don't know nothing for about five minutes. I fell right in the snow. I fell, and he kicked me, he says, 'Get up!'—I don't know nothing about it, but after I wake myself up, he says, 'Get up, you son of a ——, get up. Go to work.'—'Go to work?' I says. 'I can't work,' I says. 'I don't want to work with my arm like this. I got enough,' I says, and I start back to tent-colony. But on the way there is another soldier, and he says, 'Where you going?'—I says, 'I am going to the tent-colony.'—'Why don't you go to work?' he says.—'I don't have to go to work,' I says. 'You don't make me go to work.' And he come over and kicked me. He says, 'Go on.' And I can't say nothing—he got a rifle and a gun, and a pistol and a bayonet—everything. What's a poor man going to do—with his poor arm—with his smashed arm?"

The wound was a serious one, the flesh being torn loose from the bone for a considerable distance; the doctor declared that it would be slow to heal; Mike was an old man, and it might be a year before he was well. At which the old Slovak began to weep in despair. Yes, he was an old man—fifty-eight years—and what was he going to do, a cripple? What could a miner do with only one arm? He appealed to the doctor and the crowd of sympathetic onlookers. "Is it right? Shall a young soldier hit an old man like that? Is it good to smash an old man's arm because he goes to get his mail?"

[16]

That was the climax for Hal. He went back to his tent, and spent the evening writing a letter to Lucy May. He put the letter on the midnight train, with a special delivery stamp; and next morning he himself took the train to Pedro.

He went to union headquarters to see Tim Rafferty, and there was a Slav woman with a young baby in her arms, having been brought that morning from the town of Mateo, thirty-five miles away. The woman had gone out of her home with a shawl over her shoulders, her destination being the corner grocery, when three militiamen had stopped her and told her she was "wanted". They had put her in an automobile and driven her thirty-five miles over the mountains in mid-winter, and taken her before the "military commission", only to discover that they had made a mistake. At least, that appeared to be the situation; they turned her loose with no explanation, without money enough to get home, and here she was at union headquarters, sobbing and moaning, her lips blue with cold, her baby half frozen.

Hal heard the story through an interpreter, and then phoned Billy Keating, who was in Pedro getting material for the "Gazette's" campaign against the "Czarism" of the Adjutant-General. Billy came, and told Hal the story of his adventures—the "Czar's" headquarters were in the American Hotel, and Billy, hanging about the lobby, had been too successful in picking up gossip about the "military commission" and its doings; so yesterday a guard had been set for the purpose of keeping him away. At least the guard had no other purpose that any man could discover; he stood all day at the entrance to the hotel, and other people came and went freely—all save the reporter of the Western City "Gazette"!

They laughed; but then Hal became suddenly grave. "Billy," said he, "I have a statement to make—something I want to go into the 'Gazette'."

"Our columns are at your service," said Billy. "What's happened?"

"Sit down," Hal said; and the other, who knew shorthand, got his note-book and pencil ready.

Hal began: "With regard to my arrest, I would state—"

Billy stopped and looked up. "Your *arrest*?" he said.

"Go on," said Hal; "a stenographer shouldn't interrupt." He repeated, "With regard to my arrest, I would state—" And he went on to give an account of the sights he had witnessed, and of what he knew about the torturing of prisoners in the county-jail.

"Good stuff!" commented Billy, when he was through. "But how about the arrest?"

Hal answered, "Is it so very difficult to get arrested in this part of the world?"

"What are you gong to do?"

"I'm going to pay a call on General Wrightman. I've given you my last will and testament. And now let me tell you what to do. You watch the hotel, and when you see me arrested, go to the nearest telephone and get Mrs. Edward S. Warner, Junior, in Western City. I want it for a test, so do it without delay."

"What shall I say?"

"I've written her a letter, which she'll no doubt receive this morning. I told her what the jails are like, and how they're treating the prisoners, so I feel reasonably sure she'll do her part, if only you get word to her."

"Don't worry about that," said Billy. "I'll get word to her. What else?"

"Nothing—only give that statement to the 'Gazette'. Give it to the other papers, too, if you're capable of that much altruism."

"I'm capable of enough altruism to give it," said Billy, dryly; "but you'll find that they aren't capable of enough to take it."

And so Hal went over to the American Hotel and sent in his card to Adjutant-General Wrightman.

[17]

Hal had met the eminent eye-doctor once before, and their interview had been pleasant enough; but that had been in the "era of good feeling", before the General had realized what a desperate character Hal was. Now he found the great man sitting up straight and stiff in his desk-chair, glaring at him from under bushy eye-brows, his long mustaches making Hal think of a big walrus with white tusks.

"Well, young man?" said he.

"General Wrightman," began Hal, "I happened to be in union headquarters just now—"

"You had no business to be there!" snapped the General.

Hal was disconcerted for a moment; then he went on, "I met a Slav woman, Mrs. Bobek. She had been brought over from Mateo in an automobile by some of your soldiers. Possibly you know about the case."

"I know something about it," said the General.

"She had nothing on but a calico house-dress and a shawl. As a result of this joy-ride, her baby is dying."

"Who says the baby is dying?"

"A doctor who has just examined it."

"One of those union doctors? I wouldn't believe him on oath!"

"Well," said Hal, "perhaps you will believe the coroner."

There was a pause.

"General, you may know why that woman was arrested. Certainly *she* doesn't. To have brought her in that condition, and without a wrap, was nothing but wanton cruelty."

"I have ordered an investigation into that matter," said General Wrightman, and his tone conveyed the words, "That settles it. What else?"

"This morning," said Hal, "I was in Horton, where an old miner, Mike Sikoria, whom I have known for a year and a half, and whom I can certify to be a decent, hard-working man, was brutally assaulted and seriously injured by a militiaman."

"Did you witness the assault?"

"I did not. But I saw the man immediately afterwards, and I heard his account—"

"I am hearing accounts all the time," said the General. "They are always lies."

"This man was going into the Horton post-office—"

"The Horton post-office is in the railroad-station, isn't it?"

"Yes, of course—"

"Well, then, he had no business there."

"But it's the post-office, General—"

"When a post-office is in a railroad-station, it's not a post-office for my purposes, it's a railroad-station. Strikers are not permitted near it; they know it, and you know it."

"General, this man was struck without warning, and brutally struck—"

"Do you know the name of the militiaman?"

"Yes." Hal gave the name.

"Very well. I will investigate the matter."

Hal was tempted to smile, so many times had he heard that word "investigate". But he had come for more than words, so he pursued his course. "General, two days ago I was talking with Joe Prince, the negro. You know him?"

"I've heard of him."

"He had just come out of the county-jail. He had been there nineteen days, and had not even had a bath."

"Well," said the General, "did it occur to you to ask if he had had a bath in the nineteen days before he was taken there?"

Hal decided to "pass up" that question.

"At least he had been accustomed to have more food than they gave him in the jail. He had lost twenty-five pounds. And it seems that his cell was next to Louie the Greek. Now, I know Louie. He's a man of peace, if there ever was one in this world. He's an educated fellow, generous and fine-spirited—.

"I know all about you fine-spirited men, you trouble-makers," broke in the General. "I've heard your phrases before! You teach what you call humanity to this foreign scum, but when it comes out at my end of the machine, it's murder and assassination. I've got my own information about your 'Louie the Greek', and it's from people whose judgment I have more respect for than I have for yours, young man. What else?"

"General, they are torturing those men to try to make them confess something. They are not allowing them to sleep—"

"Stuff and rubbish!" cried the other. "I'm tired of hearing such tales!"

"Joe Prince is the seventh man who has told me what is going on in that prison. Their stories all agree, and I have talked with them before they had a chance to get together and compare. That man Prince was arrested without any charge whatsoever. He's a decent, hard-working negro—used to be a stable-boss at Pine Creek. Apparently the only purpose your office had in holding him was to frighten him into declaring that Johann Hartman had given him a gun."

Hal was watching the General, and saw the blood mounting into his face.

"That's enough, young man!" he cried. "I'll not have you come here to slander and insult my officers."

"They are torturing those prisoners—"

And the General brought his fist down on the desk with a bang. "That's enough, I say! Have you come here to be insolent?"

"It depends upon what you call insolent," Hal answered. "The purpose for which I have come is to pay a debt which I owe to my own self-respect."

[18]

The General was puzzled by this remark—enough to let Hal go on; which was what Hal intended. "For seven weeks now I have lived in the Horton tent-colony, devoting myself to helping those people, to bringing a little order and decency into their lives. I have been sanitary officer and chief of police and international diplomat and captain of a baseball team. And for a month not a day has passed that I have not witnessed outrages. I have seen peaceable and decent-hearted peasant people driven about like cattle, deprived of every right guaranteed by our laws and our constitution—"

The General brought down his fist again. The "constitution" was a red rag to him. "That's enough, I say! I'll not listen to lectures from you."

"Day after day I have seen these things, and sat impotent—"

"Young man!" roared the General. "Don't make me lose my temper!"

Hal might have thought this funny, if he had not had such a desperate purpose in mind. "General Wrightman," he answered, with the utmost solemnity, "I am perfectly willing that you should lose your temper!"

There was something so unexpected about this, that again the old man was disconcerted. "General," persisted Hal, "will you permit me to ask you a question?"

"What is it?"

"Why have you never arrested me?"

"*What?*"

"You have arrested practically everybody else who has been active in the Horton tent-colony. Louie the Greek, for example—he has done exactly what I have done, no more and no less. Why is he being tortured in prison and threatened with death, while I am left at liberty?"

The General was like the master of a sail-boat who finds the wind blowing unexpected gusts, so that he does not know which way to put his helm. "Anybody would think, young man, that you had come here in order to be arrested!"

"A little over a year ago," said Hal, "I was working as a 'buddy' in the North Valley mine. That's how I come to know more about this strike than you do, General. While I was there, Jeff Cotton—you remember, he was camp-marshal at North Valley—threatened to send me to jail. I told him there were some people in the state who *could* not be sent to jail. He saw the point. Do you see it, General?"

The General saw it, beyond doubt; the blood rushed into his face, and he thrust a trembling finger out at Hal. "You impudent young puppy!" he cried.

"But General, that is not answering my question. And if there's any other answer than the one I suggest, I want to know it. I have been watching the thing with wonder. I see your gunmen and troopers scowling at me as they pass, I hear them cursing me under their breath—but they let me alone. I go my way, I do what I please; I am protected by a mystic spell!"

The General's shaking finger suddenly became a clenched fist. "I'll show you!" he roared. "You think because you've got a rich father I'm afraid of you! You think you can come here and insult me to my face! But you're mistaken—by Jesus Christ, you'll find out that you're mistaken!"

Hal had been prepared for such language; it was what everybody reported about these interviews. "General Wrightman," he said, earnestly, "you must realize that it will be wiser to put me in jail. I am a very bad example to the people at Horton."

"I have realized that, sir."

"It is contrary to the whole spirit of our institutions that there should be one law for the rich and another for the poor. If you torture Louie the Greek, while you let the son of Edward Warner do what he pleases, you are simply giving arguments to anarchists; you are causing everybody to believe that the commanding general of the militia is a coward and a sycophant, who oppresses the poor and helpless, but cringes before his masters, the well-to-do and prominent—"

"Hold your tongue, young man!" roared the infuriated old walrus. "You want to go to jail, I'll send you there—and I'll see if I can't take some of the starch out of you while I'm about it!"

The General pressed a button at his desk, and instantly the door was opened by a sergeant. "Take this man to the city-jail!"

The sergeant saluted; and Hal, smiling his most irritating smile, remarked, "General Wrightman, I'll make a wager with you—for the drinks. It's ten o'clock now; I'll wager you can't keep me in jail until two!" And so leaving his victim on the verge of apoplexy, Hal followed the sergeant, and rode to the jail in a coal-company automobile.

[19]

Of this city-jail Hal had heard many terrifying accounts; it was in some ways worse than the county-jail—without ventilation, and crowded like the Black Hole of Calcutta. But still more prisoners came, until they were having to use the storage vaults underneath the court-house!

"By God, another!" said the head-jailer, when Hal was brought to his office. He was a burly Irishman, and sat scratching his head. "Whatever am I going to do with this one?"

"You needn't worry, Kerrigan," said Hal, promptly. "You're only going to keep me a few hours."

The other looked at him in surprise. "Sure, and how do you know that?"

"It stands to reason," said Hal. "My father's a millionaire."

"Go on!" said Kerrigan, staring. "Whatcher givin' us?"

"Sure thing!" laughed Hal. "Did you never hear of Edward S. Warner?"

"I have," said Kerrigan. "You tell me he's your father?"

Hal took out his card-case, and handed a card to the jailer, who looked at it. He looked at Hal, and then at the two militiamen, who were standing by; but neither of them seemed to know. Could it really be that this genially impudent young person was the son of the coal-man? Or was he daring to mock them?

Kerrigan was a Catholic, and perhaps had heard of the famous argument of Pascal. It is the course of prudence to believe the teachings of the church, for there is no penalty to be feared if you believe them and they should turn out to be false, but, on the other hand, if you refuse to believe them, and they should turn out to be true, you are damned. "Sure now," said Kerrigan, "if they'll be lettin' you out in a few hours, there's no use puttin' you in with them bums."

Hal had not foreseen this possibility. "Oh, but I want to go in!"

"What?" cried the other. "What for?" This looked decidedly as if the chances were in favor of the church!

"Why, you see," said Hal, "I've never been in a jail."

"Sure, this is no jail," remarked Kerrigan, "it's a menagerie. It smells like the monkey-house. 'Tis no place for a decent person."

He signed the receipt for Hal and gave it to the two militiamen. "All right," he said. "I'll take care of him." And after they had gone, he turned to his charge again. "Now," said he, "and what has a young gentleman like you been doin'?"

"I've been telling the Adjutant-General what I thought of him."

The other's eyes showed a flash. "I'd like to been there!" said he.

To which Hal answered, "If you want me to, I'll do the same thing to you."

"How do you mean?"

"First tell me—are you the fellow who knocked down Louie the Greek when he came in here?"

So the geniality of the head-jailer suddenly dried up; he perceived that he had a dangerous man to deal with, and took him to one of the cells and shut him in with a bang.

The size of this cell was eight feet by seven, and there were five men already in it—three of them sitting on the floor. There were six such cells in a tier, and two tiers, one on top of the other, and every cell crowded in exactly the same way; in addition, boards had been laid on top of the second tier of cells, where the hoboes and drunks were packed. Hal found that Jailer Kerrigan had been correct in his description of the odor of the place.

All five of the men in the cell with Hal were strikers; four of them having been arrested for picketing in the early days. Hal introduced himself, and they shook hands all around; then, standing in the middle of the floor, Hal suddenly began to sing at the top of his voice:

> "We'll win the fight today, boys,
> We'll win the fight today,
> Shouting the battle-cry of union!"

The song was taken up in chorus, and seemed to shake the very walls of the building. For a minute or two the city-jail was a glorious place to be in. But then one of the jailers, having marked the source of the trouble, rushed up with a bucket of ice cold water and hurled it into Hal's face, and the young man's revolutionary fervor was ingloriously extinguished!

[20]

It had been ten-twenty when Hal gave up his watch to the head-jailer; it was one-three by the same watch when it was returned to him. "However did you do it?" inquired Kerrigan, as he counted back Hal's money to him.

Then he added, somewhat nervously, "I hope you'll understand, Mr. Warner, it ain't my fault if this place is crowded."

"You could resign," remarked Hal.

"What good would that do?"

"It would give you a chance to protest and let the public know what is going on."

"I'm sorry," said Kerrigan, "but I've got a family."

"Children?"

"Six of them."

"And you choose to feed them on the blood of these prisoners?" And so, leaving the head-jailer with his jaw fallen down, Hal went out with his guards, and returned in the automobile to the American Hotel.

This time, however, it was not to the commanding officer, but to the Judge-Advocate. Major Cassels sat at his desk, tapping nervously with his pencil. "Well, Warner, you've had a little misadventure," he remarked, with what was meant to be a genial smile. He was a foppish person with an affected accent, and his black-rimmed eye-glasses looked strange with his military uniform.

As Hal made no answer, he indicated a chair. "Have a seat."

"I prefer to stand," answered Hal.

"Now Warner," began the other, "I want to have a talk with you, and see if we can't come to an understanding."

"Barry Cassels," demanded Hal, abruptly, "have you been inside that city-jail?"

"Now—"

"You are willing to pile men in there—to starve and suffocate them—men who you know are innocent of any crime—"

"I have something to say to you, Warner—"

"I have something to say to you, Barry Cassels, and mine is the more urgent. In the beginning, I was puzzled, I could hardly believe it. You are what the world calls a gentleman—a college graduate, a lawyer who has sworn an an oath; and you are setting out deliberately to 'frame up' men whom you know to be innocent–"

The foppish Major's patience was beginning to wear through. "That's strong talk, young man!"

"Not half as strong as the situation deserves. I've been to the bottom of this matter, Cassels, and I know the knavery of what you and Wrightman are doing. I've read the statements you've given the press–one series of lies after another–"

The Major clenched his fist. "Be careful!" he cried.

"I say lies, Barry Cassels–lies! You were lying when you gave out that interview with Johann Hartman. You were lying when you wrote that statement for Joe Prince to sign. You were lying in everything you said about Louie the Greek. You were lying when you quoted the deputy commissioner of labor–I know, because he told me so. You were lying in the statements you made about Tim Rafferty. You were lying when you said there was typhoid at the Horton tent-colony–"

The Judge-Advocate had become livid with rage. "By God! I'll make you sweat for this!"

Hal laughed at him. "Come off, Cassels!" he said. "Don't I know that if you could have kept me in prison you'd have done it? You've sent for me to let me go, so hurry up!"

For a minute Cassels glared in silence. Then with an extreme effort he controlled himself. "We wish to give you another chance; that is provided–"

"Cut it out!" broke in Hal. "I'll not make terms with you."

"You must understand, you are not to return to this district–"

"No, I'll not have to return–I'm not going away."

"You're to be sent out on the next train, young man!"

"Oho! Like Mother Mary! You honor me, Cassels! Will you send your whole army to escort me, as you did for her?"

Major Cassels rang a bell. "Judson," he said, to the soldier who answered, "take this man and put him on the train for Western City."

"You'd better tell him to call out the artillery," taunted Hal. "I'm popular with the strikers, you know."

Evidently the Major agreed with Hal's judgment, for a detachment of thirty men marched down to the depot to see him off. And of course that was notice to the strikers that something important was happening. A crowd gathered, and when they saw who was being shipped away, they sang the union song and cheered vociferously. It warmed Hal's heart; it was his reward for the discomforts he had undergone!

[21]

Hal had refused to buy himself a railroad ticket. When he explained matters to the conductor, that official was vastly amused, and agreed to put him off at Sheridan for non-payment of fare. But evidently a warning had been sent ahead, for there was a squad of militiamen on hand at Sheridan, and they had conceded the point of paying Hal's fare up to Western City.

Hal was content to go, for now was the time to reach the public. The newspapers would have an account of his arrest and deportation—the "Gazette" would force them to that. So the channels of publicity would be open to him; he could tell the public a little of why he was behaving in this disgraceful fashion. Such is the pass to which things have come in our land of freedom—the only way to let the public know about strike-outrages is to get one's self in jail!

Hal's train got in at seven o'clock in the morning, and he went directly to his brother's home. Edward was shaving, a rite which must not be interrupted; but Lucy May ran to welcome her wayward brother-in-law, clad in an embroidered pink dressing-gown. She caught him by the hands, and there were two little pearls of tears, one in the corner of each eye; it was evident that the Philadelphia lady had been in a state of tremendous excitement. "Oh, Hal! How terrible!" she exclaimed.

"It was jolly!" laughed Hal—"Just long enough not to be monotonous! How did you manage it?"

So Lucy May told her thrilling little story. She had got Hal's letter, with the dreadful account of the jail, and had read it aloud to Edward at the breakfast-table; and then, two or three hours later, at the dressmaker's, had come Billy Keating's telephone-message. Lucy May had gone nearly beside herself; she had called up her husband on the phone, and they had nearly broken up matrimony on the spot. For Edward had refused to do anything, declaring that the best thing for his brother would be to stay in jail and cool off!

So Lucy May had jumped into her limousine and sped to Hal's father, and read him the letter and told him the news. "Hal!" she exclaimed. "It was wonderful! Just as it used to be before he was ill!"

"Dear old Dad!" cried Hal.

"He was so angry! I'd never seen him so angry! He didn't stop to think that somebody might cut off his credit, and compel him to stop work on the new mine! He said, "That boy's coming out of jail!'—I said, 'What will you do, Dad? See the Governor?'—'What?' said he. 'That nincompoop? Not much! I'll see the head of the firm!' And he called up Peter Harrigan!"

And Lucy May stopped. "What did he say?" asked Hal.

"I can't tell it, Hal—he used such bad language!"

The other laughed. "I'll remember the quotation marks. Go ahead!"

"It was dreadful, you know. I remembered what the doctor had said, about Dad's not getting excited; and really, he was terribly excited. And I couldn't forget it was Old Peter he was talking to. Afterwards Dad told me what he'd said."

"Tell *me!*" said Hal.

"Dad went right for him. 'I understand your tin soldiers down at Pedro have put my boy in jail!'

" 'Well,' said he, 'why don't you keep your boy out of my coal-mines?'

" 'When my boy was in your coal-mines, he worked,' said Dad. 'He earned his wages, and profits for you besides. Since then, if I understand the matter, he's been the guest of some people who pay rent for their land, and have a right to be there. Anyhow, I've called you up to tell you that my boy comes out of jail, and comes out quick!'

" 'Well,' said he, 'I've got nothing to do with it–'

" 'Don't talk that rot to me! I want that boy out of jail!'

"Then Old Peter swore for a while; but finally he came down to business. 'I won't have him making trouble down there in that strike. If I get him out, will you see that he keeps out of the district?'

" 'I'll make no promises,' said Dad. 'I've tried my best to control the boy, but he's seen too much of the way you treat your working-people. And understand me, Peter Harrigan–you can abuse your Dagos and Hunkies, but by God, when you put Edward Warner's son in a cell you've gone too far! I'll give you just one hour to get that boy out, and if you don't I tell you right now there'll be trouble.'

" 'What will you do?' he asked; because, of course, he's not used to having men talk to him like that.

" 'I've got the letters my boy has been writing me,' said Dad. 'I've got the inside story of that strike, and I'll send for the newspaper reporters and give out an interview that'll blow you and your tin soldiers to kingdom come! And understand, Harrigan, if I go on the war-path, I'll stay on. I see there's going to be a mass-meeting at the Auditorium tomorrow night–if my boy isn't up here in Western City before that, I'll go there and give them a talk that this city won't forget in a hundred years. You think it over now, and get busy!' And then he hung up the receiver."

Hal chuckled with delight. "That's the way to talk to Old Peter!" he cried. "You're a jewel, Lucy May! I knew you'd get me out!"

And she put out her hands to him imploringly. "You aren't going back to the dreadful place, Hal!"

At which he became instantly grave. "Think of it, Lucy May! All the poor devils whose fathers don't happen to be rich, and who have no way to frighten Old Peter, and have to stay down there in that hell and starve and suffocate!

[22]

After this Hal had the customary argument with his brother. There was news in the morning paper which gave material for controversy—a "confession" by Dinardo, the Italian who was accused of having shot Pete Hanun. Here was the whole conspiracy revealed—an elaborate account of the shooting, how Rovetta had got him the gun, how Minetti and Hartman had paid him for the deed. These were Hal's friends from North Valley—the very people he had tried to persuade his brother to meet! A bunch of conspirators and assassins!

"It's an obvious frame-up," declared Hal; but how far would that get him with Edward? How far would it get him with any of the friends he hoped to influence? The statement of Dinardo was published in full all over the state, and did its intended work of alienating sympathy from the strikers. When Dinardo came out of prison at the end of a couple of months, a broken man, he repudiated the so-called "confession", declaring that when he had signed it he had been so nearly insane from lack of sleep that he had had no idea of what he was doing. But that, of course, was after the public had lost interest in the Hanun case, so the papers did not consider it "news".

Hal went to see his father, to thank him for what he had done—and to have his heart torn with fresh grief. The old gentleman had disobeyed the warning of his doctors, and now his hands were trembling so that he could not hold a glass of water. Of course he pleaded with Hal to promise not to go back to that dreadful strike-country; and Hal had no way to meet his plea save to tell about the sights of horror he had seen.

It was a painful situation; Hal realized that there might be deeper complications than he could see. He was fighting one group of coal-companies, with money derived from another group. Could he expect the world to regard that as an altruistic proceeding? The Warner mines were in what was called the "Northern field", and were union properties. But what, precisely, did that mean—how did Edward work it? Just now the Warner Company was "in clover", as the phrase has it, because its big rivals were tied up in a strike; but suppose it were to occur to Old Peter to have the "Northern field" tied up as well! The Schultz Detective Agency would know how to arrange it, turning loose some "radical" agitators, telling the workers of the Warner Company that their union leaders were a lot of grafters, standing in with the bosses, and that now was the time for them to join their fellows in the South and get their full rights! A "sympathetic" strike! Edward wanted to know what would be Hal's attitude in such an event; the Harrigans would want to know also, the newspapers would want to know!

Hal went to pay his call on Jessie Arthur. It was the first time he had seen her since their parting in London; she was thinner and paler, and evidently suffering intensely. She seemed to him exquisite, yet at the same time fragile—costly, artificial, like some rare flower that blooms indoors, and that a breath of rough wind might destroy. He took her in his arms and kissed her gently, and discovered that tears were running down her cheeks.

It was not merely that she was moved at seeing him after so long; she went on sobbing, until he asked, "What is the matter?"

"Oh, Hal! I've been hearing such dreadful things about you! You have been risking your life!"

"Yes, I suppose so," he said.

He wanted to tell his story; but she could not wait to listen. "It's Papa!" she exclaimed. "What are we going to do about him?"

Old Mr. Arthur was taking Hal's conduct as a personal insult, it appeared. He had made the cause of the operators his own; it was the banking-house of Robert Arthur and Sons which had taken the lead in advancing the money to preserve "order" in the coal-country; and now the head of the house had read in yesterday afternoon's paper about the arrest of his future son-in-law! He had read it, not in the "Gazette", with Hal's statement, but in the "Herald", one of the interest-controlled papers, with a statement of Major Cassels, to the effect that young Warner had made himself a menace to peace in the strike-district, giving encouragement to rioters and assassins. And on top of it, this very morning had come the confession of Dinardo, involving Hal's friends and intimates, the father of the Dago mine-urchin whom he had brought into the Arthur home!

"Hal, he's wild!" cried Jessie. "What in the world are we going to do?"

"I don't know, dearest."

"He's forbidden me to see you! If he finds you here—"

"I'll go, if you think best," Hal said; but the suggestion came too late. There stood the old gentleman in the doorway!

[23]

The old gentleman was really as "wild" as Jessie had described; he was as "wild" as anybody could possibly have described. When he saw Hal, he gave a jump, and stood with his fists clenched and his cheeks swelling. He brought his fists down, and cried, "Well, sir!" Three times he brought them up and down, as if pumping up his rage, and each time he cried, "Well, sir! Well, sir! Well, sir!" Then he began, quite literally, to romp up and down the room; he would walk a dozen steps one way, and shake one furious fist at Hal, then he would wheel about and walk as many steps the other way, and shake the other furious fist at Hal. "So you've got out of jail, sir! You condescend to honor us with a visit, sir! Did you escape? Or have you served your term out? How does it happen they failed to shave your head, sir?"

Hal answered nothing. Jessie made an effort to interpose—"Papa!"—but she only succeeded in diverting the storm to herself for a moment. "Hold your tongue! I'll have order in my home, even if it's nowhere else in the state!"

And again the old gentleman turned upon Hal. "How dare you show your face in a respectable home? To bring your shame to sully my daughter's pure name? Look at that, sir—look at that!" And one of the trembling hands indicated the library-table, where lay a copy of the morning paper. "Murderers and assassins! Italian black-hand conspirators—your own associates, convicted out of their own mouths! And you introduced them to my daughter, you brought them to my home! Go back to your nest of criminals, sir—your *tent-colony*, as you call it!"

Four times the old gentleman had raced up and down, and his color had deepened to a fiery purple with the unaccustomed violent exercise. Now suddenly, as he turned, he saw through the open doorway the figure of the butler passing by. "Horridge!" he shouted, and the black-clad, elderly person appeared, trying to keep his correct, impassive face. "Horridge, I want the servants up! Bring them all! Instantly! You understand?"

"Yes, Mr. Arthur."

"Bring the cook! Bring Thomas! Bring the gardener and his boy! Bring Jane and Ellen and Kate! I want them all—every one of them!"

"Papa!" cried the horrified Jessie; but the old gentleman shouted at Horridge, "Go on! Don't stand there gaping at me!" And as the butler disappeared, the exercising up and down the room continued.

Presently, out of the torrent of indignation Hal gathered the meaning of this new turn of the scene. A couple of weeks previously a reporter of the "Herald" had interviewed Hal and printed a few sentences of what he had to say about the strike. "Are you a Socialist?" the reporter had asked; and Hal, being a new hand and not seeing the trap, had answered, "I suppose I may be—after a fashion." The reporter had made this the heart of the interview: "Coal-magnate's son a Socialist after a fashion!" And so old Mr. Arthur had got the phrase stuck in his crop. A Socialist after a fashion! An enemy of law and decency after a fashion! An incendiary and assassin after a fashion!

The servants came; frightened, yet curious, of course—knowing there was a "scene". They stood in the doorway, each trying to keep behind the others; the old gentleman, turning in his mad career and seeing them, rushed up to them. "Come in! Come in! Howdy do, Yung?"—this to the fat and grinning Chinese cook. "Come in, Thomas, and you, Jones!"—this to the kitchen-man and the gardener. "Good morning, Kate! Howdy do, Jane? Walk in, Ellen! Walk in—don't be afraid!" And he took them by the shoulders and pushed them into the room—the footman and the gardener's boy, the chauffeur and the upstairs girls—eleven of them altogether, all that Horridge had been able to gather in the sudden emergency.

"Welcome! Welcome all!" cried the head of the banking-house of Robert Arthur and Sons. "I've called you up to introduce you to Comrade Warner. A Socialist after a fashion, you see—he'll be charmed to make your acquaintance! Comrade Yung, shake hands with Comrade Warner!

Thomas, you're a Socialist after a fashion too, I believe—shake hands, shake hands, all of you!" For some reason Comrade Yung and Comrade Thomas hung back shyly, which did not please the old gentleman. "I want you to sit down and make yourselves at home! I mean what I say—sit down, sit down! We're all social equals now, there are no more classes, and I'm going to divide up my money and give you all a share, and we'll do the dirty work together. Comrade Kate, one of our women comrades, shake hands with Mr. Warner—you have pretty red cheeks and he'll be interested in you. We're going to be free-lovers now, you know—"

"Papa!" screamed Jessie.

And the old gentleman whirled upon her. "Yes, indeed! Didn't you know that? We're all free-lovers—after a fashion—"

"Papa, you shan't talk that way!"

"Oh, but you must get used to it! He has one woman down in the tent-colony, I hear, and you'll be the next. And Kate here—"

So far Hal had not said a word; but he thought it was time to speak now. "Mr. Arthur," he said, "I realize that I made a mistake in coming to your home—"

"Yes, sir, you did indeed, sir!"

"So I think now, if you don't mind, I'll ask to be excused."

"Very well, sir! Go back to your assassins and free-lovers—your Socialists after a fashion!"

Hal had turned, and started to the door; but he heard Jessie rushing after him, and she flung her arms about him, shrieking, "No, no! You shan't go!" And she turned upon her father. "How dare you say such things? How *dare* you insult your daughter?" She whirled upon the horrified crowd of servants. "Get out! Get out!" And she waved her hand with a gesture that made the group fairly reel.

There are times when discreet servants understand that the ladies of the household take precedence over the gentlemen, and this was one of the times. Yung and Thomas and

Jones, Kate and Jane and Ellen, the gardener's boy and the footman, the chauffeur and the upstairs girl—they backed precipitately out of the room, and the last of them decorously closed the door.

Meantime Jessie was rushing on, a little virago, possessed by sudden unguessed demons. "How *dare* you say such things? I think you are *horrid!* I think you are *wicked!* I'll never speak to you again! You shan't drive Hal away—or if you do, I'll go with him! I'll go down to the coal-country with him, I'll join in the strike with him, I'll go to jail with him, you'll never see me again!"

The head of the banking-house of Robert Arthur and Sons quailed before this terrific blast. What was the world coming to, if a respectable father of a family could not lose his temper without his children being possessed by unguessed demons? He stood for a minute or two, pumping his hands up and down, puffing his cheeks and gasping like a cat-fish in the bottom of a boat; then he flung out his arms in a gesture of abandonment. "All right! All right! Have your own way! I wash my hands of the two of you! Go away with him if you wish—let him make you into a free-lover and an assassin!" And with this last cry of an elder and perishing generation, Mr. Arthur turned and rushed out of the room.

[24]

Jessie Arthur stood weeping in Hal's arms. Of course it took but a minute for the storm of her rage to pass, and then she was horrified at what she had done; she had never even thought a disrespectful thought about her father before—and now she had told him he was horrid! Did not Hal see the misery he was causing her, bringing all this dissension and distress into her life? How was she to stand it—all her relatives scolding her—brothers and sisters, cousins and uncles and aunts!

"Sweetheart," he answered, "I see your trouble, but what can I do? I have a duty—"

"You have a duty to *me!*" she cried. "I need you, Hal!"

"Dearest, you don't need me as the strikers do. If you only knew what is happening to them!" And again he began to recite the cruel story: Mrs. Bobek with her poor, half-frozen little baby; Old Mike with his mutilated arm; the jails with their half-crazed inmates. Jessie had never heard of such horrors in her life, and she gazed at him aghast, the tears running down her cheeks.

"Jessie," he said, "you want me to help *you*; but why can't you help *me*?"

"What could I *do*, Hal?"

"You might come down there and support those people." Seeing her look of dismay, he added, "You threatened to, just now."

"Yes, I know. But I couldn't really, Hal!"

"Why not?"

"Girls don't do such things."

"Some girls don't," he answered.

"What could I accomplish?" she asked, catching the note of bitterness in his voice.

"You might comfort people who are in distress; you might be the means of making others hear what was happening. The newspapers, you see—" Then suddenly he stopped, thinking what the newspapers would do if a daughter of the banking-house of Robert Arthur and Sons were to join the coal-strikers!

And she saw why he had stopped. "Mamma and Papa would go crazy!"

"I didn't suppose Mamma and Papa would enjoy it, Jessie."

"They would lock me up, Hal—if they knew I was even thinking of such a thing!"

"There's a way you can prevent that, sweetheart. We can go and get married; then people would expect you to come with me."

Such things have been done in the world's history, but to look at Jessie's face you would not have thought so. "Hal!" she whispered. "Papa would never see me again!"

"Papas have threatened that," he said—"and changed their minds later on."

He was looking at her. She wore a costly house-dress, exquisite, fragile, with colors chosen to match her eyes and hair. A maid had tended her soft hands, arranged the last strand of her golden-brown hair. Her little slippers were of cream silk, and would probably not be worn a dozen times before they were cast away. Hal, seeing them, had a sudden vision of the thick red mud at the Horton tent-colony!

Perhaps if he were to urge her, she would take the plunge. And it was a temptation, for he loved her, and when he was with her his senses were intoxicated. But his reason said no. If she came, it must be of her own impulse; it must not be with the idea, conscious or unconscious, that she could draw him back into the old life. Hal's mind had become clear on that point. He would not go back; he had enlisted for the war.

Then too, there was a doubt about his sweetheart, gnawing like a worm in his heart. How *could* she show so little effective response to the thing that was dear to him? Was there something lacking in her? He made the excuse that she was so young, but he had to admit that she was not so very young; she was nineteen—and surely that is old enough for a woman to discover that the jewels she wears are the crystallized agony of other people. Seeing that she did not discover it, he pointed it out to her, many times and in many ways. He waited for her to show that she cared about it; but all she showed was that she cared about *him*!

There was something Jessie was now trying to say to him; blushing and hesitating, not meeting his eyes. That dreadful story—that thing her father had referred to—

"What about it?" he asked, coldly. He would not help her on that matter.

"I want you to tell me, Hal!"

"I am not going to discuss scandals, Jessie. All I have to say is that the tale they are telling about me is false. You will have to believe that."

"I believe it, Hal. But then—we have to think how things look to other people. Isn't that girl at the tent-colony?"

"She is at the tent-colony—because her young brother is lying there with his foot maimed by a bullet. Would you have her anywhere else under the circumstances?"

"No—I suppose not." Then, after a pause, "Is she in love with you, Hal?"

"I don't know, Jessie," he answered. He had no right to tell her about Mary Burke's affairs, and he would not take the chance of her relatives asking her questions.

He went away from the interview, leaving her unsatisfied and miserable. She would take his word that there was nothing dishonorable in the affair; but she was sure that Mary Burke wanted Hal—how could any woman fail to want Hal? And among those "dreadful people", as she called them, anything might happen!

[25]

Hal went for consolation to his friend Adelaide Wyatt. Adelaide had proposed a resolution in the Tuesday Afternoon Club, calling for an investigation of coal-strike outrages, and she told how she was being "cut" on account of this bold action. The ladies of Western City "society" were ablaze with anger against the strikers, who were interfering with the business of their men-folk.

"Mrs. John Curtis came to see me," said Adelaide.

"Ah?" said Hal. This was the lady he had appealed to in Percy Harrigan's car.

"She came on a delicate errand," added the other. "She wanted to know if I was aware of the report that you had tricked me into employing your mistress in my home."

"By God!" cried Hal. "You don't mean it!"

"Didn't I warn you of it?"

"What did you say to her?"

"I defended you, of course, but I'm not sure if I convinced her. She may think you're deceiving me—or she may even think I'm abetting you."

Hal had come to Adelaide with a wonderful scheme. He wanted her to visit the strike-country. She could help the strikers enormously, for the reporters would flock to her, and would print anything she said. But Adelaide made him see how impossible his project was. She was a woman living apart from her husband; in a few days Mrs. Curtis would be hearing a report that Hal had two mistresses instead of one in the tent-colony!

Yes, that was their way of fighting. If you lifted your voice in opposition to their greed and oppression, they crept upon you in the dark and shot you through with a poisoned arrow. And because you knew this, you kept silence, you shut yourself up in your own private affairs, and let your life be ruled by fear.

"I wonder why it is," said Hal. "There seems to be so much of that nasty element in our Western City politics."

Adelaide answered—it was one of the curious and unforeseen consequences of woman suffrage, or rather of woman suffrage granted too early, without the women having had to work for it, and develop intelligence and public spirit. "Men don't pay much attention to scandals," she said, "but when you're dealing with women voters, there's nothing pays so well as a nasty story."

She went on to explain that the "interests" which ruled in the city had a regular factory for that sort of campaign material. There was a certain Dr. Anna Carlton, who had what was supposed to be a medical-office, but who received a regular salary, with an expense account, out of which she paid spies and agents to seek out all manner of scandals in the lives of persons whom it was desired to threaten—politicians who would not vote as ordered, journalists who dared to be independent, labor leaders who called strikes. If there was no scandal in existence, the doctor would make one; fitting it in so carefully to the known facts of the person's character and circumstances that it could with difficulty be denied. She would start this story through a score of underground channels, and in a few days it would be everywhere whispered and believed.

Later on, Hal saw much of this bureau's operations. When magazine-writers came from the East to investigate and write up this strike, there was a scandal got ready for each of them within a week of his arrival. As Dr. Carlton could not find out much about these people, and was pushed for time, she was forced to pair off men and women indiscriminately, without any regard to their tastes. This got to be a joke among the victims; but even as one laughed over the joke, one thought of all the good, earnest people who took it all for gospel, and were thereby led to withhold their help from the strikers in their pitiful distress!

No, Adelaide could not visit the tent-colony. But there was another plan in Hal's mind—and before he told of it he went to the doors of the drawing-room, and looked outside, and then closed them carefully. "Suppose," said he, "there were something you could do in secret?"

"What, Hal?"

"When I went down to that strike, I had my mind made up that I would not countenance violence. But now—well, I see the soldiers closing in on us, and I've had to revise my program. I don't mean to stand by and see those tent-colonies wiped out!"

"But you can't fight the state militia!"

"We're going to *have* to fight them, Adelaide!"

"But Hal, that's absurd! You'd stand no chance!"

"I'm not so sure. We outnumber them ten to one—"

"But the arms, Hal! The ammunition!"

"That's what I'm talking to you about!" There was a pause; then Hal continued, "You saw in the papers yesterday that General Wrightman has issued an order forbidding gun-stores anywhere in the state to sell arms for the strike-field."

"Yes, I saw that."

"Well, then, if we're going to buy on a big scale, it will have to be outside the state. And we must have somebody we can trust, and whom the enemy would not suspect."

Adelaide sat with her eyes fixed on Hal; at last she answered, quietly, "All right; when you need me for that, let me know!"

[26]

In New York and Chicago and other big cities to the East, it was becoming quite a fad for the sons and daughters of the idle rich to get themselves arrested in strike troubles. And in Western City they strove diligently to keep up with Eastern fashions. As Hal walked down the street, the members of the "younger set" whom he encountered were keen with curiosity. They "kidded" him, of course; it was a lark to greet the son of Edward S. Warner as an ex-convict and jail-bird; but then they wanted to know everything that had happened, and what it had felt like; they imagined themselves acquiring this wild and perilous kind of distinction. These members of the "younger set" went hunting in the mountains and clamored up perilous peaks; they drove racing-cars and broke the necks of themselves and others; they rode wild horses, and fought professional boxers, and ran away with chorus-girls; but here was something brand-new, the ne-plus-ultra of fashionability—to beard an old military walrus with white mustaches, and to be locked up in jail, and come to town next day and see your name on the front page of both morning papers!

There were older people, of course; and the old are conservative by nature, and cannot understand the need of youth for new sensations. A couple of ladies who met Hal on the street showed their good breeding by making no mention whatever of the shame he had brought upon his family; while others mentioned it in tones of grave reproof—old Dr. Penniman, for example, whose duty in life it was to discuss other people's conduct and morals, and who had an especial right to discuss Hal Warner's, because the young man was a member of his congregation, and had disgraced the hallowed name of St. George's.

Dr. Penniman knew all about the strike—he had read the details of it every day in the newspapers. There were fierce foreign criminals, with anarchistic ideas in their heads and daggers and bombs in their hands; there were gallant sons of good families, preserving the supremacy of the law and of the flag at deadly peril to their lives; and here was Hal Warner, showing what became of young men who no longer come to church regularly, but take up with modern infidelity and sedition. There was a fierce argument on the street—until Dr. Penniman noticed that people were staring at him, and remembered his dignity and hurried away, leaving Hal swearing at the bourgeois world and its prejudices.

More than anything else it was the newspapers! Twice a day people read these class-owned sheets, and it was as if they breathed poisoned gas. These papers had made a vulgar sensation, a scandal, out of Hal's arrest; they had suppressed his statement, his explanation—and that, of course, was as if they had cut out the brains of his action!

Suddenly, as Hal walked on, brooding, he lifted his eyes, and before him towered a great building of brown stone, with a sign across the top of it: "The Western City Herald". Underneath this sign was a second line of words—big, so that you could not miss them, graven in stone, so that they would last forever:

"Justice, when expelled from other habitations, make this thy dwelling-place!"

Hal had seen these words many times before, but never in his present mood, with his present knowledge. They came to him as something new and startling, incredible. He read them over and over—staring like a countryman who comes to town for the first time in his life, and is moved to awe by the great sights of the city. Suddenly an impulse laid hold of Hal Warner, and he went across the street and entered the elevator of the building.

"*Office of the Publisher*", read the sign on one of a long row of doors; and Hal walked in, and handed his card to a clerk, and the clerk disappeared, and came back and said, "This way please"—all quite suddenly, before Hal had time to realize the consequences of this mad impulse which had seized him. He went along an inner corridor, past several doors with names on them, until the clerk opened one with the name, "Mr. Anthony Lacking".

It was a big room, with big windows which looked out over the roof-tops of the city. Between the windows was a big desk, and at this desk sat a big man. He had been a fighting man in the early days, this "Tony" Lacking— he had come to town as a "tin-horn" gambler, and started a paper in a place where everybody had scandals, and he had shoved his way among them, black-mailing, brazening, blustering in huge headlines black and red. First he had plundered individuals, then he had plundered corporations—until at last he had discovered the final destiny of the great newspaper, which is to assist the corporations in their plundering of the public. So now this ex-gambler had a big building and an odor of prosperity, and was all for law and order and the sacred rights of privilege. But he was still the same "Tony" Lacking to all the city—a florid face and a florid mustache, a diamond on his finger, and a voice that had been made in the mountains. "Hello, Kid!" said he, when he saw Hal.

The other was not disconcerted, for though he had never happened to meet Mr. Lacking, he knew him by sight, and had heard the voice of the mountains. He took the greeting as it was meant, in good fellowship, and remarked, "Justice has accepted your invitation, Mr. Lacking."

The other looked at him. "What the devil?" said he.

"I hope you won't fail to recognize her!" And without being invited Hal went and took a chair beside Mr. Lacking's desk. "Justice comes to make an appeal for Mrs. Bobek, a Slav woman, whose baby is dying." And Hal continued, simply and earnestly, with the story of this poor mother and her "joy-ride". Then, "Justice comes to make an appeal for Mike Sikoria, Mr. Lacking." And Hal told the story of the old Slovak—not merely how he had had his arm crippled for trying to get his mail, but how he had wandered about in the coal-camps of the state, being robbed and mistreated—

"Say, Kid!" interrupted "Tony". "Why all this sob-stuff?"

"For justice, Mr. Lacking. She has been expelled from other habitations, and she comes to make this her dwelling-place."

The publisher of the Western City "Herald" leaned back in his chair and laughed his mountainous laugh. He knew "one on him" when he heard it! But when Hal went on to protest that he was in earnest, Mr. Lacking told him that that was not a natural condition for one of his age. When Hal started to tell what had happened to Louie the Greek—"Cut it out, sonny!" said "Tony". "You can't wring my hard heart! I know you idle youngsters who have to blow off steam, but you can't use this paper for your devilment, and you might as well get that clear at the start. The 'Herald' is a moral paper."

But Hal persisted in being serious, and in trying to argue; so it became necessary for "Tony" to become emphatic. "Listen, Kid," said he. "I sit in this room every day, and all sorts of people come to me, thinking they can pull the wool over my eyes. They want publicity for this and that, they have their grafts of one kind or another—and I have to sort them out. Your turn came sometime ago, and I looked you over and put you in the waste-basket—d'ye get me? Any time you get yourself arrested or hung, you can have your name in the 'Herald'. I'd keep it out even then if I could, for the sake of your old father, who's a gentleman; but I can't do it—business wouldn't let me. But one thing I can see to, and that is that you don't use my columns for the defense of riot and assassination. So forget it, Kid—and run along and get yourself a girl somewhere, and drop this scheme of turning the world upside-down!"

[27]

The State Federation of Labor had called a convention to discuss the outrages in the strike-field, and this convention was to open its proceedings with a mass-meeting in the Auditorium. Hal besought his friends to attend the meeting—so that he might do his explaining wholesale, instead of here and there on street-corners! He kept the telephone busy, and he heard what several people thought of him before the day was by.

His father would come; in spite of the doctors, he would show Old Peter he was standing by his boy. But Brother Edward would not come—not even for a pretense at open-mindedness. On the other hand, Lucy May was coming—more breaking up of the home! And then came another spiritual wrestling-match in Will Wilmerding's study. It was a desperate proposition to put up to Peter Harrigan's assistant rector; but how could he say no—he who was a man of conscience, professionally that! One had a right to demand things of him which one could not demand of a business-man like Edward, or a man about town like Bob Creston!

"You see," said Hal, "you happen to have a 'call' here. Other people can say: 'Such things can't happen!' But you know they have happened—because I come and say to you, 'I have seen them!' "

And that was true. This boy was a product of Will Wilmerding's moral work-shop, so to speak; he had the Wilmerding trade-mark on him. He was being put to strange uses, but the material was sound, and the maker knew it.

"If you'll only come and see, Uncle Will!" For Hal believed that if he could once get his friend to know the strikers, the problem would be solved. "Uncle Will" really was one of the most democratic people alive. He would labor over the soul of a poor boy exactly as hard as over the soul of a rich boy; he would welcome the poor boy to the church, and try to mix him up with the rich boy. And he was entirely naive about it—when they wouldn't mix, when he found the church becoming a church of rich people, with a mission on the side for the poor—it really distressed him, he did not know what to make of it. He did not see that the world outside was organized on exactly the opposite principle from his church; that he was trying to teach men to be brothers on Sunday, while all the rest of the week the world was teaching them to be wolves.

"Or rather, you see it," argued Hal. "You admit that in social life and business, in the newspapers and even in the police-court, the rich boy stands upon a different basis from the poor boy. What blocks your mind is your idea of heaven. The things of this wicked world don't really matter—"

"Not at all! Not at all!" cried the clergyman; for he was what is called a "muscular Christian", desiring righteousness on earth. "But I see the boy I love going into strange movements, being drawn away from his faith in his Redeemer—"

"You're turning things exactly about, Uncle Will! It was because I gave up the old dogmas that I set out to try to realize something here and now. You can tell yourself that Jesus was a god, and you can't blame men if they don't act like Him. But to me Jesus is a workingman suffering injustice; a proletarian leader, exactly like John Edstrom or Louie the Greek, whom you'll meet down in Horton. As for Peter Harrigan, who comes to worship Jesus in your church—he is the Jewish Pharisee and Roman plunderer who had Jesus nailed on a cross!"

Such is the new theology; like all theologies, it sounds blasphemous when first revealed. It did not convince the assistant rector of St. George's, and Hal was about to give up in despair, when a funny thing happened. He chanced to mention his interview with the publisher of the Western City "Herald", and how "Tony" had advised him to "go and get himself a girl".

A startled look came on Wilmerding's face. "Tony Lacking said that!"

"Yes, of course." And at once Hal realized how this remark must hit his friend. The clergyman was devoting his life's energies to combatting "the sinful lusts of the flesh"; the precise point on which he was keenest was that young men should not get themselves "girls"! Hal saw his hands clench, and his face grow ruddier than he had ever seen it before. Yes, there could be no question about it, there was wickedness in high places in Western City! And who was to rebuke this wickedness if not one ordained in the line of the prophets and apostles? Because of that infamous jeer of the publisher of the "Herald", the assistant rector of St. George's would face the scandal of being seen at a trade-union meeting!

[28]

The meeting came off, and a couple of thousand people heard Jim Moylan and John Harmon tell what they had seen and experienced in the strike-field. Also Mother Mary, who had been shipped out of the strike-district, delivered one of her "tirades", as the papers never failed to describe them. And then came the turn of Hal Warner.

It had never occurred to Hal that he was an orator, or could become one; but he had the main essential of true oratory—something to say, something he cared so much about that he had no time to think how he said it. If he could make this audience see and feel what was happening in the strike-country, some of them might get busy and save Minetti and Rovetta from torture! He told the story, with all the fervor and earnestness he could muster; he told it so that women wept, and men shouted with indignation. When he had finished they crowded about him to shake his hand and assure him of their sympathy and support.

So Hal went home, thinking he had really accomplished something; but next morning he got the papers, and some of the conceit was taken out of him. The papers featured the "tirade" of Mother Mary—for the obvious reason that the prejudice against her discounted what she said. One paper made no mention of Hal whatever, while in "Tony" Lacking's organ he had four lines, as follows: "Another speaker was Harold Warner, recently arrested in Pedro for disorderly conduct. Young Warner denounced the militia authorities in unmeasured terms, declaring that General Wrightman had the manners and ideals of a bandit-chief." Now, Hal had said those words, and considered them well-chosen; but he was forced to realize that, standing alone, they were not convincing to the casual reader!

Next morning the convention gathered, and Hal lured his group of friends still farther into the "lime-light". "Among those who watched the proceedings from the gallery were Edward S. Warner, the coal-operator, Mrs. Edward S. Warner, Junior, Mrs. Adelaide Wyatt, and the Reverend William Wilmerding."

There were five hundred delegates to this gathering, representing three hundred labor organizations, with a total of more than fifty thousand members. Things began to boil from the first moment, for Johann Hartman had succeeded in smuggling out a letter from the county-jail, and there was the story of its horrors in accents of unmistakeable truth—the starvation, the vermin, the cold, the torturing of men with bayonets and cold water. The reading of this letter almost broke up the proceedings at the start, for Mother Mary sprang to her feet and called upon the delegates to march forthwith to the Capitol to demand that the Governor should stop these outrages. There was vehement discussion, back and forth, and when the politicians among the labor men voted down the motion, Jim Moylan and Mother Mary and some fifty of the delegates went out and marched for themselves!

The rest stayed to listen to the deputy state labor commissioner, who told of his efforts to investigate peonage in the camps; also to Louie the Greek, who had been let out of jail the day before; also to George Tareski, an old Croatian miner whom Louie had brought with him, to tell the delegates how the militia had arrested him and set him to digging his own grave.

Tareski had worked for twenty years in the mines of this district; for ten years he had been an American citizen, and had a wife and three children on a little ranch which he had bought with the savings of his labor. He had written a letter to a fellow-countryman in the Hazleton camp, asking him not to work as a strike-breaker, but to come to the tent-colony, where he would be taken good care of. This letter had fallen into the hands of the mine-superintendent, and poor Tareski, setting out in the morning with his pick and shovel to hunt rabbits, was seized by two militiamen, taken to their camp, and locked in a cellar for three days without being allowed to communicate with anyone.

"The place was dirty, full of lice," said the old Croatian. "I told the soldier up there, 'I can not sleep there, I got to stand up all night.' Then some officer come in, and on Sunday morning he took me out of jail and he got my pick and shovel, and two soldiers took me away from the house and showed me the ground, marked already, and he says, 'Here is your job. Two and a half feet wide, six feet long, eight feet deep. You dig this.' And while I was digging for a while the soldiers came around and they asked the guard, 'What is that going to be, a toilet?' And he said, 'We got a toilet.' And the soldier says, 'That looks to me like somebody going to get buried in there.' And the guard says, 'There's the man, who is digging it.' And after that some of them came around and say, 'What are you going to use, a blanket or a coffin?' Some of them say blanket and some of them coffin, and the officers come and they put me in front of the line and told me to be ready. I think they are going to shoot me. They go to a place on the hill, a little bit level, and then a man come to me and talk to me in Polish language. He ask how I am, and I say, 'Not much.' And he say, 'Friend, I am sorry to tell you, but you are digging your own grave, you are going to be shot tomorrow morning.' He told me that in my own language. And I say, 'Have you heard about it? What for?' He say, 'I don't know. You must have done something very bad.' 'You are sure?' I say. 'Yes, everybody knows it already. They are sure going to kill you tomorrow morning. About three men going to shoot you.' And then I fell down in the hole, because he make me believe. I go to see officer, to send for my wife and children before they kill me, and he say, 'Nothing doing!' I was crying. I say, 'Give me a piece of paper. I want to write to them something.' And he say, 'Nothing doing!' He say, 'Dig and hurry up. If you don't dig I raise hell with you. If you don't dig I shoot you before time.' But I feel so bad I can't do any more, because he make me believe. And when I am so weak I can't dig, the officer put me in the ground in the cellar and then shut the door. And after a while another officer come and he ask, 'What's the matter?' And they say to him, 'I guess that fellow is pretty near crazy. You better do something for him.' And he say, 'Now I turn you loose. You go home and stay with your wife and children, and don't write any more. Don't talk anything, and don't go any place from the place where you are living, because if you do I get you again.' When I come to my house my woman was pretty near crazy. She sent her brother to telephone to soldiers, if she couldn't come over there and see me or something, and they wouldn't let her come at all. And when I got home I dressed myself, I clean myself, I get some of them lice off me that I got in the jail, and then I come to union headquarters."

Such stories as this, told with the simple pathos of the poor, sent Lucy May home to make her husband's life a burden; they sent Will Wilmerding home to his study to pray. After his prayer—possibly because of it—he paid a visit to the office of Judge Vagleman, chief counsel for the General Fuel Company, and one of the trustees of St. George's. For an hour that eminent lawyer argued with his assistant rector, pointing out the other side of the complex social problem. What did Wilmerding know about this Tareski, for example? About the letter he had written to his fellow-countryman? Suppose it had been a threat of murder, instead of a polite invitation to quit work and come to the tent-colony! Here was a letter just received from Bernard Vagleman, the judge's son in Pedro, telling of his efforts to convict strikers who had ambushed four mine-guards in an automobile and shot them. The host of perjured witnesses the strikers were bringing! Was it not a fact that the rank and file of these poor wretches were being intimidated by agitators? It would surely be a calamity if a man of influence like the assistant rector should let himself be drawn into such a controversy, to give public encouragement to law-breakers!

After the clergyman had left, Judge Vagleman called up Peter Harrigan; and Peter Harrigan called up Dr. Penniman, and there was a terrible scene between Uncle Will and his rector, about which Hal did not hear until long afterwards. Dr. Penniman was unsparing in his indignation. How could Wilmerding feel that he had the right to go over his rector's head, to compromise the church, to bring scandal upon its hallowed name? It was not merely an act of folly, it was insubordination, rank presumption! So Wilmerding went back to his study to pray some more!

[29]

Meantime the convention was continuing its sessions. A few of the delegates wished to call a general strike; others wished to order an investigation. Matters were brought to a head by Mother Mary, who made another speech, urging her idea that the convention should march in a body and demand an interview with Governor Barstow. So it was voted, and on the following morning occurred a march of five hundred delegates, and perhaps a thousand sympathizers, to the white marble State House on the hill. At the head marched Mother Mary and Louie the Greek, carrying an American flag, and behind them came the delegates singing the union song. They went quietly, but there was resolution in every face. Their mood had been voiced by the little old woman: "When you go to the Governor, don't say Your Honor, for he has no honor. Don't call him Your Excellency, for there is nothing excellent about him. Go up there and demand your rights and see that you get them!"

The little cowboy Governor was of course in a panic. At first he refused to meet the delegates at all, but finally he consented, provided their questions should be submitted in writing, and provided they should not ask any questions he did not want to answer. With a burly detective on each side of him, he came to the legislative chamber and listened to plain talk from the miners' leaders.

A more pitiful exhibition of futility could hardly have been given by a public man. First he backed down squarely from the position he had permitted General Wrightman to take; it had never been the intention of the military commission to "try" the strikers, it was purely a commission for "investigation". But what was the difference, demanded the delegates—so long as the General could put men in jail and keep them *incommunicado* for as long a time as he wished? The Governor declared that the civil authorities of Pedro County had refused to prosecute the cases: in answer to which the strikers presented in a couple of hours a telegram from the District Attorney, declaring that he had never refused to prosecute a case, and never would refuse. The Governor repeated his usual meaningless platitudes, that he wished to "see justice done", that he was "determined to enforce the law". The strikers answered by presenting a telegram from Pedro, to the effect that General

Wrightman had that morning turned out his troops and cleared all the principal streets of the town, shutting up the inhabitants in their homes while a crowd of two hundred strike-breakers were brought through to the coal-camps!

But the convention adjourned without ordering a state-wide strike—to the intense disgust of Jim Moylan, Hal Warner, and others of the younger spirits. The forces of fear and conservatism—possibly more sinister forces, who could tell?—held the delegates back from the one move which might really have brought deliverance to the miners. "It's all a fizzle! All a farce!" cried Moylan, in despair. "We're just where we were when we started!"

The Governor had declared that he would appoint a committee to go down to the strike-field and investigate the charges against the militia. "But he'll back down even on that!" declared the young Irishman. And sure enough, when they went next day to ask about the committee, they found that Peter Harrigan had been ahead of them. Let the labor men name their own committee, and do their own investigating!

There was no way to hold the Governor to his word, so John Harmon set out to find some men who would command public confidence, and were willing to give their time and energy to investigating the charges against the militia. The deputy state commissioner of labor would serve, but his statements would be discounted because he was a labor man; the same was true of the president of the carpenters' union, and the secretary of the brewers. What was needed were doctors, lawyers, merchants, clergymen, professors—but these were sought in vain. Harmon went from one person to another; Hal did the same; but not one man or woman could be found who could spare the time, or who was willing to incur the public odium involved in investigating complaints of coal-strikers.

So Hal was driven again to Will Wilmerding. He made an appointment, and took John Harmon with him to the clergyman's study, and for an hour the three of them fought it out.

"You complain," Hal said, "how hard it is to be sure about these charges. I tell you one thing, Vagleman tells you another! Well, now, we ask you to come and see for yourself! How can you refuse that?"

Looking at his friend, Hal thought that he was several pounds lighter since the coming of this crisis into his life. He took a long time to answer—so long that it might have been another prayer. "Mr. Harmon," he said, at last, "I have been profoundly shocked by the charges made at your convention. But I am not at all sure that I am the sort of person you want for your investigating committee. If I should go, it would not be as a member of any partisan commission, submitting to any sort of partisan control."

"Of course not," said Harmon. "We had no idea of such a thing."

"But I might find that your charges are exaggerated, Mr. Harmon. I might find one side as much to blame as the other. If so, I should consider it my duty to say so."

"Of course," said Harmon, again.

But apparently the other could not believe that Harmon meant what he said. "I have an idea of the sort of report that you and Hal want written, and I can't imagine myself being willing to sign such a report. If you invite me, you must realize that you are running the chance of having a minority report."

"I understand that, Mr. Wilmerding."

"You wish me to go, on the terms that I shall use my own judgment and tell the truth as I see it?"

"You will *go* on those terms, Mr. Wilmerding?"

"Yes, I will go."

And Hal jumped up from his chair with a cry of triumph. He felt at that moment as if he had won the strike!

[30]

So anxious was the assistant rector of St. George's to make clear his non-partisan attitude that he refused to travel to the strike-country with a disturber of the peace like Hal Warner. As a favor to his friend, Hal agreed to stay in Western City until the work of investigating was done! The five members of the committee—John Harmon was the fifth, for lack of anyone else—went to see Governor Barstow and obtained credentials, and then, journeying to Pedro, presented themselves before the Adjutant-General of Militia.

Hal was told about this interview soon afterwards, and it was exceedingly funny. Here was Will Wilmerding, with his intense conscientiousness, his desperate determination to be impartial; and here was General Wrightman, with his "manners and ideals of a bandit-chief"! At the very outset he put his foot down. There existed a state of war in this strike-district, and he would not have civilians meddling with military discipline, asking questions of either officers or privates of the militia.

The General announced this decision, and expected that to settle it; but to his surprise the clergyman attempted to argue the point. "How can we get the truth, General, if we only question one side?"

"It will be entirely agreeable to me, sir, if you question neither side!"

"But General, the Governor told us—"

"Do you expect me to take your word for what the Governor told you?"

"No, as it happens, you do not have to. The Governor has written here—will you kindly read it?"

"I don't care to read it, sir!"

"Well, then, I will read it to you." And the clergyman read—impressively, as a clergyman learns to read: " 'You will please give this committee every assistance within your power, to the end that they may secure what information they desire. Please have them furnished with any information you may have, or direct that anyone who has information shall give it to them.' Surely, General Wrightman, there is no misunderstanding those words!"

The General was tapping with a pencil on his desk—always a danger-sign. "I have given you my decision, sir."

"I understand, then, that you refuse to obey the orders of the Governor?"

A silence.

"I am informed that in a military sense the Governor is your superior officer. You refuse to obey his orders—and yet you talk about military discipline?"

"Sir!" cried the Adjutant-General, "you are insolent!"

Now when one has been for many years accustomed to standing in a pulpit and having a congregation listen in helpless respect to whatever the Lord may inspire one to say, one acquires a certain notion of one's own importance. So now there was a lively scene. When the General shook his finger under the nose of the clergyman, the clergyman shook his finger under the nose of the General. "I know what is the matter with you, sir—I can see at once that you are a victim of military megalomania!"

Apparently General Wrightman had never heard of this disease. He was taken aback, and the other rushed on: "Yes, sir! You dress yourself up in epaulets and gold-lace, and have a wholly exaggerated idea of your position, and of the deference that is due to you. You talk grandiloquently about 'a state of war' and 'the duties and responsibilities of a soldier'. Well, sir, when I accepted this responsibility I took the trouble to consult an authority about constitutional law, and I would inform you that when you talk about 'a state of war' in this district, you are talking nonsense. There is no war here; there is no insurrection, there is not even a riot. You are doing police duty, and your position is that of a police-officer, in every way subordinate to the courts and the civil authorities of the state."

"All right, sir!" said General Wrightman. "If that is your notion, you are welcome to it. All I can tell you is that if you attempt to act upon it in this district, you will regret it."

"*What*, sir?" cried the other. "You threaten me? When I have come to this district with credentials from the Governor? Let me tell you, sir, that I am here for an important public purpose, quite as important as your own, and I will permit no military autocrat to intimidate me!"

So the matter was left; the contestants drew apart, growling like two dogs which have been interrupted in the middle of their fight. The Reverend Wilmerding was so indignant that he was now willing for Hal Warner to come and assist the committee. John Harmon telephoned up, and Hal took the next train to Pedro.

Meantime the clergyman had already got to work. He would be impartial, in spite of the Adjutant-General's efforts to prevent him! He would ferret out all the dark secrets which the strike-leaders must be hiding from the public! Before anybody in Pedro knew who he was, he dodged away from the rest of the committee and began strolling about the streets, talking with the rank and file of strikers and with neutral citizens. Judge Vagleman had assured him so earnestly that the strikers were terrorized by agitators; now it was his duty to discover and expose this terrorizing!

Sometime afterwards there was a Federal investigation, and Wilmerding told under oath what befell him on these expeditions. He picked out a likely-looking striker and began: "You don't care for the union, do you?"

"Care for the union!" was the reply. "What can we do without the union? We want our freedom."

"Well, but what do you mean?"

"I mean that we can't ask for anything without the union to back us."

"What do you want to ask for?"

"My weights, to begin with."

"Your weights? How is that?"

"Why, I mine coal and I don't get it."

"You don't tell me that?"

"Sure I tell you that! You can't know much about coal-mining."

"How do you know you don't get the weight?"

"I know it because it's my business. Don't you suppose a miner can tell how much coal he loads? Where do you come from, anyway?"

"From Western City."

"What's your work?"

"I'm a clergyman."

"Well, when you've read the Bible through enough times, you know what's in it, don't you? And when I've loaded a few thousand cars, I know what a car weighs. A miner measures a car with his fore-arm; he loads her level or loads her high, and he can tell you within a hundred pounds what she'll go. He won't miss one time in a thousand. And here we work year after year, knowing that we've got to load three thousand pounds to get paid for a ton."

"I honestly didn't believe it," said Wilmerding. "But when I found that man after man—and not poor Slavs or Greeks, but Englishmen and Scotchmen, men with names as good as any of our names—told me time after time that there were short weights, then I was compelled to believe in their grievances."

In the same way he was compelled to believe in other grievances. He saw the drunken militiamen, carousing in the streets with their prostitutes; and one day, walking with John Harmon, he saw the mine-guard Dirkett assault and beat a young Polish boy. He wanted to interfere; he quite naively proposed to ask Gus Dirkett why he was beating the boy—thinking that Dirkett must have some profound philosophic reason which he would be pleased to explain to one ordained in the line of the prophets and apostles. He only desisted because Harmon assured him that people who asked questions of Dirkett met the fate of Tom Olson; and Harmon could give plausibility to his statement by pointing out the shape of a revolver in the mine-guard's overcoat-pocket!

[31]

The committee got to work. The Reverend Wilmerding sat in a hotel-room, and hour after hour, day after day, witnesses came before him and gave testimony. Louie the Greek came, and Joe Prince, the negro, and Rovetta, just released after fifty-five days in hell. A few days later Johann Hartman was released—and Wilmerding took pains to get hold of this man before anyone else had a chance to tell him what to say! Tim Rafferty came, and Kowalsky and Mike Sikoria; Rosa Minetti and Mary Burke and Mrs. Bobek. Not only strikers—there came a Socialist lecturer and his wife, who had been thrown into jail, and an undertaker and his wife whose home had been invaded and wrecked in a search for arms; also two clergymen of Pedro, and the postmistress of Horton, and the District Attorney and his assistant. So on through a long list—a hundred and sixty-three witnesses in all, over four hundred thousand words of testimony.

Hal would sit and watch the face of his "Uncle Will". The clergyman could hardly contain his emotions; he was far more deeply stirred than the labor men, who had learned lessons of patience in the long, slow, heart-breaking struggle of their class. To a man like John Harmon, for example, these strike-outrages were an old story. He was used to reading labor-journals and convention proceedings and committee reports, in which such events were told over and over. In this one state the militia had been called out in labor-troubles no less than thirty-four times in eleven years; and it had always been the same militia, called out for the same purpose!

But while Harmon was living in this world of realities, the clergyman had been living in abstractions, in comfortable maxims and formulas, religious, moral and legal. He had consulted "an authority upon constitutional law"—

and had really and sincerely believed that the principles this authority had explained to him had some relationship to modern life. For instance, the maxim that a man's home is his castle; he thought that applied in America—even to Hunkies and Dagos and wops who come to America to look for jobs! In the same way, it was a principle of law that a man could not be imprisoned to compel his testimony; Wilmerding had no idea that it is a universal practice of police-officials in America, not merely to imprison men to compel them to testify, but to inflict tortures upon them, in order to compel them to testify what the police-officials desire.

The committee visited the Horton tent-colony. It happened to choose a day when the occupants of the tents were lined up under the guns of the soldiers, so that a search for arms might be made; and Wilmerding, strolling about the camp, climbed a little ridge to where Lieutenant Stangholz stood patting his machine-gun—his "baby", as he called it—turning the weapon and sighting it, showing how from this place he commanded the whole colony. The clergyman started a chat with this mine-guard with the face of a bull-dog, who had seen cruel adventures in many lands, and delighted to tell about them. Wanton beyond belief was Stangholz—so foul-mouthed that a person of decent instincts writhed to listen to him. He hated the strikers with a fierce and blazing hatred; he poured out oaths and obscenities upon them in the fashion that his "baby" poured out steel bullets.

While the committee was still in Horton, it chanced that a group of Stangholz's troopers were riding down the road at night, and one of the horses stumbled over a strand of barbed wire. The rider was carried into the depot at Horton, and a few minutes later the Lieutenant rushed in. Without any investigation, he knew what had happened—some of those God-damned red-necks had set a trap for his men! It happened that Louie the Greek was standing in the station, having a pass, and the Lieutenant leaped upon him with furious curses, seized him by the throat and beat his head against the wall. He drew his revolver and was about to brain the man, when two other officers interfered and dragged him off. A minute or two later he broke loose and attacked a Greek boy upon the station-platform, beating him ferociously in the presence of many witnesses.

Now the truth about this wire was definitely settled next day, when John Harmon went to the spot with one of the militia officers, and showed him that the wire had been cut quite innocently, in order to leave a short cut across a field. The wire was not long enough to reach across the road, nor even into the road; the horse had been thrown because it was off the road, on the prairie—and later the injured man admitted this had been the case. But these tedious explanations had no interest for Stangholz, who proceeded to teach the strikers a lesson by dragging down all the wire surrounding the tent-colony, rolling it up, together with the posts, in one tangled mass, and dropping it into the well from which the strikers obtained their water-supply.

A couple of days later it chanced that Mrs. Olson, the school-teacher, was going to the train, and set out with her cousin, a young lad who was not a striker and had nothing to do with the colony. They were stopped and turned back, and so the young woman went to the militia-captain and obtained a pass. But when she presented this paper to Lieutenant Stangholz, he refused to honor it, and abused the woman; turning upon the boy, and shaking his fist in his face, he shouted in furious anger, "I am Jesus Christ! All my men on horses are Jesus Christs, and must be obeyed!" After that, the strikers had a nick-name for Stangholz; a name which compelled the Reverend Wilmerding to bow his head with embarrassing frequency!

The Governor had declared that he was ready to act promptly upon presentation of adequate evidence; the clergyman was now so indignant that he consented to the committee sending a long telegram, detailing the Lieutenant's behavior on these and other occasions, and asking that he be relieved from duty, pending a trial upon charges of unfitness. But the Governor did not take this action, nor any other action. He said that he would not condemn anyone in advance of a trial!

[32]

The committee remained on the scene for a couple of weeks, and then went back to Western City and made its report, which was a scathing indictment of the militia, substantiating every charge the strikers had made. The curious thing about it was that Wilmerding, who had been so dubious and full of cautions—he was the one who wrote the report and put in all the "brave words"!

He arraigned the militiamen for arbitrary conduct; for drunkenness and debauchery; for abuse of women; for the torture of prisoners; for the invasion of homes; for the assaulting and robbing of strikers; for the maintaining of peonage in the camps. He spared neither the rank and file nor the officers. "Khaki and even gold lace and epaulets can not make a soldier," he declared. "Think of a man really fit to be Adjutant-General of a state's national guard angrily shaking his fist in the face of a grey-haired widow, whose offense was the singing of the union-song in her home, which she had owned for over twenty-five years!"

"Some of the testimony," continued the report, "relating to the meaner and more cruel forms of oppression, would be almost incredible were it not corroborated both directly and by circumstances, and by the appearance and conduct of the witnesses; and were it not, moreover, confirmed by the commanding officer's assertion of his privilege to infringe the most fundamental and sacred rights secured to men under Anglo-American law." And again, "Whether a robber is drunk with liquor or with power, the effect on the person robbed of his liberty or his property is the same. Lawlessness begets lawlessness, and when subordinates of all ranks witness the violation by their superior officers of great underlying laws of civil society, they will naturally gratify their own low desires and get themselves what they can of the spoils of war."

One of the clergymen of Pedro who appeared before the committee had told of the public conduct of the soldiers with their prostitutes, of their "rushing the can" in the town-hall, of their drunkenness on sentry-duty. He told how he had protested to General Wrightman, who had answered by calling the accusations "lies", and accusing the clergyman of "besmirching the uniform of the soldier". "Robberies and holdups by militiamen," continued the report, "the General disposes of in the same way; but the instances of this sort of valorous conduct are far too numerous, too varied in circumstances and scattered over too wide a territory to be so simply gotten rid of. They range from a forced loan of twenty-five cents; or whiskey 'for the captain'; or a compulsory gift of three dollars; or whiskey, gin, cigars and champagne; or a ton of coal—to the downright robbery of three hundred dollars, and other considerable sums of money, with watches and other small pieces of property."

This report was transmitted to the Governor, who politely acknowledged the receipt of it and did nothing about it. It was briefly mentioned in the newspapers, and printed as a pamphlet and circulated by the miners. Hal Warner and Mary Burke and Mrs. Olson sat up nights for a couple of weeks mailing out marked copies all over the country.

And meantime the Reverend Wilmerding was having scenes with his rector, and with the trustees of his church, and the host of his parishioners and friends. He was so desperate about the sights he had seen—he actually proposed to stand up in the pulpit of St. George's and preach about conditions in the coal-country! In the presence of Peter Harrigan and Judge Vagleman, and other directors and leading stockholders of the General Fuel Company! As a result, he was not allowed to preach at all; and when he went about the city, seeking other ways to reach the public, he found that all ways had suddenly closed tight. Everywhere word had gone about that the assistant rector of St. George's, hitherto such a favorite at Y.M.C.A. entertainments and church sewing-clubs and charity theatricals and Chamber of Commerce banquets and civic improvement conferences—that the said assistant rector had become possessed by devils, and turned suddenly into a firebrand and fanatic, an offender of good taste, a menace to law and order. Hearing about his plight, Hal wrote to his college-friend, Morris Lipinsky, who got up a Socialist mass-meeting, at which the clergyman was free to vent his insane emotions. And of course that finished the process of his social and clerical downfall; when decent people opened their newspapers and read that the once-popular assistant rector had delivered a harangue at a gathering of revolutionists and incendiaries—well, they set to work to let him know what it really meant to be ordained in the line of the prophets and apostles, who had been stoned and scourged and fed to lions and thrown into kettles of boiling oil!

[33]

At this time there fell in the coal-country what the strikers came to know as the "big snow". There must have been two feet of it—the tents were buried, and men had to go out every hour or two while it was falling, and clear the roofs to keep them from being borne down. And this of course laid everybody helpless; the pickets could not get about on the roads, and their amazon auxiliaries could not get down to the depots. So General Wrightman saw his opportunity. He was still supposed to be enforcing the Governor's "policy" of keeping out strike-breakers from the mines; but it was humiliating to a military commander to have to make pretences to strikers; so now the General announced that Governor Barstow had "modified" his orders over the telephone. When the perplexed Governor stated that he did not know of having done anything of the sort, the General said nothing, but went ahead with his own "policy", which was to see that all men who wished to go to work were protected in their constitutional rights—the constitutional rights of men who did *not* wish to work being meantime suspended. That was not the way the General put it, of course; he merely said that if the strikers continued to resist his orders, he would herd them all into a stockade, and not let anyone out without a military pass.

From that time on, the militia was frankly an agency for the breaking of the strike. The soldiers met all trains regularly and escorted them up to the coal-camps. All over the country hordes of ignorant foreign-speaking men were being hired, under the grossest misrepresentations, and brought to the mines and held for "debt". So many escaped, so many circumstantial stories were sworn to, that there could no longer be any doubt about it. The deputies of the state commissioner of labor, endeavoring to protect these victims, were several times arrested, and finally barred from all the camps in the district.

Such proceedings could of course not fail to have a weakening effect upon the strike. Everywhere the mines were filling up with workers, the newspapers were publishing statements to the effect that the strike was broken and the mines in operation. The strikers' claims could with difficulty be presented, for most of their leaders were in prison, their headquarters had been raided, their papers suppressed. Under the circumstances, it was not surprising that a few of the weaker men lost heart and gave up.

One of these was old Patrick Burke. As liquor was banned from the tent-colony, Patrick was accustomed to disappear each week when he got his union benefits, and not return until he had spent them. But one week he did not return at all, and a few days later came the report that he had gone back to North Valley a scab. It was a terrible humiliation to his three children; but there was nothing they could do about it, no way they could get to the old man to dissuade him from his shame. The sheep in paradise and the goats in purgatory were separated by no wider chasm than the strikers at Horton and the strike-breakers at North Valley!

Mary Burke shed tears of shame in secret; but she went on with her work, trying to make up for the treachery of her father by extra diligence on behalf of the sufferers. There was a constant stream of wounded men provided by the militia, and Mary assisted these at all hours of the day and night. For weeks she would not leave the tent-colony at all; she had no pleasure outside, on account of the militiamen, who made her life a burden. It had come to be the generally accepted idea among these men that "Red Mary" was Hal's "girl"; and they did not see why she should be so exclusive with her favors. They made advances to her, and when she spurned them, they never lost an opportunity to insult and terrify her. They would walk behind her, discussing the flaming splendor of her hair and the shape of her healthy body. They would speculate aloud as to her price, and make her offers. Several times they caught hold of her and kissed her, and once three drunken ruffians in khaki seized her and tried to drag her into an alley.

[34]

And then still more trouble fell upon Mary. By way of protest against the violence of the soldiers, it was decided to have a procession of the strikers' women in Pedro. A number from Horton went to attend, among them Mary and her young sister. A parade started, with American flags and printed signs; but when it reached the post-office, there was cavalry barring the way, and when the paraders refused to turn back, the General rode in at the head of his troopers, driving women and children pell-mell before him. Some were trampled by the horses; one woman had her cheek cut open by a sabre, another had an ear slashed off. In the midst of the excitement, the General fell off his horse and landed on his back on the side-walk; and little Jennie Burke laughed aloud—would not any child laugh aloud to see an old red walrus fall off a horse? The General, wild with rage, climbed upon his steed again, rode into the crowd, and as the child was trying to back away, kicked her savagely in the breast.

Now "Red Mary" was Irish, and had an Irish tongue, and in that moment of stress she made use of it. "Arrest that woman!" cried the General; and the woman was dragged away, and herded into an alley-way with eighteen others—among them Mrs. Jack David.

The little Welshwoman had had nothing to do with the parade, so she claimed. She had her two children with her, one three years old and the other four, and they had been standing in the doorway of a store, watching the soldiers driving the women down the street. "Move on!" one of the troopers had commanded her, and she replied—somewhat indiscreetly, perhaps—"I don't have to." Whereupon he seized her, twisted her arm behind her back, and beat her with his fist. "Shame! Shame!" cried the spectators; and Mrs. David assailed him with a new and ferocious weapon, her muff.

Now came General Wrightman, riding up to inspect the round-up.

"That's Mrs. Jack David," said one of his subordinates.

"Oh, indeed!" said Wrightman. Perhaps she had been too free in public discussion of "tin willies" and "scab-herders"; or perhaps her husband's title of "general" was regarded as a challenge. "Take her to jail," said Wrightman. "That red-headed one too!" And so Mary and Mrs. David and the two children were "military prisoners"!

And prisoners they remained, day after day. There was no way to tell how long they would remain, nor even what was the charge against them. It was at this time that Major Cassels, in a habeas corpus proceeding in Judge Denton's court, explained the theory upon which he was proceeding. "The question of the guilt or innocence of these people is a matter of no importance. It was deemed necessary, for purposes known to General Wrightman, to lock them up, and they will remain locked up until General Wrightman orders them released."

But if the General thought that he could tame the spirit of either Mary Burke or Mrs. David, he miscalculated. The freedom of woman's tongue is an institution which has managed to survive many different kinds of despotism. They were in the second story of the jail, in adjoining cells, and they would stand at the windows, and when they saw strikers passing, they would sing the union song in chorus. Mrs. David clamored to see her husband, and finally this permission was granted, on condition that a militiaman should be present during the interview. It gave the little woman keen delight to circumvent this order by conversing with her husband in Welsh!

Mrs. Olson made an effort to visit Mary, but was turned back. As for Hal, he knew that he could do nothing; he would only add fuel to the flame of slander—and possibly get himself shipped out of the district again. He went to Captain Harding, to try to persuade that officer to intervene—but in vain. There was a vehement dispute between the two, for Hal told his cousin exactly what he thought of him, and the opinion was not flattering.

Eleven days passed, and Mary came out, with her Irish complexion faded, but her Irish tongue as sharp as ever. It was a sorrowful home-coming, for little Jennie's breast was swollen, with a big black bruise, corresponding to the shape of the toe of an Adjutant-General's boot. The doctor took it seriously—such things often turned into cancer, he said. Poor Mary almost fainted when she heard this; for she had seen old John Edstrom's wife die of cancer, up in North Valley, and in her mind was the memory of a ghastly stench.

Yes, these were black days for the Burke family. Mary had worked so hard for her young brother and sister—

pinching and scraping, sewing, scrubbing, scheming. She would pull them through, and when they grew up, her life would be free from its heaviest burden. But now they were both of them to be cripples! Tommie was up again, but had to have a crutch, and would never be able to run and play like other boys.

Hal's heart was wrung with pity; and the "poetry-books", as Mary called them, record the psychological subtlety that "pity moves the soul to love". The clouds of despair which overhung Mary Burke's life would be shot through with sunshine—if only Hal were to follow the impulse which was now seldom quiet in him. He would find himself alone in a tent with the girl, and would want to put his arms about her, and let her cry out her grief upon his shoulder. Why should he not do it? Why should he pay heed to warning voices which could be only survivals of old prejudices, old cruel instincts of caste?

The world outside, Hal's world of culture and "refinement", would say that he was slipping into a pit, that he was becoming demoralized, here in this intimacy with low people. But what did he care what his world thought about him? Was he not bracing himself in a furious struggle against it? How long would it be before he was ready to cast the die, to burn his bridges behind him, to take the final plunge—to do whatever mixture of metaphors might express the awfulness of the temptation which beset him—to clasp to his heart in the intimacy of love a low-caste woman, the daughter of a drunken miner!

It was a fire smouldering inside his heart, and threatening to burst into conflagration. He would speculate about Mary, what she was thinking and feeling. Did she smell the smoke? What would she do if suddenly the flames were to leap out and seize her? He would have an impulse to go and find out; but instead he would turn in a fire-alarm, and a whole department would respond with hurry and clamor—the great water-tower of Brother Edward, with its powerful stream of worldly counsel, and the chemical engine of religious asceticism driven by the Reverend William Wilmerding, and the hook and ladder rescue-truck with Jessie Arthur lashing the horses to a gallop!

But this rescue-truck was far away, and might not always arrive on time. Letters from Jessie came to him frequently, pitiful and touching—the protests carefully veiled, nothing expressed but gentle pleading. If they failed of their effect upon Hal, it was because of the distrust of his own world which was storing itself up day by day in his heart. He was coming to believe in nothing in this world any more. He could not ever see it as it wanted to be seen, a place of ease and graciousness and charm; he saw it only as it appeared to striking coal-miners—a club that came down on one's head, a bayonet that was plunged into one's vitals.

[35]

What saved Hal from these passionate allurements was hard work: the burdens that kept falling upon his shoulders, the new efforts that clamored to be made. No sooner was Wilmerding's pamphlet printed and sent out than John Harmon came with a new plan. In their campaign to break down the conspiracy of silence of the news agencies, the miners sought to persuade Congress to appoint a committee to investigate the strike. Would Hal go to Washington and try his luck in the role of lobbyist? It was unfortunate, but true, that members of Congress would be more impressed by one member of the leisure class than by any number of working-people.

So Hal took the three-day journey, and appeared before the "House Committee on Mines and Mining". Partly as a result of his testimony, and partly of the statements which he read from Wilmerding's report, the resolution was carried, and a sub-committee of five members of Congress set out for the West.

Here was a great opportunity! In Western City, Wilmerding got busy and persuaded a friend of his, a professor at Harrigan, to represent the miners as counsel. This was Professor Purdue, the "authority on constitutional law" whom the clergyman had been so bold as to cite to General Wrightman. He came to Pedro, arriving only the day before the congressmen, so that he had to put his witnesses on the stand without having a chance to talk with them in advance: which any lawyer must admit was a severe test for the most learned "authority upon constitutional law"!

The coal-operators were represented by Bernard Vagleman and Judge Evans, an eminent statesman of Pedro. At the beginning of the strike Hal had found himself horrified at the idea that gentlemen who had been to college, and held high rank in the honored legal profession, could be guilty of "framing up" evidence against low-down, pitiful coal-strikers. But now Hal was on the inside, where he could see Messrs. Vagleman and Evans at work day by day. And when he had got through with this investigation-battle, nobody could ever talk to him about the ethics of corporation-attorneys!

The battle was hottest about the question of "peonage" in the camps. The American public does not care so much what is done to strikers, but the "scab" is a sacred personage; has not a revered college-president hailed him as the "true American hero"? So now the strike-leaders wished to prove the fact that hundreds of these true American heroes from Russia and Bohemia and Italy and Greece were being kept by force in coal-camp stockades, and worked practically as slaves. But the difficulty with these heroes, as you found when you came to deal with them, was that they were terrified by their heroic experiences; they were hard to get hold of, and still harder to keep. Also, alas, they sometimes consented to take bribes, and to repudiate their affidavits at a moment's notice; even when they were sincere, they were so ignorant that they could hardly make themselves understood.

The strikers presented a Roumanian who had been "shanghaied" in Pittsburg, brought to North Valley in a locked steel car with an armed guard, and put to work at the point of a revolver in the hands of Hal's old pit-boss, Alec Stone. The man had worked for two months, and his total credit for this time had amounted to twenty-two dollars and eighty-seven cents—applied onto his transportation from Pittsburg! Four times he had attempted to make his escape, and been driven back by militiamen.

And now came Vagleman and Evans, to discredit this testimony. They put on the stand a Russian strike-breaker who swore that Tim Rafferty, Johann Hartman and two of the strikers had met him in a room at union headquarters and paid him money to give false testimony in this matter of "peonage". It happened that at the time the testimony was given all four of the accused men were in the room; Tim Rafferty was seated on the platform within ten or fifteen feet of the witness, and he signalled to Hal, who saw the opportunity, and tipped off Professor Purdue. The Professor announced to the committee that the accused men were present, and suggested that the witness should confirm his story by identifying them.

Vagleman of course made objection, but the chairman of the committee insisted upon the test, and Hal stepped across the room and deliberately placed himself in front of the lawyer, making it impossible for him to give signals to the witness. So the poor fellow sat there, gazing blankly at the rows of faces in the room, unable to pick out a single one of those he accused.

"I protest against this!" cried Vagleman, excitedly. "The lights are bad, the witness cannot see the people."

"Very well," said Professor Purdue. "Let the lights be raised."

While this was being done, Vagleman moved a little to one side; but Hal followed suit, standing within three feet of the lawyer, and looking at him. The lawyer knew quite well what he was there for, and was white with rage. "This is a farce!" he cried. "The witness can't see the men! They're too far away."

"Let him move towards the audience, if he wishes to," said Professor Purdue. "The men are in plain sight. The committee can see every one of them. We shall point them out afterwards."

The little drama went on for several minutes. Another of the operators' attorneys attempted to signal to the witness, but Professor Purdue appealed to the chairman, and there was a lively scene, at the end of which the accused attorney sat still. The result was that the "frame-up" collapsed completely, and the committee declared the witness discredited.

[36]

So the eminent lawyers had to arrange something else. At the request of Professor Purdue, the committee was keeping the "peonage" witnesses in a separate room, and the counsel for the operators brought down the camp interpreter from North Valley, and smuggled him into the room where the committee's witnesses were kept. It happened that Hal got a glimpse of the man and recognized him—his old enemy Jake Predovich! He told Professor Purdue about it, and the Professor brought the matter to the attention of the committee, and there was a great to-do. At first Vagleman attempted to deny all knowledge of it, then he pleaded that no harm had been intended. The chairman expressed his indignation in no uncertain terms, and they paid Vagleman back by putting Predovich on the witness-stand, and forcing him to testify concerning his treatment of strike-breakers. Whenever these charges of "peonage" were brought, the operators were always ready to prove that the victim had signed a paper declaring that he came with the full knowledge that he was to work as a strike-breaker. It had been noticed that these signatures were somewhat alike, and now the Galician interpreter was forced to admit that he had written over fifty signatures himself!

Then again, there came testimony concerning a group of strikers who had flagged a train. A Schultz detective had witnessed this episode, but had not identified the man who led the strikers, and he now tried to get this information from Jim Moylan, coming up to him in the course of the session and starting a conversation. Moylan, who knew the man, saw what he was up to, and mentioned quite casually that Vink Santifonti, the Italian organizer, had "handled that job". So the detective went away in triumph. At last the operators had something big!

They took the trouble to notify the editor of Vagleman's paper, the Pedro "Star", to be on hand for that session, as an important revelation was coming, and he would wish to describe the scene. The Schultz detective took the witness-stand, and, under the skillful guidance of Vagleman, told a detailed story about the flagging of the train and the dragging off of strike-breakers. He pointed out Vink Santifonti in the audience, declaring that he was the man who had led the raid. A paid organizer, a responsible official of the United Mine Workers! There was a stir in the committee, and a chuckle of delight from the strikers; Vink Santifonti took the stand and presented a copy of the proceedings of the recent annual convention of the Mine Workers, proving that he had made a speech in Indianapolis on the day the train was held up!

Nor was this the end of the troubles of Vagleman and Evans. The latter eminent citizen was fighting valiantly for his employers, but they had only recently employed him, and previous to that he had been their political opponent. Now came witnesses to testify to evil doings of the coal-companies in Pedro County, and Judge Evans would riddle these witnesses with the arrows of his sarcasm. Professor

Purdue let him go on in this way for a week, and then produced a stenographic record of a political speech which the Judge had made two years before, describing in detail all the conditions of which the miners complained! He had told how political conventions were "fixed", the delegates being selected and sent down by the superintendents of the coal-camps; how the political leaders got together in a back room of the Palace Hotel up in Western City and named the "slate". "'We will take for county clerk So-and-so; he is a good man for the purpose.' Some other man says, 'But I think within the last eight or ten months he has had trouble with some pit-boss.' He isn't right with the company, and they don't want him; he goes off the slate. And so it is from bottom to top—the candidates are selected, not with a view to their fitness, not with a view to their ability to discharge their duty, not with a view to their integrity, but 'are they satisfactory to the company?' If they are, that settles it."

And when Judge Evans was cross-questioned about this speech, he admitted he had made it, but explained that he had not said the companies controlled the officials—he had only said they controlled the *nomination* of the officials! When he had stated that lawyers did not dare to bring damage-suits against the companies, because they were afraid the companies would "black-list them and be against them politically and in every other way"—what he had desired his audience to understand was that this was the fault of the lawyers, not of the companies! The lawyers feared these things, but their fears were groundless, the companies never really did such things at all!

[37]

You might have thought that in the face of evidence such as this, the investigating committee would have considered itself bound to do something. But if you thought that, you would be showing yourself ignorant of the ways of investigating committees in America. For ours is a government of "checks and balances", and we pride ourselves upon the difficulties we throw in the way of getting anything done by our authorities. We believe in "individual initiative"—which means the power of great wealth to get done what it wishes for itself.

The "Subcommittee of the Committee on Mines and Mining" set a date for its return to Washington; and on the day before its departure, Hal sought out Representative Simmons of Indiana, a farmer-legislator who in spite of his prejudice against unionism had shown some human feeling throughout the proceedings. "You're going away," Hal said, "and you see our situation; these militiamen have been like a pack of hounds held in leash, and the moment you go, they'll be turned loose on the people who dared to appear before you."

"It's hard to see what we can do about that," said the legislator.

"One or two of you ought to stay here," argued Hal. "You surely owe it to your witnesses to protect their lives!"

"We can't stay, Mr. Warner. We have other matters pressing for attention."

Hal could understand that. It had been evident towards the end that the legislators were chafing to get back to Washington; they had a thousand errands to run for their constituents, their share of "pork" to secure, their "fences" to keep in repair. And here, two thousand miles away from the centre of events, they were compelled to sit and listen to the complaints of coal-miners who had no votes to give.

"You see," explained Mr. Simmons, "ours is a government of divided powers. The task of this committee was to investigate a possible need of new legislation."

"Will you recommend any new legislation?"

"It's difficult to think of any law that would remedy this present situation."

"Let me suggest a law, Mr. Simmons—a law that would end this struggle at once. Make membership in a union compulsory in all coal-mines."

It was evident from the look on Mr. Simmons' face that he was not going to recommend any law like that! "It would seem to me," he said, "that the problem is to get the present laws enforced. And that is the duty of the Governor of your state."

"But he won't do his duty. We've spent five miserable months proving that! So what next?"

The congressman was not in haste to reply to this. "You have an appeal to the President, of course—"

"But the President is a long way off, and has many things pressing for his attention. He's always reluctant to interfere in the affairs of a state, and he doesn't know much about the labor question anyway—he never had to work with his hands, and it's hard for him to imagine life in a mountain-fortress of the 'G.F.C.' So he won't do anything until matters get worse. Isn't that about it, Mr. Simmons?"

The other indicated by silence that that was about it.

"All right then. I just want to get the situation clear. You are a representative of the government, and you've come out here and seen what is happening to us. You saw Major Cassels take the witness-stand at the outset and make the announcement that he would let the strikers air their grievances, and then riddle their case. But now we're done, and he has no evidence to present at all! General Wrightman refuses even to take the stand—unless you'll agree that our attorneys shall not cross-question him! He refuses to produce any of his 'military prisoners', and your committee hasn't had the nerve to insist. Now you're going away again, and you have no hope to hold out to us—nothing to tell us, except that we must continue to suffer, until we make so much trouble that the President will be compelled to think about us."

"Now!" protested the congressman. "I didn't say quite that."

"No, you wouldn't put it in plain language; but isn't that what it comes to? You are going away, and certainly you must know that hell is going to break loose when you're out of the district. What is there left for us but to get ourselves arms and fight the militia of our state?"

"Young man! Don't talk like that!" exclaimed the congressman.

"I talk like that, Mr. Simmons, and it's not just talk. I have sources of information—people who come and tell me what Wrightman and Cassels are planning. I know it is their intention to break up these tent-colonies; also I know that the miners don't intend to stand it, and for my part, I don't intend to advise them to stand it. So we're going to get ready, and when it comes, it'll be civil war in this coal-country—nothing less than that!"

The farmer-legislator laid his knobby brown hand on Hal's shoulder. He liked this clear-eyed, desperate youngster, who for weeks had been drumming up witnesses and throwing them at his legislative head. "My boy," said he, "take my advice, and don't do anything foolish!" And that was the end of the interview—that was what American government had to say on the subject of the class-war!

BOOK FOUR

CIVIL WAR

[1]

Springtime came late in these high mountain regions; it was the hope deferred which "maketh the heart sick". The sun would shine, and pent-up women and children would come out of the tents; but then would come more rain and cold, and they would go back into their crowded quarters. There seemed an endless number of tiny snowstorms hidden back over the mountains, and the roads were red and sticky swamps. One night there came a furious gale, which blew down seventy of the tents. The shivering strikers found what consolation they could in the fact that the elements were impartial, and took the militia-tents as well.

Through all these troubles Hal Warner stuck by the Horton people, sharing their hardships, their hopes and fears. If you read the newspapers, what he was doing seemed the abyss of futility, for the strike was pictured as lost; strike-breakers were pouring into the district, and the mines were working at full capacity. But the old heads among the miners did not let themselves be cowed by such pronouncements. They knew the character of these strike-breakers, they knew that coal-mining on a big scale could not be done by such riff-raff. So they bided their time, keeping themselves warm as best they could in the tents, and occupying themselves with music and card-games and reading, and with baseball and football when the sun shone. The children went to school, and the grown-ups went to English classes, and to a class in economics, where John Edstrom and Mrs. Olson taught them the meaning of their effort, and of the whole effort of labor throughout the world.

At first this class had met one evening a week, but now it met every other evening, so intense was the interest. It was a sort of open forum, where the strikers threshed out their problems, and sought to make intellectual profit out of their physical sufferings. Some times the wind beat upon the big school-tent so that it was hard to hear the speaker; men sat in their overcoats, and could see their breath in the cold. But still they stayed on, questioning and disputing—just as in the "social study club" at Harrigan! The foreign-speaking leaders—men such as Louie the Greek, Wresmak and Rovetta and Kowalewsky—would carry the ideas they gained to groups of their compatriots. Also the plan had been taken up in other tent-colonies; so the light spread, and seeing it, Hal could keep his courage and faith, even in the moments of bitterest wrong and suffering.

During such long evenings in the tent-colony Hal had a chance for heart-to-heart talks with many of the people. He heard their stories, and came to know their inmost thoughts. There was John Harmon, for example—a truly great man, whom it was a heartening thing to have encountered. Harmon did not know he was a great man, he was really without thought of himself; he displayed an annoying inability to tell his own story, to realize that there was anything picturesque or romantic about his struggle for an education and his life-long efforts for the welfare of his people. If you wished to draw him out, you must get him to talking about these people, the wrongs he had seen them suffer, the dreams he knew they were dreaming.

"John," said Hal, one time, "do you know whom you make me think of?"

"No," said Harmon. "Who?"

"George Washington."

The other looked at him, thinking that he must be joking; then he began to laugh. He laughed a long time, in his slow, quiet way. It seemed to him immensely funny.

"I've read his letters," explained Hal. "I've an idea that he was about your sort of a man. He set free a nation, and you are trying to set free an industry. Of course it will depend upon whether you succeed—what history has to say about you."

"Well," answered the other, "it never occurred to me that history would be interested in me. But you can set this down for certain, Hal—we are going to succeed. Maybe not this time, maybe not in our generation; but sooner or later, labor will be free. That's what I'd like to say to our masters, if I could ever get them to listen; there's only one thing they have to say about the matter—shall it be peaceably, or in a revolution, with a lot of bloodshed."

"And which way do you think it will be?"

"I don't know. If I could have my way, there'd never be a law broken; but as you see, I'm not having my way. Peter Harrigan is having his—and teaching our people to hate the law."

In his most desperate moods, Hal got the habit of going to John Harmon for comfort; for no outrage, no matter how terrible, ever was able to shake his steadfast faith. "I know how you feel," he would say. "This is your first strike, it seems to you that to lose it will be more than you could bear. But take it from me—I've lived through many strikes, and I've never got all I asked for, nor even all I hoped for, but all the time I've seen the movement growing. We are educating our people and disciplining them, we are training leaders for them, we are working out our tactics and studying our campaign. We are slow, but we move like a glacier, there's nothing can bar our path forever. Before we get through, we're going to give the world a new democracy—and one that is good every day in the year, and not merely on election-day!"

[2]

And then there was Mother Mary. When this old lady first came to Horton, the General sent a squadron of cavalry to put her on a train. When she came back, he arrested her and put her in a hospital. It sounded better to say she was a patient in a hospital, rather than a prisoner in a jail, though the difference was not very great, since she was kept in one room, and was not allowed to have a newspaper or a book, or to be spoken to either by the nurses, or by the five soldiers who were always on guard. When she fell ill, and the nurses brought in a physician, he was ejected by the troopers; and so Wrightman kept her for five weeks—until he learned that her attorneys were about to make application to the Supreme Court for a writ of habeas corpus. Then he put her on the train again and shipped her out. But she turned round and came a third time, so he had her taken off the train at Sheridan and put in a damp cell in the basement of the court-house—the same cell which had caused the death of a Greek prisoner from rheumatism of the heart. This old woman of eighty-two years he kept in this place for twenty-four days. And when he learned that the Supreme Court was about to hand down a decision in her favor, he turned her loose, and after the court had adjourned, he re-arrested her—thus playing, as it were, a game of tag with the most august tribunal of the state!

But in the end Mother Mary's persistence won out; she was set free and came to Horton, where she was welcomed with a procession, and with cheers and singing. During the week or two that she was resting in the colony, Hal spent many an entertaining hour listening to her pungent wit. For the old lady differed from John Harmon in this respect, she knew the part she played, and had a sense of its picturesqueness. She would tell endless stories about her adventures: about strikes she had led and speeches she had made; about interviews with presidents and governors and captains of industry; about jails and convict camps, and a "bull-pen" full of small-pox patients. All over the country she had roamed, and wherever she went the flame of protest had leaped up in the hearts of men; her story was a veritable Odyssey of revolt.

Adelaide Wyatt had sent to Hal a copy of "Simple Simon", the weekly scandal-sheet which was helping in the war on the strikers. It contained an article about the old "agitator", which did not trouble to employ the usual method of innuendo, but set forth explicitly how "Mother Mary" had been the "madame" of a number of houses of prostitution, giving the street-addresses of the houses and the names of different business partners. "I wonder if any of it's true," wrote Adelaide.

Hal asked the old lady, who told him the story of her early days. She had supported herself by sewing corsets, and had been this far depraved, that she was willing to work for anyone who would pay her wages. In this way she had come to know a poor girl of the town, who died of consumption, and begged with her dying breath to be buried by the church. Mother Mary applied to one priest after another, and all refused, and for this she denounced them in a letter to the newspapers.

"That was my first appearance in public," she said; "and, of course, it got me placed. Afterwards, when I began to speak up for the workers, the Pinkertons went to work, and they've made me the 'madame' of every place where I ever did sewing!"

The old lady took a great fancy to Mary Burke, who of course regarded her as a kind of divinity, and hung upon her every word. They sat one night in the headquarters tent, the two women with John Harmon and Hal, talking of the labor movement, its hope of freedom, its vision of brotherhood.

"Mary," said the older woman, "you're a fine girl, and there's many a fight coming where you'll be needed. So keep your head up—don't let them get you with pretty clothes and an easy life, the things they use to break down our girls. I'm getting old, Mary—there'll have to be some of you to take my place."

She laid her hand on the younger one's arm, and Hal, watching them from the corner where he sat, saw the girl gazing in front of her with a rapt look on her face. "I'll try, Mother, I'll try," she said; and Hal realized what this little scene meant to her—it was like Samuel anointing King David, Elisha laying his mantle upon the shoulders of the new prophet.

"Keep your heart clean," said the old woman; "like John Harmon, here. You see, he can go about, not caring what lies they tell about him, what plots they frame up on him—because he knows he's never done wrong to man or woman in his life."

"Now, now, mother!" said Harmon, breaking into the conversation. "You're putting it on thick!"

But Mother Mary would not be diverted. "You can see it in his face," she said. "He looks it, doesn't he?"

And Harmon glanced at Hal, with a twinkle in his eyes. "He looks like a statue of George Washington!"

[3]

Jerry Minetti came out of jail. The authorities were proceeding to try Dinardo for the murder of Pete Hanun, but they had given up their idea of proving a "conspiracy", and let the last of the supposed-to-be conspirators go. Jerry came to Horton unannounced, driving from the station in a hack, and Hal saw him coming up the street of the tent-colony, walking very slowly, as if dazed; Hal stared, recognizing him, yet unable to believe his eyes. The Milanese had the horrible pallor of prison, grey and ghastly;

he was thin, and so weak that he walked like an old man. There was left no trace of the verve and energy which had made Jerry notable among miners; the spring was gone from his step, the fire from his eyes.

Hal was so deeply moved that he caught the poor fellow in his arms; and then came others, including Rosa and the children, crying out in a mixture of delight and fear. They led Jerry to his tent, and the neighbors crowded inside, while he sat upon the bed, his wife kneeling at his feet, pressing his hands, sobbing, asking questions. Jerry answered feebly, and not always to the point; before long they realized that he did not hear well–he had got a blow over one ear which had caused what the doctor called an abscess. The doctor did not seem to know just what to do about it, and Jerry suffered agonies; he had horrible dreams–he would begin screaming and struggling in his sleep, and would have to be waked up and soothed like a baby half a dozen times in the night.

And here in the tent stood Little Jerry, staring at this wreck of a man who was his father. Hal saw the terror on the face of the Dago mine-urchin; he went and put his arms about him, and the little fellow burst into weeping on Hal's shoulder.

Such things as this put a strain upon the self-control of an amateur sociologist; they made it difficult to find contentment in observing and classifying the facts of the labor problem. Hal Warner lost the last trace of his light-heartedness, the spirit of adventure in which, at the beginning, he had gone up to North Valley. This grim struggle had got its roots into the very deeps of his being. It was the spectacle which has driven good men mad through the ages–"truth forever on the scaffold, wrong forever on the throne". It was the strangling of justice, the exalting of greed and violence–and he set his teeth together and flung himself into the war against it. Day and night he brooded over the situation, he made new efforts, he launched new campaigns–and in the process he became, without realizing it, a grown man. There were lines of care in his face, a grim, desperate look about the corners of his mouth.

And of course all this drew him nearer to Mary Burke. Mary was going through the same experience, only still more intensely, because her occupation kept her in close contact with physical suffering. When someone's head was broken, it was Mary's task to wash away the blood and stand by while the doctor sewed up the cut. If a man had an eye mashed by a spent bullet, or a woman lost an ear from a cavalryman's sabre, Mary Burke saw the raw, quivering flesh, and heard the moans of suffering. She held little orphaned children in her arms, stilling their sobbings and rocking them to sleep. She nursed and fed the men and women who came out from General Wrightman's dungeons of torture.

So Mary was one of those of whom the poet sings—"whose youth in the fires of anguish hath died". She was thin; the roses seemed gone forever from her cheeks; her lips were set and her look that of battle. How she hated the militiamen! Oh, how she hated them! Sometimes she would curse them; but there seemed nothing of the mining-camp about this cursing, it was not shocking to hear such language from beautiful feminine lips. It was a pious cursing, a religious rite!

"Someone ought to kill them!" she declared. "To begin at the top and shoot them down, one by one!"

John Harmon happened to be in the tent when she expressed this sentiment, and Hal was interested in the argument between them. "Tut, tut, Mary!" said Harmon. "Don't let yourself talk like that. Someone might say you meant it."

"I do mean it!" cried the girl, and went on to tell what was in her mind. Groping about for a way of deliverance, she had come upon the thought of a secret revolutionary tribunal to pass sentence of death upon the tormentors of labor! Let it once be known that men who committed such crimes would be killed, and one would see the end of them.

"Has it never occurred to you," said Harmon, "that the other side might take up the same idea?"

"They've taken it up now!" cried Mary. "Aren't they killing people?"

"Some, Mary, but not so many as they'd like to. They'd like to kill you and me, and Joe Smith here, but you see they haven't dared."

"We'd win in the end," she exlaimed. "They've more to lose than we have, they're more afraid to die."

"I'm not so sure of that," answered the other. "You can be sure that if we put them in the right, they'd be just as determined as anybody. Don't forget that, Mary—if we keep the right on our side, help will come in the end."

"I can't stand it!" she cried.

"I know how you feel, Mary. I had it out with myself when I was your age. It's the great temptation we have to face. If you give way to it, you're no good all the rest of your life—no good to the labor movement, I mean. Don't you see that?"

"No," she answered. "I don't."

"We're working for organization, Mary; and that has to be done in the open. We have to say exactly what we're doing, and why. If we resort to conspiracy, we are lost, because the enemy will have spies among us, and we can't know whom to trust. So our organization goes to pieces, our work comes to nothing."

"That's all very nice for talk," argued Mary. "But are we going to tell these militiamen they can do whatever they please to us? How do we get the world outside even to hear about us—except by raising a row?"

"That's true, Mary, that's true," answered the other. His face was grave, for it was a grave problem. But Hal made them both laugh by telling of something Tom Olson had said to him. "No, Joe, no violence! No violence—until you've got enough of it!"

[4]

The prophesy which Hal had made to Congressman Simmons had been fulfilled. The militiamen resumed their old tactics, and with especial vindictiveness towards those who had dared to testify against them. It seemed as if they must have a list—and every day someone on this list encountered a painful fate. There had come a number of witnesses from the Harvey's Run colony; and the very day the congressmen took their departure, the troopers fell upon this settlement. The body of a colored man was found upon the railroad-track, having been run over by a train, and the militiamen pretended to believe that the strikers had murdered him. By way of punishing the crime, they tore down thirty-eight tents and scattered the contents about in the snow, giving the inhabitants forty-eight hours in which to get out of the district. One family happening to be away, they wrecked the home and left a six-year-old child crying in the midst of the ruins. Two infants died from exposure as a result of the raid.

Another of the strikers who testified before the Congressional Committee was John Edstrom, who had been made treasurer of the Horton colony, and was one of the pillars of the union. One morning as the old man was coming up from the village, in company with nine other strikers, the party was stopped by soldiers and marched up to the North Valley mine, beaten all the way. Upon their arrival a party of drunken militiamen and mine-guards proceeded to amuse themselves by standing them up against a stone wall and going through all the preliminaries of an execution, bringing out a cannon and training it upon them, compelling the terrified men to stand motionless while the cannon swung back and forth before their faces.

This amusement went on at intervals through the day; tiring of it at last, some of the troopers got rawhide whips and drove their victims out of the camp on a run, following them on horseback and lashing them back to the tent-colony. Old Edstrom being unable to run far, the soldiers fell upon him and beat him to insensibility; he crawled into the tent-colony that night on his hands and knees.

Hal happened to be in the hospital-tent when his friend was brought in; he went that same night to see his cousin, and there was one last argument, ending in a bitter quarrel. It was Hal's contention that by remaining in the militia, Appie was making himself responsible, lending the prestige of his name and social position to those who were committing outrages.

"There is Stangholz! You admit that he's a ruffian—you have said that the proper place for him would be a home for the criminal insane. But what do you do about it?"

"You talk as if I had the power to turn him out!" exclaimed the other.

"You have!" insisted Hal. "Write a letter to the Governor, offering your resignation and stating your reason. If you were to give that letter to the press, it would have an effect, wouldn't it?"

"Not very much."

"At least it would clear your conscience—it would be a duty done. The public has no suspicion of these conditions, no idea how the militia is being prostituted. And you are a lawyer, with a duty to society—you've no right to think only of your career."

"I resent your insinuation," replied the Captain, haughtily.

"Well," said Hal, "you admit the facts, but you refuse to do anything. There must be some reason."

"It's not for you to judge the reason."

Appie wished to stand upon his dignity; but Hal had the advantage of him, and pressed it, and presently the Captain was goaded into revealing that he had already gone to Cassels and Wrightman with a threat of resignation. They had tried to brow-beat him, but had been badly frightened; finally they had asked for time to consider the matter, and that day they had given him an answer—in the form of a notice that the militia was to be withdrawn from the field.

"But what does that mean?" cried Hal. "You know they won't leave these mines without protection!"

But Harding could not answer this. Wrightman and Cassels had not gone into their plans.

"It's evident enough what they're up to!" said Hal. "They're going to get you out of the way without a scandal—you and your fellow-officers who will no longer stand for their deviltry! You're offered the chance to go back to your law-practice, and you're going to pocket the bribe and hold your tongue!"

It was not many days before the meaning of this new move had become clear to everybody in the strike-country. One militia company after another was put on the train and shipped out—all save a part of their membership, the ex-mine-guards and other coal-company employes. These were put into a new organization, under the command of Major Curran, the saloon-keeper politician from Western City, with Lieutenant Stangholz as his right-hand man.

"Troop E" was the name of this organization, and the greatest secrecy was maintained concerning it; there were plenty of rumors, but definite facts were not to be obtained. General Wrightman himself made false statements about this troop, not merely at this time, but throughout the subsequent troubles. He refused to furnish the roster of the troop, even when the legislature of the state demanded it; until at last the legislature appointed a committee to wait upon the Adjutant-General of Militia at his office, with instructions to stay until he yielded the point. So at last the truth was exposed—of the hundred and thirty members of this troop, a hundred and twenty-two were coal-company employes, remaining upon coal-company pay-rolls at the same time that they were paid as members of the state militia!

The news about this new troop which filtered in from the coal-camps broke down even John Harmon's resolve for peace. There was a conference in the headquarters tent, and at the end of it Harmon and Hal Warner took the midnight train for Western City. Hal had the thrills of a real revolutionist now, for the enemy must have had a pretty good idea what his trip signified; two coal-company detectives went on the same train, and two others met the party when they arrived in the morning. Hal took a cab and drove to Perham's Emporium, and after buying some neckties and collars, he lost himself in the crowd at a "bargain-day" sale, and so managed to throw his pursuers off the track and escape by another door. After which he went to the nearest telephone-booth and got Adelaide Wyatt on the phone, instructing her to wire immediately for a varied assortment of musical instruments, to be expressed to her home from the East. There was to be a surprising amount of musical activity in the tent-colonies, it appeared; the shipment would include two "pianos"—which in the code adopted signified machine-guns.

[5]

Jessie Arthur had written that she must see Hal the next time he came to the city. Her father had forbidden her to meet him; a whole week of hysterics had failed to break down this command; so now for the first time in her life Jessie would commit an act of disobedience. Hal must have some other person telephone to her home, and she would drive to the park in her electric, and pick him up at a certain unfrequented place. More revolutionary thrills!

Jessie had sunk into the background of Hal's consciousness; but when he met her, the spell was renewed, as poignant and intense as ever. The smile upon her lips, the light in her eyes, the very odor of the perfume she used, the touch of her soft garments—all these were intoxicating to his senses, and threw his mind into confusion. What was this mad thing he had been doing—casting away the gracious things of life, and going down into a bottomless pit of sordidness?

He controlled himself—listening meantime to what Jessie was saying. She was so unhappy; she had waited for him to come to her, but the weeks dragged by, and the months. How much longer was this dreadful strike going to last?

"I would have to be a wise man to tell you that, Jessie."

"And you mean to stay on, no matter how long it drags out?"

"I couldn't possibly do anything else."

"But then—what am *I* to do, Hal?"

So he told her of the decision he had come to. "I don't think there's any possibility of our making each other happy. You will never approve what I am doing, never be interested in it. We should only be tearing each other to pieces, and we ought to realize it, before it's too late."

"*Hal*!" she exclaimed. Her voice was stricken with fear. She could not go on.

"That's it," he said, taking up her unspoken thought. "We ought to part, Jessie."

She drew up the electric by the side of the roadway; for it is not safe to run an automobile when one's eyes are blinded with tears, and one's hands trembling. "Surely, the strike can't last forever, Hal!"

"Not this one; but there will be others—there is a class-war, which will last longer than your life-time or mine."

"And you'll always be mixed up in it?"

"Always."

"You're never going to work, Hal? I mean—like other people?"

"You mean at making money? But my brother runs the business; he wouldn't let me have anything to do with it if I wanted to."

"But so many other things you might do, Hal—besides being a labor agitator!"

"I might study, Jessie, and write about these things; I might take to editing a paper, or even go into politics; but it wouldn't make a bit of difference—I should still seem dreadful to you and to all your world. If I didn't, I would know that I was on the wrong track—that I wasn't accomplishing anything."

He paused; realizing how perverse his last statement must sound to her, and being moved by her grief, he began once more trying to explain—the old, wearisome propaganda, to which she listened because he forced her to.

"But what can I *do*, Hal?"

"There's nothing you can do, dear—because you don't want to do it."

"But if I *wanted* to, what would it be?"

"The words are in the Bible—leave all and follow me. Break with your family and friends, everything you consider decent, and come with your mind made up to help me."

There was another long pause; at last Jessie spoke, in a whisper. "Hal!"

"Well?"

"Isn't there somebody else?"

"How do you mean?"

"I mean—some other woman?"

"No, Jessie—I'm not thinking about Mary Burke, if that's your idea. Can't you understand, I'm fighting for something dearer than life, and I'm under the shadow of destruction! I've seen such things in that tent-colony—there's no use telling you about them, there's no use telling what I fear, or why I've come to town at this moment. It would only drive you to distraction. I've thought it out, and I see that I can never be what you want me to be, I can never make you happy, I can never do anything but tear you to pieces. So we ought to have it over with, once for all."

She sat twisting her hands together in distress. "Oh, Hal, it's so dreadful!"

Now it is a habit of women to suffer, and break men down. Sometimes they do it by instinct; sometimes, having found that it can be done, they do it deliberately. Being young, but on the way to maturity, Jessie was now doing it a little of both ways. She might have succeeded, had there not come an interruption–an automobile passing by, driven by a girl who recognized them, and waved her hand.

Instantly Jessie flew into a panic. That was Estelle Edmonds, and she would tell people she had seen Jessie and Hal, and the news would come to Jessie's father! Oh, what miserable luck! Jessie seized the lever of the electric. She must catch Estelle and pledge her to secrecy!

"You'd better let me out first," said Hal, quickly.

"Oh, but I want to talk to you some more!"

"Yes, but meantime we'll meet other people. You can't drive me out there on the main avenue!" He started to get out.

"But Hal, I can't part from you like this. I can't! I can't!" And she wrung her hands in excitement and distress. "I've got to think it over!"

"All right," said Hal, "think it over, and write me your decision." And he stepped out of the limousine. "Hurry up, or you won't catch her."

Jessie gazed after the disappearing car, and then, in anguish, at her lover. "Oh, please wait for me here! I must see you some more! I've so much to say!"

An instinct told Hal that it was better to make his escape while he could. What was the use of suffering to no purpose? What was the use of talking on and on, and getting nowhere? This very episode was proof where Jessie's real interests lay–in the world of obedience and propriety. "No," he said, "others would surely see us. Go on, and tell Estelle, and write me what you have to say."

"But you'll wait to hear from me, Hal? You won't do anything irrevocable! Promise me that—promise me!"

To which he answered, "If anything irrevocable is done, it will be by General Wrightman and his soldiers."

So Jessie was reassured. She was not nearly so much afraid of General Wrightman and his soldiers as she was of a wild rose in a mining-camp! Hal stepped back, and she started the electric, disappearing down the driveway at a pace which promised trouble with the first traffic-policeman on the way.

[6]

From this interview Hal went to get the latest reports on the history of contemporary Christian martyrdom. Having been denied opportunity to convert the congregation wholesale, the Reverend Wilmerding had begun insidious efforts at private proselytizing. He had persuaded a number of ladies of St. George's to read his incendiary pamphlet, and had had secret conferences with them, attended by no one knew what dark proceedings; he had even gone so far as to persuade one of them to turn a church sewing-circle at her home into an opportunity for him to unsettle the minds and disturb the home-life of some twenty innocent and hitherto blameless females. These underhanded procedures having come to the ear of Dr. Penniman, there were further clashes between them—clashes so open and shocking that Wilmerding could no longer conceal them from Hal.

Poor Uncle Will! The very foundations of his soul-life were crumbling beneath him. He had loved Dr. Penniman as a child loves a father, he had reverenced him as a deputy of heavenly powers. And here, because his assistant presumed to differ from him on political and economic questions, Dr. Penniman was proceeding to suspect that assistant of the basest and most inconceivable motives! Motives, not merely of insubordination and presumption, but of common jealousy, of vain-glory, of greed for attention—things of which the Reverend Wilmerding was no more capable than he was capable of sitting on a broomstick and flying to Walpurgisnacht. But his rector had attributed these things to him, in tones of shrill rage; so that Wilmerding had stood with tears of shame and grief in his eyes.

Hal, who had gone deeper into these questions than his friend, endeavored to give him comfort. He must realize that Dr. Penniman could not take the class-war as a purely political and economic question; Dr. Penniman had definitely enlisted himself and his church on one side—his social reputation, his intellectual prestige, his very moral sanctions. He had built his church on the prevailing system—made it a place of privilege, a school of comfort to the rich; he had had rich men appointed to the positions of trust in it, so that they stood before the world as the church itself. And now in his old age it was to be changed overnight—and at the behest of a humble assistant whom Dr. Penniman had trained up!

Hal had given his friend books to read, expositions of a heresy which masqueraded as "Christian Socialism"; and Wilmerding had absorbed these, and had made the mistake of quoting them to Dr. Penniman—being so far unbalanced as not to realize how their very titles must terrify a respectable rector: "The Call of the Carpenter", "The Carpenter and the Rich Man", "Christianity and the Social Crisis"!

What a subtle and cunning fiend was Satan! When he wished to rend and destroy a stately religious institution, to terrify and scatter a fashionable congregation—did he burst through the floor of the edifice with a glare of flame and an odor of brimstone and sulphur? Not he! He took upon himself the form of a serpent of cunning and plausible new thought, and crept thus into the minds of members of that fashionable congregation, setting them to seething with strange ideas—with motives, not merely of insubordination and presumption, but of common jealousy, of vainglory, of greed for attention! So in a short while, the stately religious institution was become as it were a nest of scorpions, stinging one another; a place of ugly hates, base suspicions, cowardly fears. So that on Sunday mornings the words of the Litany ascended to heaven like a wail of despair: "From envy, hatred and malice and all uncharitableness, we beseech thee to deliver us, good Lord!"

[7]

Next Hal went to the "Gazette" office, to see Keating and Pringle. The former had got word from Perry White of another conference of the secret organization of the coal-operators, in which a balance-sheet had been submitted, showing that efforts to operate coal-mines with "scab" labor were leading to bankruptcy. So a new fund had been subscribed. What was to be done with it was a secret not entrusted to the rank and file of the membership, but Perry White had given so definite an idea that Keating was returning to the "field" again, to be ready for the next move of "Troop E". He was to take the night train, and Hal would go with him.

Meantime Hal went away by himself, and the shadow of destruction looming over him drove him to a desperate course. He sought Lucy May and poured out his soul to her. He was going back to Horton, it might be to his death; so now, if ever, the lady from Philadelphia must summon her nerve and her aristocratic tradition, and *act*. "What do you want?" she asked; and Hal told her, and she stared at him aghast.

Then he saw her eyes go to the wall of her reception-room, where in two large gilt frames hung a stately lady in furbelows and a stern-looking gentleman in ruffles, each with yellowish-brown complexions cracked with age. These were ancestors, it appeared; of the names and doings of such the little lady's mind was a store-house. Hal would tease her about them, declaring that she used these portraits as ikons, or shrines; when someone failed to invite her to a dinner-party, or when a rival got more votes for president of the Tuesday Afternoon Club—then Lucy May would come to this holy place to remind herself that she was a daughter of colonial governors and of duchesses from over seas!

So Hal had jested; and Lucy May replied with words which made him stare at her—the little witch! No, he did not care anything about his ancestors, the men of two hundred years ago were dead and buried to him; but what about the men of two hundred years from now? Did he never make appeal to *them?* And Hal realized that this was exactly the custom he had adopted, in the stress of this cruel struggle. When the whole world flouted him and humiliated him—when General Wrightman sent him to jail, or when Tony Lacking threw him into the waste-paper basket—he would make his appeal to the future, and cheer himself with its imagined applause!

Pretty soon Brother Edward came home to dinner, and Lucy May announced that she and Hal were going that evening to hear Mrs. John Curtis expound her plan for a home for destitute cats and dogs. For God's sake, said Edward, what were they coming to now? But of course, when one got started fooling with crank ideas, there was no telling what would come next! No, thank you—Edward would not be roped into a rich woman's scheme for self-advertising! He would stay at home and read the latest adventure of Sherlock Holmes. So, at eight in the evening, Hal and his sister-in-law, in their gladdest rags, entered the latter's electric and set out upon a journey.

It was the same road by which Hal had taken Little Jerry to the New Year's party at the home of "Mr. Otter".

But this time they went farther yet—to a place where a spur came down from the foothills and spread itself into a lofty table-land, a mile wide and two or three miles long, overlooking the landscape for enormous distances. The road went round the foot of it, and all the way was a great fence of iron-railings, twelve feet high. If you were riding on the "rubber-neck wagon", which every day came out to behold these sights, you would hear the man with the big horn explaining how many thousands of tons of metal were in this fence, and how if the palings were set end on end, they would reach to Omaha, or to Tokio, or to the moon, or wherever it was.

At one place in the top of the ridge was a bend in the fence, where a road turned in and stopped in front of massive gates, with a brown-stone keeper's lodge and guard-house on either side, but so hidden with trees and bushes that you could get no glimpse of them, even by day. In front of these gates Lucy May's bold little limousine came to a halt, and rang its tinkly bell; and a man came out of a side-gate, and took the card which was held out to him. "Will you please to wait," he said, and without stopping to hear whether the lady would please or not, he went into the lodge to the telephone. A new calling-custom, introduced since the strike, it seemed. Uneasy lies the head that wears a crown!

As all believers in fairy-stories know, there are magic words which have power to cause the most massive gates to swing back upon their hinges. It is the modern custom to have these magic words engraved upon little pieces of card-board, which the fortunate possessor has only to present. Silently and majestically the gates gave way, and the bold little electric rang its tinkly bell again, and glided up a long winding avenue lined with pine-trees, coming to a great yellow building trimmed with white, and stopping under the shelter of a porte-cochere with columns big as the pillars of the temples of Karnak.

A man-servant opened the door of the electric, and took it in charge; another man-servant opened the door of the house, and a third stood by to remove the wraps of the guests. Before them went up a broad stair-case, with a carpet of royal purple, woven all in one piece; on one side was the mantel of white marble, six feet high, from a palace in Ferrara, and on the other side were tapestries woven for the popes at Avignon. You might read the prices of these art-treasures ever so often in the society columns of the Western City "Herald"; you might hear malicious doubts of their genuineness in any Western City drawing-room.

"Mr. Warner will wait for me," said Lucy May, to the servant; and so Hal was free to wander about and improve his artistic sensibilities, while the little lady followed the lackey down a broad corridor, and through a side door, into a passage comparatively low, almost a tunnel. There was another door ahead, and as they approached it the knees of the little lady were trembling, the hands of the little lady were clenched, and the little white teeth were biting the little red lips so hard that the blood almost came. Deep within the little lady soul were voices crying out: "You are not afraid! You are not afraid! You are a daughter of colonial governors and of duchesses from over seas! He is nothing but a pack-peddler—he is common! Common! Common as dirt!"

The servant opened the door, and the little lady went through. But it proved a false alarm, for the corridor continued to another door—it took several such barriers to protect an American man of business from the society-doings of the women-folk of his family! Again the little lady clenched her hands and bit her lip; again the voices spoke, and the servant opened and bowed, and Lucy May passed through, and into the presence of the Coal King.

[8]

It was a common-place looking room, its ceiling low, its walls covered with book-cases and filing-cabinets. At one end was a library-table with a shaded lamp, and a couple of worn leather arm-chairs. Between that and the door stood a chess-table, and before it sat two men. Neither of them moved when the door opened; until the servant's voice broke the silence: "Mrs. Warner, sir."

One of the men turned and got up—slowly, with what seemed reluctance. He was a heavy man, with a powerful frame, stoop-shouldered, sluggish; his head hung forward, so that when he turned it to look at you, you thought of some animal, swinging from side to side in a menagerie cage. By the eyes, and the folds of the skin, it might have been a hippopotamos; by the set of the jaw and the protruding lip, a bulldog. The old man's hair was grey, and straggly, because he ran his hands in it when he thought. He had on a brown smoking-jacket shiny at the elbows, a pair of baggy black trousers, and ragged green carpet-slippers which might have been a wedding-gift.

The heavy-lidded eyes fixed themselves upon the unexpected caller, a lady in a brocaded opera-cloak of blue silk, with a tiny diadem of a hat on top of soft brown hair; a lady petite, exquisite, with fine, sensitive features, now wearing a look of ineffable serenity, of proud assurance. You might have thought it Queen Titania, come to command some surly old Caliban-monster.

"Good evening, Mr. Harrigan," she said; and her voice was musical, as poets tell about the voices of mountain-streamlets.

"Good evening, ma'am," said the old Caliban-monster.

"Mr. Harrigan, I have to talk to you about something personal."

"Yes, ma'am?" said the monster. (One of the monstrous things about him was that all the ladies of his family together had not been able to break him of that plebian form of address.)

"I shall have to see you alone," said the little lady; and she looked at the other man, who sat at the chess-table, not having lifted his eyes. He was white-haired, thin and old, with the pale face of a student. Lucy May knew all about him—who in Western City did not know about Jacob Apfel, who had been a dealer in second-hand clothing, and had "staked" Old Peter in his early days, and now lived with the Coal King, collecting bugs and postage-stamps and snuff-boxes by day, and in the evening playing chess with his crony!

"Alone, Mr. Harrigan," said the little lady, in the voice of the duchesses from over seas.

And the head of Peter Harrigan swung round. "Get out, Jake!" So the old man rose from the chess-table, and without a sound or glance went through a door at the other side of the room.

"Now, ma'am?" said Old Peter.

"Sit down, please, Mr. Harrigan," said Lucy May. She realized perfectly well that he did not want to sit down, nor to have her sit down; but she was mistress of ceremonies wherever she went, and he, who hated snobbery, cursed it with all the force of his forceful being, was yet bound by it, helpless.

"Mr. Harrigan," said Lucy May, having put herself at ease, "I have come on a strange errand."

"Yes, ma'am?" said the uneasy old monster.

"I will begin by mentioning some things I have *not* come for. I don't want any money from you; I am not raising funds for anything."

"Yes, ma'am," said the other again. He did not smile, but there was a trace more of humanness in his voice.

"In the next place, I have nothing to do with the affairs of my husband. I gather that you are not on cordial terms with him. I wish you to know that I am not either. I am unhappy about his business—almost as unhappy as you are about yours."

Not by so much as a flicker of an eye-lid did Old Peter betray his attitude to that remark; but the little lady had that strange sense which enables ladies to feel their way—as bats fly in darkness, dodging a thousand obstructions. She went on, the smile going out of her face, and her voice becoming grave. "Circumstances, Mr. Harrigan, have willed it that your fate and mine are bound together. You don't know it, possibly, but I know it, and you will know it before long. You may not wish this, but it is a fact that I have to play a part in your life, whether for good or evil. So it seemed the sensible thing to come and talk matters out with you."

Again there was a silence.

"You, Mr. Harrigan, are a lonely and unhappy old man. Life has played a trick on you. You have worked hard, and made a success, but you haven't found happiness—you are embittered, desolate, shut up in yourself. You sit here now and peer out at me, and you're saying: 'What new kind of dodge is this? What's this one after?' You're saying that, aren't you, Mr. Harrigan?"

"Yes, ma'am," said Old Peter.

"You've got something that everybody wants—that's money; everybody is putting his wits to work, thinking up some way to get at you, some new scheme, some pretence, some 'spiel', as they call it. There is no end to their devices—charities, reform movements, causes, religions, arts, sciences; a new kind of healer to make you well, a new kind of prophet to save your soul, a new kind of power or dignity or learning to impress you—to separate you from your money! They gather like flies about a honey-pot—you have to shut yourself up in a cage, with servants and guards outside, and miles of iron fence all about you, to keep yourself safe. And even then they break through!"

"Yes, ma'am," said Old Peter.

[9]

There had come the trace of a smile to the eyes of the old Caliban-monster, and the little lady saw it, and renewed her courage. "You have an impulse towards humanness, Mr. Harrigan—isn't it so? But then the old habit reasserts itself. You say: 'Watch out now! This is a smooth one, this is the smartest yet—but all the more dangerous for that! She has her graft!' Now listen, Mr. Harrigan—get this clear in your mind. I don't want anything from you, there's nothing you could offer me that I would take. I have money of my own, and as you know, I have my social position—I don't have to come to the Harrigans."

The little lady was sitting very erect, looking into the eyes of Old Peter with a haughtiness beyond the power of words to convey. "You understand me?"

And she waited for an answer. "Yes, ma'am," said Old Peter, obediently.

"All right. And then let me mention one other thing about myself—that I'm a woman who can keep her own counsel. I shall not be overpowered by the fact that I have had a private interview with the Coal King. I read somewhere about a Greek philosopher who laid down a receipt for practicing self-control: to walk ten miles on a hot, dusty road, to come to a spring of water, and fill your mouth, and then spit it out without swallowing any; and finally—here's the point—'Go thy way and tell no man'!"

For the first time there came a really human look upon the old monster's face. But it was only for a moment. "I know what you're after!" said he. "It's that brother-in-law of yours."

"You are mistaken, Mr. Harrigan. Even though I brought him with me—"

"*What?*" And Old Peter's hands clenched and his lower lip was thrust out farther. "You brought that fellow to my home?"

"You needn't be afraid, Mr. Harrigan–"

"Afraid? Who says afraid?" The old rhinoceros with the scaly hide and the deadly horns glared about him, snorting, scenting his enemy, ready to rush him, trample him to the earth, gore and mangle him!

"Morally afraid of him, Mr. Harrigan," explained Lucy May, gently.

Old Peter sat up. "I'll not go on with this, ma'am. I'll not discuss that young puppy!"

"I haven't come to discuss him, Mr. Harrigan. Upon my word."

"What *have* you come to discuss?"

"It's a little hard to make it clear. I might say to discuss my three children."

"What do you mean by that?"

"My children, and other children–"

"Strikers' children, I suppose!"

"No, Mr. Harrigan, the children of the rich. I have three, and I sit and watch them while they sleep, they're so innocent and lovely, and I think about the world outside–how they have to grow up and go out into it, and there's something there that poisons them, destroys them. I look at my own little ones, and ask, Is it going to happen to them–in spite of everything I can do? Am I going to find myself a heart-broken parent? Shall I have a son who gets drunk, and throws a brick through the window of my home when I refuse to honor his checks–"

Old Peter's face turned scarlet; and the woman stretched out her hands to him imploringly.

"Ah, Mr. Harrigan, we don't change the facts by refusing to mention them! You must know that I know these things–you have publicly cast off one of your sons, and you call the other a nincompoop. And your daughters–they haven't disgraced you, but you've no love for them, nor they for you; they spend your money in things you despise, in empty snobbery. This palace, and the art-junk in it, price-marked and advertised in the papers—"

"It's their mother!" cried the old man, with a gesture of fury. "She's a damned fool!"

"Was she a fool when you married her, Mr. Harrigan?"

"How do I know? What does a man know about the woman he marries?"

The little lady smiled, appreciating an epigram. But Old Peter did not smile; he was on a topic that stirred him to the deeps—fool women, and the fool notions they put into the heads of the young, and the idle, dissolute, addle-pated jackasses they made of them!

[10]

So Lucy May had got her probe into the soul of this lonely, embittered old man, and could ask him why he wanted so much money, which brought him no happiness; why he persisted in forcing his will upon tens of thousands of his fellow-creatures, filling them with terror and hatred, never by any chance with gratitude or affection.

It was the same with industry as with women and children, said Old Peter. You gave jobs and homes to working-men, but they were incapable of gratitude or understanding, they were self-willed, lazy, turbulent—a prey to self-seekers with glib tongues, to suggestions of jealousy, suspicion and greed. So there was nothing to do but drive them, to keep a strong hand upon them. And he, Peter Harrigan, would do it—they would find they could not cow him, by God—"I beg pardon, ma'am," said the old rhinoceros, stopping in the middle of his charge.

"That's all right," said Lucy May. "If we are really going to talk things out, you must be at liberty to swear. But can you honestly be sure, Mr. Harrigan, that there are no real grievances for these agitators to work upon? You see the difference that would make–if there were really something that ought to be changed, and you wouldn't change it, or even hear about it."

"I've no doubt, ma'am, there's plenty of things they'd like changed. They'd like to get double wages–they'd like to run the industry–"

"They say they don't get their weights, Mr. Harrigan."

"Well, they lie!"

"You're really sure of that?"

"What rot–to imagine that a concern the size of the General Fuel Company could be run on a system of cheating!"

"But mightn't the bosses be doing it, Mr. Harrigan?"

"I'll not deny that might happen now and then. We have seventy-three properties to run, and many departments in each. I can't get angels for all of them."

"No, not angels–just men, with men's weaknesses and vices. And you pit them one against another–compare their returns each month, making them compete?"

"Of course, ma'am; there's not a big concern in the country that's not run on that principle."

"But then–mayn't that drive some of the bosses to take advantage of the men?"

"If it does, ma'am, nobody wants to know it more than I do. If the men would bring me proof–"

"But can they get to you, Mr. Harrigan?"

"Of course they can! I'll see any employe of mine at any time."

"But it's a long way to Western City, Mr. Harrigan."

"Let them write me, then. I read all my letters."

"You have detectives among your men, to see if they're organizing, don't you?"

"I do, ma'am. I've been driven to it."

"But have you ever had detectives among the bosses, to find out if they are cheating the men?"

For a moment Old Peter was at a loss. Then a flush of anger came into his face. "I know what you're doing!" he cried. "You've been listening to that brother-in-law of yours! An idle young scape-grace, with more money than is good for him, mixing himself up with labor agitators—playing with dynamite! Yes, ma'am, that's what it comes to! Going into my properties to stir up strife—"

"Oh, but that's not what he went for at all, Mr. Harrigan!"

"Maybe that's not what he tells *you*—"

"But I know all about it! I talked with him before he went! He had no idea in the world but to understand the industry, the labor-problem especially. That's one of the things you must realize, Mr. Harrigan—so that you won't feel so much bitterness."

And Lucy May rushed on, passionately, swiftly, bearing down interruptions, trying to make Old Peter hear the story of Hal's experiences. Old Peter did not want to hear, of course; it was maddening to him, so that he got up and tore about, and swore frightfully, and without apologizing. But he did not put the little lady out; she had made that much impression on him—he had to defend himself, even while he shook his finger at her and cried that he cared nothing about her opinion, that she was a fool woman, who could not understand men's affairs, and had no business messing with them! It was a world of realities, the coal-industry, and pretty ladies with their fine feelings had better not know about it. The ignorance of those low-down foreigners, their turbulence, their criminal impulses, that had to be so sternly repressed—

"Why do you employ such people, Mr. Harrigan?"

"Because they're all I can get. Do you suppose I wouldn't hire white men, if they were to be had?"

"But if you made conditions better—if you paid better wages—"

"I pay the wages I can afford to pay. Do you think I made this industry out of my head? I have rivals, and I have to meet their prices. That's what I tell you about business—it ain't for you women—"

"Plenty of women have understood business, Mr. Harrigan; I think I understand all you've told me. I'm not here to blame you, but to see what we can do. You must know the present situation is impossible—"

"I know one thing that's impossible, ma'am—that's these loud-mouthed agitators, that go in and stir up trouble in the camps! Men with no sense of responsibility—"

"Oh, but I know that's not so, Mr. Harrigan! I went to the convention and met a number of them. I had a talk with John Harmon, and he's really a straight man—a man one could deal with—"

"Harmon is better than the run of them, I've no doubt; but he hunts with the pack—if he didn't they'd turn on him and eat him up. It's the principle of the thing I'm fighting—my right to control my own business that I've built up. There's a lot of things I'd do for my workers, if it would help; but it wouldn't, because of these agitators, who'd rather see the industry ruined than have the men take a favor from the company!"

So they were launched upon a mighty ocean, an ocean of controversy which the wisest economists and social philosophers find difficult to navigate. And here was one little lady, pleading for kindness and mercy—like a cockle-boat tossed about in a tempest of hatred and prejudice and fear!

[11]

They argued until Lucy May had convinced herself that she was not getting where she wanted; and then she said: "Mr. Harrigan, I don't want to hurt your feelings, or to make you angry; but on the other hand I don't want to let you have a contempt for my intelligence, so I must tell you that I understand you're not telling me the truth. You *know* there are evils in your industry—monstrous evils that have grown up like weeds, while you've been meeting your rivals' prices and keeping your credit good. It's an old story to me, because I've had to wrangle it out with my husband—"

And Old Peter burst out: "Why the hell don't you stay at home and reform your husband?"

"It happens to be you that have the strike, Mr. Harrigan—"

"So that's it? Well, maybe if you want one for your husband, we can accommodate you!"

"What does *that* mean, Mr. Harrigan?"

"Never mind!" growled the old monster—realizing that perhaps he had gone too far.

"In plain words, you mean that while you denounce labor agitators, you might be willing to use them to get an advantage over your rivals."

"They've done it to me, ma'am—not once, but many a time."

"There you are giving me the true answer, Mr. Harrigan. Every wrong you do—you have to do it first! So you go into politics and buy officials—"

"A bunch of skunk politicians, black-mailers that want to skin me alive—"

"One more device to get at your money, Mr. Harrigan! Enemies around you, everywhere you turn—hating you, plotting against you! And you shut up here, lonely, afraid—not even daring to indulge in the luxury of saying what you think!"

There came such a queer look upon the face of the old monster that Lucy May could not keep from laughing. "I know, Mr. Harrigan—it sounds strange to you! You think that by swearing and blustering, you can pass for a plain-spoken, blunt old martinet. And all the time you're hiding from the facts, cowering in a corner, like a frightened child in the dark."

"So that's the way I seem to you!"

"You see this dreadful situation heaping up—you're confronting it aghast. You know something has to be done, but you don't know what it is; and people are trying to force you, and before you'll be forced, you'll let yourself be torn limb from limb. That's your main trouble, Mr. Harrigan—you're so stubborn, a perfect mountain of obstinacy and self-will. You've set your teeth together and shut your eyes, you're driving yourself to destruction. Oh, I'm so sorry for you—"

"You can save your pity, ma'am!"

"Ah, don't make that cheap answer! When a person comes to offer you friendship—a person who really wants to help, who hasn't a selfish motive—"

"What do you want?" shouted Old Peter. He had started up, and stood glaring at the tormenting little witch-lady. "You ask me to give up to these agitators? To let them get in and ride my business to ruin? Is that what you ask?"

"What I ask, Mr. Harrigan, is for you to meet me fairly—to talk to me straight. And I ask you to hear my brother-in-law—what he's seen with his own eyes, what he knows about your working-people, about the bosses you put over them."

"I'll not hear him!" raged Old Peter. "You had no business bringing the young scoundrel to my house!"

"You're using words without meaning, Mr. Harrigan. You can't possibly think that Hal's a scoundrel."

"He's one of your fine phrase-makers. I know them—I've read their stuff, they're more dangerous than the anarchists. They're going to make the world over in a night, and we're to go to school to them—we old fellows that have spent our lives learning our job. No, ma'am, I've got my hands full, solving this problem—"

"But that's the trouble, Mr. Harrigan—you're *not* solving it!"

"Who says I'm not?"

"You've had half a year to solve it, and all you can do is to bring on a civil war."

"I'm going to break that strike!" stormed the old man. "I'm going to drive those agitators out of my camps! I'm going to have law and order in that coal-country! They've put their tent-colonies down on the roads to my properties, they're terrorizing honest workingmen that want to earn their livings, and there's a white-livered little son-of-a-fool up here in the State House that won't use the power of the law and move them—"

"Mr. Harrigan, you must know what that would mean—wholesale murder!"

"Murder? Who says murder? If the law had been enforced from the first day there'd have been no talk of murder."

"What's the use of talking that newspaper talk to me, Mr. Harrigan? The law can't be enforced, because you've broken its back; you turned over the government of that coal-country to a private detective agency, and if you don't know what they did, you're the only person concerned that's ignorant. Haven't you even looked at Mr. Wilmerding's report?"

[12]

The mention of the Reverend Wilmerding was like a spark to a powder-magazine. That ecclesiastical viper that Old Peter had nourished in his bosom! After all he had done for St. George's—the mission he had built for it, the ten thousand dollar organ he had donated! Old Peter clenched and unclenched his hands, his face turned from purple to white and back again, he said things truly shocking for a daughter of colonial governors to hear.

"But Mr. Harrigan," she persisted, "you didn't think all these things about Mr. Wilmerding until he opposed you."

"I didn't know them about him!"

"You don't know them now, Mr. Harrigan—you don't know a single thing, except that he's opposed you. You've given him no chance to defend himself—any more than you've given Hal Warner. It's just one more proof of what I tell you—that you've set your teeth and shut your eyes, and you're driving yourself to destruction. You're driving me, along with you—"

"*You?* What am I doing to you?"

"Mr. Harrigan, do you suppose I can see my brother-in-law risking his life—"

"Then get the damned young fool out of my coal-camps!"

"I tried to do that; my husband went there and tried, and failed. I have come to see the reason—the boy's moral sense is involved, he has seen things there that are not to be tolerated. If you and your associates won't see, won't care—why, somebody has to throw his life away to *make* you care—"

"I'll not care, even so!" shouted Old Peter.

"Mr. Harrigan," said the little lady, "you try your best to seem a wicked old man, but you aren't in the least—you're just pitiful and tragic. You're trying to run away from your conscience—and you're perplexed and bewildered, because wherever you run, your conscience is there waiting! You're fighting me as hard as you can—and all the time knowing that I'm right, that you'd like to win my regard—to talk out your heart—"

Old Peter sank back in his chair. "There's no use arguing with you, ma'am—you know more about me than I know about myself!"

"I know that every man likes to have a woman to tell his troubles to. And since I've met you, I'd so much rather be your friend than your enemy. And of course, I have to be one or the other. Things have got to a pass where I can't let Hal fight alone; if I can't move you, why then, I've got to go out in the world and use my tact and my social position to raise up an insurrection of women against you.

—Yes, I see you get ready for another fight! Let all the women in the state come on, you'll meet them! But you won't have to, Mr. Harrigan—because I'm going to help you."

Old Peter sank back again. "Yes, ma'am," said he. "What will you be pleased to do, ma'am? Grant the men's demands tonight?"

"I'm going to ask you to press a button and call a servant."

"I'll not see that boy, I tell you!"

"He's waited so long, Mr. Harrigan—"

"You can take him away, ma'am. I'm not keeping him."

"He can be so valuable to you, Mr. Harrigan! Just think of it—he has seen your industry from the other side, the side you're not allowed to see—"

"I'd not believe a word he told me!"

"All you'd have to do, Mr. Harrigan, is to meet him, and you'd realize that he's an absolutely truthful man. It would no more occur to you that he was lying than it would that *I* am lying."

"I'll not see him, I tell you!"

"Is it because you're afraid of the truth, Mr. Harrigan? Afraid of the hours you have to spend alone—"

Old Peter had got up from his chair and begun to pace the room. Suddenly he turned upon his tormentor. "I've had enough of this!" he exclaimed. "You've no business forcing yourself in here, meddling in things you don't understand."

"You mean, Mr. Harrigan, that I understand too well!"

"There's no use arguing any more! I'll not give way! I'll not see him!"

"Mr. Harrigan, I want you to hear how he was treated when he asked for a check-weighman—"

"He didn't ask for a check-weighman! He started a conspiracy in my camp."

"How do you know, Mr. Harrigan?"

"I know the whole story. I had a full report on it."

"In other words, you had your subordinates write out what they wanted you to believe. Now come and find out whether your subordinates told you the truth!"

[13]

They wrangled on: check-weighman and the check-weighman law, camp-marshals and those who assassinated them, superintendents who were mayors of coal-company towns and enforced their will and called it "law". Each time Lucy May would come back to her demand that Hal should be brought in to tell the hidden facts; and each time Old Peter would shout, No, no! He would not see the young puppy! He would have the young puppy put out of his house! He would not permit a woman to intrude into his home and browbeat him. It was an outrage—and the old Caliban-monster would do his best to blow fire and smoke. But Lucy May only smiled patiently, and told him that she was the voice of his conscience, of his better self, his true and real self. She had come for one purpose—to persuade him to hear her brother-in-law's story; she would stay there all night, if necessary, until she had accomplished that purpose. Mr. Harrigan must understand that a woman was an obstinate creature; that when she had got her head set—

"Look here, ma'am," said Old Peter, "I've stood a lot from you—more than I've ever stood from anybody in my life. I know you're a lady, and that your emotions have run away with you; but I've had enough now, and you've got to quit."

"I can't, Mr. Harrigan—things are in such an awful way! Just think, down in those tent-colonies—"

"I tell you to go, ma'am!"

"There's going to be more fighting, Mr. Harrigan—"

"I've got a lot on my mind these days, ma'am. I have to be at my office at seven every morning, so I keep regular hours—my doctor's orders—"

"But it would take such a little time, Mr. Harrigan! If you will let Hal come—"

"I give you fair notice, ma'am. I propose to go to bed."

"Only half an hour, Mr. Harrigan—"

"You'll have to excuse me. I'm going to undress."

"Let me call him myself. Please! *Please!*"

"You're in my private room, ma'am, and you refuse to get out. So—" And Old Peter finished the sentence by beginning to untie his neck-tie.

Lucy May took one glance; then she sprang to her feet. "Mr. Harrigan!" she cried, wildly.

"I'm sorry, ma'am, but I've told you it's my bed-time." And off came the long string tie.

"Don't forget yourself!" exclaimed the little lady.

"I'm sorry, ma'am, but it's the time I always do it." And he unfastened his low collar.

"You are no gentleman, sir!"

"I never pretended to be, ma'am. I'm just a rough old fellow—started life as a pack-peddler, you know." And he began to slip out of the worn brown smoking-jacket, exposing a shameless spread of white shirt.

Tears of rage came into the eyes of the little lady. She saw that she was going to be beaten—and oh, how she hated to be beaten—she, the daughter of colonial governors and of duchesses from over seas! "Mr. Harrigan, *stop!*" she cried.

But Old Peter had sat down in his chair, slipped off the ragged green carpet-slippers, and begun to tug at one sock. "Any time you want to go, ma'am—" said he.

Lucy May stood clenching and unclenching her hands. Such an indignity! Such a grotesque, an unthinkable ending to her desperate emprise! And while she stood agonizing, the sock came off, exposing a horrible old pink foot, veined with purple and covered with black hairs, suggestive of who shall say what other Caliban-monstrosities? The little lady gave a scream, and turned her back, and stood choking back a sob, while the old monster took off the other sock, and stood up to unbutton his shirt. "I'm truly sorry, ma'am," he said; "but there was no other way." And he added, with delicate consideration, "I hope you won't turn around again, ma'am."

The little lady did not turn round. She stalked to the door; but before she opened it, she stopped, and stood until she had controlled herself and found her voice. Then, in a tone of withering scorn she spoke: "I might have known what would happen. You are a low, vile, *common* man!" And with that she opened the door and went out, shutting it behind her with as much decision as is permitted to a lady of her social position.

[14]

Hal Warner returned to Horton, and learned that on the previous evening, while he had been inspecting the marble mantel from Ferrara and the tapestries of the popes at Avignon, the militiamen had been making another "search" of the tent-colony. A woman had come to the militia-camp, complaining that her husband, a strike-breaker, had been assaulted; Lieutenant Stangholz, guardian of order, had come over with a dozen of his men to seek the culprit, and had ordered out all the strikers from the tents. Most of them had been in bed, but no matter—dressed or undressed, they had to come. They were herded out into a field with blows and curses, and compelled to file one by one past the Lieutenant and the woman who had made the complaint; and all this for nothing—the man they sought had never been near the tent-colony! And meantime others of the militiamen were prowling about in the tents, insulting women and young girls, helping themselves to anything of value they could find. If they were caught breaking open a trunk they had always an answer—they were "searching for arms".

The militia were now making these "searches" at any time the fancy took them; they would carry off money, food, cigars, clothing. Many of the strikers had now got arms, and so they were more reluctant to have these searches made. So the tension became greater each time; the hatred which had come to exist seemed a thing un-human, belonging to a world of devils. You would see a group of militiamen come walking down the road, and where they went every activity would cease, every voice would fall silent, every eye would follow, glaring. You could imagine that the strikers' hair rose up, as on the back of a snarling dog.

Hal got another bit of news—old Patrick Burke had run away from North Valley, and was again in the colony. He had not found life as a strike-breaker the delightful thing it had been pictured; they had not been interested in keeping him drunk, but had insisted upon his getting out coal. He had developed a tooth-ache, and got permission to come down to Pedro—and so had made his escape. The strikers had welcomed him like the hundredth sheep which has gone astray; all save Mary—who had taken advantage of his humiliation to arrange that his three dollars a week should be paid to her! She would see that it went for food instead of for liquor, and there would be less chance of her father's being lured away again.

Hal talked with the old fellow and heard his description of North Valley, which was become a place of terror. Alec Stone, the pit-boss, had been turned by the harassments of the strike into a sort of demon. He was declaring that he would kill any man who tried to get away owing money to the company; and of course everyone owed money to the company. Also the guards told harrowing stories about the mishandling of runaways by the strikers; so that between two fears the men stayed on and worked for nothing. At Reminitsky's they were living like pigs in a pen; they would reach onto the tables and grab pieces of meat with their hands, and if you were not one of the first grabbers, you got no meat at all. Yes, Old Patrick had had enough of "scabbing". They might trust him from now on—and so would Joe Smith try to get him at least a part of his three dollars? That girl of his was getting to be so hard, there was no doing anything with her anymore!

[15]

Hal sat that evening talking with Old Patrick's girl in the door-way of the hospital-tent. Inside was John Edstrom, still ill from his beating; Kerzik, a Slavish fellow who had got a bayonet through his arm trying to defend his trunk from a "search"; Zanelli, a young Italian with a broken scalp, and Haggard, an English miner, dying of tuberculosis. Such were Mary's charges, and sitting by the tent-door in the moonlight, Hal could see that she was worn almost to exhaustion.

Yet she was happy that evening. Hal discovered the reason—because she could hold her head up, could look her fellow-strikers in the eye. Lieutenant Stangholz might insult her, the militiamen might call her any names they pleased—but at least nobody could say that her father was a "scab"!

Hal told her the news from Western City; about the "pianos", and about Mr. Wilmerding's martyrdom, and about the interview with the Coal King. Mary listened awe-stricken to this last narrative; for of course to her Peter Harrigan was a far-off, mythological monster, and the idea that "Joe Smith" had been in his home on the previous evening endowed him with superhuman qualities. So Hal got a painful sense of his own worldly greatness—painful to one who had sworn to root snobbery out of his soul, to have nothing to do with the glamor of wealth, but to be in all things a democrat!

Once more there was strife in his soul, a war of impulses. He was sorry for Mary, who looked so pitiful, sitting in the moonlight with her hands in her lap and her shoulders sunk, her grey eyes dark with fatigue. He had vividly in mind the picture of Jessie Arthur in her stately electric—that most royal of devices, which, if you watch closely, you may see evolving a costume and manner, even a cast of countenance for its users! Jessie had been pitiful too, but she had not failed to remain lovely; Hal realized that no matter how deeply she might suffer because of his neglect, she would never lose that loveliness—she would never fail to take herself through the processes necessary to preserve her charms.

These charms were real to him; the memory of them held him, even while his mind rebelled against them. They were costly to the point of cruelty; and their purpose was to hold his admiration, the admiration of men in general. He, who called himself a democrat, who told himself that he wished to root snobbery out of his soul—he nevertheless was in the snare! He admired and craved the woman who led an easy life, who took care of herself, who sleeked and smoothed herself like a well-fed house-cat! He recalled what Mary had said about Jessie, when she had come up to North Valley in the midst of grief and horror; she had seemed "like a smooth, sleek cat that has just eaten up a whole nestful of baby chicks, and has the blood of them all over her mouth!"

Hal had said that he was not thinking about Mary Burke; and that had been true at the time—but it was not true now. In the moonlight of this soft spring evening, he had the old impulse to put out his hand to Old Patrick's daughter, to touch her, to say words of comfort, of tenderness. And what was it that kept him from doing so? The smooth, sleek cat-women counted upon the power of snobbery, the glamor of their worldly greatness. Would it not be a worthy action to shake them out of their cynical self-assurance? A revolutionary event, indeed—if he, Hal Warner, the desired of ladies, were to give his affections to a woman who was willing to neglect her beauty and her charm to make herself haggard and thin in a struggle for other people's rights.

[16]

The hour grew late, and the moonlight waned. "Mary," said Hal, suddenly, "what have you been thinking about me?"

"About you, Joe? How?"

"I mean—about you and me."

There was a silence. When the girl spoke, her voice was low. "We weren't going to talk about that, Joe."

"I know, Mary; but I've been thinking about it—a lot. There's something I want to tell you. I had a talk with Jessie yesterday."

He saw her hands clench suddenly. It was the first time the name of Jessie had been spoken between them for perhaps a year.

"I told her, Mary, that I'd made up my mind that we could not make each other happy, and I thought we ought to part."

"*Joe!*" exclaimed the girl; her voice had sunk to a whisper.

"She asked me to wait and think it over. But I've thought, all I need to think. I'm free."

She was staring at him, her eyes wide with wonder. And the old impulse came again; his hand went out and took hers. And it was as if he had given her an electric shock. She drew it back sharply. "Joe! Ye must not do that!"

"Why not, Mary?" He leaned towards her, gazing into her eyes. "Why not?"

She was at a loss for words, because of the suddenness of his attack. "I've not let meself think about it, Joe!"

"You still care for me, Mary?"

She caught her breath, and he saw that a storm of emotion was sweeping her. He put out his hand again, touching her arm. But again she shrank away. "Don't do that way, Joe! I want to think!"

So he sat, and she sat gazing at him, as if her eyes would pierce into his soul. "Ye must not make any mistake, Joe. Ye have to be thinkin', too!"

"It's come to seem simple to me, Mary."

"Such things are never simple, Joe. It seemed that way to me in North Valley, but I found it was not. And since then, I've thought about a lot more difficulties."

"Tell me what's in your mind, Mary."

She hesitated, then began: "For a long time–more than a year–I've known ye'd never be happy with Jessie Arthur. And I've wanted ye so–it's been all I could do to hold meself in! But I said, 'No, I'll leave it to him! I'll do nothin', nothin'!' And I haven't, Joe. I can say that for sure."

"Yes, Mary," he said.

"And now, listen; there's somethin' else I got to say. A couple of years ago, when I come to know ye at North Valley, I was hopeless, desperate, ready to throw meself away. But now–now it's different; I've got somethin' to live for. To be sure, 'twas you gave it to me–but now it's mine. I'm worth somethin' to meself, and maybe to the world. So I can't throw meself away."

"Of course not, Mary!"

"Ye know what they say about us, Joe. But our knowin' it ain't true makes a difference, our friends knowin' it ain't true makes all the difference. If we–if we was to do that–we'd be puttin' ourselves in the hands of the enemy, we'd be killin' our work."

"Don't misunderstand me, Mary," he said. "I meant that we should go and get married." He said this quite firmly and decidedly. So one is swept along in a tide of revolutionary events—he wanted to take her in his arms, he wanted it desperately. But when he tried, she shrank away again. "Wait, Joe! Wait!" And she glanced about swiftly, to see if any of her patients were being disturbed by this revolutionary tumult. "Come out here!" she said, and went away from the tent a bit.

[17]

The girl stood there, looking very beautiful and pathetic in her fright, her desperate desire to keep her self-control. She waited for Hal to come close, but not too close; then, with her hands half extended in appeal, "Joe," she whispered, "I don't want to make any mistake! Help me, won't ye—let's talk it out fair!"

"I'll do what I can, Mary," he answered, gently.

It was silent all about them, for the colony had gone to rest, save for one baby that was crying somewhere in the distance. But they had to be circumspect, not knowing when some prowler might come along.

Mary waited, until she had forced her voice into calm. Then she began: "Joe, ye talk about marryin'. Have ye really thought what it would mean if ye were to marry me?"

"Would I have asked you if I hadn't, Mary?"

This was a rank evasion, of course; and it did not help him. "Ye might very easily, Joe. Ye might be havin' a sudden fit of wantin' me. Is that it?"

"I've been thinking about the matter for a long time, Mary."

"But ye don't realize, Joe! Ye can't know what ye're saying! To go and get married to your friend's parlor-maid!"

"Don't talk like that!" he exclaimed—for the words hurt.

"But it's the way everbody that knows ye would talk, Joe!"

"Well, damn them—that's all I can say."

"You'd have to be saying it, Joe—think how many years! And to how many people! Ye'd be burning your last bridge, Joe Smith. As it is, ye can go back whenever ye please and be Hal Warner; they'd forget all this craziness, they'd call it a new kind of wild oats. But if ye were to go and get yeself married to Mrs. Wyatt's parlor-maid, then ye'd be done for! There'd be no more dinner-parties, no more clubs! Go ask your people about it, Hal!"

He could not help laughing. "You're too humble, Mary. You've been reading romances about the aristocracy! They'd take you in after a bit—truly!"

"Maybe I'm not so humble as ye think, Joe. Maybe I'd not want to be 'taken in'. Maybe I'd rather stay what I am—a workin' girl."

"Well," said he, "if you can manage to forget that I was ever a gentleman, I'll agree never to ask you to be a lady."

"Don't joke about it, Joe!" she pleaded.

"It does no harm to meet one's troubles with a smile," he answered. "I assure you I'm serious enough. The fact that I've lived all my life without working, that I've travelled on the backs of you and your people, that all the culture, the power and prestige you've stood in awe of, have come out of your unrequited labor—that is something I might well be ashamed of; something you might fairly distrust me for, too!"

He paused.

"And as for your going in and learning to 'do' society—what I'd want you to do, Mary, would be to brand a sign on your forehead: 'I'm a miner's daughter!' Just as a soldier puts on a uniform—so as to give fair notice to the enemy! So, when you go and meet the grand ladies of my world, and they try to snub you, because they know the names of silks and jewels and wines, and you don't—

because they can quote French and eat peas with a fork, and you can't—well, you'll not need to give them quarter!"

"That's all very well, Joe," she answered; "but that's a man's talk, and I'm a woman. I lived a year with Mrs. Wyatt, and I watched those grand ladies, and I learned a lot. Maybe more than ye know yourself, Joe Smith!"

He began to laugh. "We'll have to compare notes some day. Did you run into the charming young matron who told me I was the one valid argument against monogamy?"

But Mary could not be got to smile. "I'm not talking about that sort, Joe; I mean the ones that are good, accordin' to their own ideas. Ye may laugh at them, but I know that a man who was brought up with them can't get away so easy. 'Tis like a poison; 'tis in the food ye eat, in the clothes ye wear, in the air ye breathe. Ye'll still have to go to their homes and sit at their tables. And there'll be an evil spell in every move of their hands, in every liftin' of their eyebrow!"

"I understand about the eyebrow," said Hal. He said it unsmilingly, as he saw she wished it. "You mean you couldn't trust me, Mary?"

"Yes, I mean that! I mean I'm afraid to love ye, Joe—that's the God's truth! For I could never set out to be like them. They have all the money in the world and all the time; they've nothing to do nor to think about but to make themselves beautiful! While I've got a job to do, Joe, and always will!" she paused, and he stood gazing into her burning eyes. It was really startling to him, for the girl was voicing the very thoughts that he himself had been wrestling with for many months! The problems of a "misalliance"! The application of revolutionary theory to the difficult field of matrimony!

[18]

He came a step nearer, and spoke gently. "You'll have to take me—you, 'Red Mary'—and make a real revolutionist out of me! You'll have to save me from the wiles of these dangerous leisure-class ladies."

"Ah, Joe!" she cried. "What a job to set a woman! To be watchin' a man—keepin' other women away from him!"

He began to laugh again—he could not help it. "You set a hard problem for me! What are you going to do—give me up to them?"

"Joe," she said, as gravely as ever, "ye remember just now, ye said, did I care for ye. Ye didn't say that ye cared for *me!*"

"You cut me off, Mary! I was going to say it—truly I was!"

"Maybe, Joe—but first ye wanted to take me in your arms. But I don't want it to be that way! I don't want ye to take me because ye're sorry for me, nor because ye happen to be near me. I don't want ye to take me because ye want to get away from Jessie Arthur, nor because ye're disgusted with your own world, and want to throw me in its face. I want ye to take me because ye know what I am, and ye really care for that."

"I see what you mean, Mary," he said. He was no longer laughing. "I can say it—I really can."

She was looking at him, still with the fear in her eyes. "What a woman wants, Joe, is for a man to love *her* as much as she loves *him*. And don't tell me that's true with *us*—because it ain't—it can't be. I had the start of ye, Joe, I loved ye when I first met ye; I loved ye so that I was ready to throw meself away."

He answered: "I'm trying to catch up with you now, Mary!" And he took another step, in this process of catching up. But again she shrank back, and put out her hands—not tempting him to come nearer, but imploring him to keep away.

"This is what I'm thinkin' about, Joe—what I been afraid of all along. If ye don't love me enough, then I couldn't make ye happy, no matter how hard I tried. 'Tis not that I'd want to be askin' pledges, 'tis not a thing a man can pledge. All I can do is to make sure ye really know me, and ye really want what I have to give."

"Mary," he answered, gently, "you make me feel very humble. If I didn't want you, I'd begin to. I don't know any way to measure love and compare it. I know quite definitely that I'm through with my world—and that I'm only at the beginning with you."

She drew a deep breath, and he thought she was going to yield to him. But the fear of him and of his world, of his prestige, his wealth, his power, had been burned deep into her soul. When he tried to take her hands, she drew them back. "Joe!" she whispered. "If ye really love me, I want ye to prove it!"

"How, Mary?"

"By waitin'! By goin' away and thinkin' it over!"

"How long, Mary?" His tone indicated that it was not thinking he was interested in.

"Just a while, Joe! We can't love each other now—with ruin hangin' over us as it is!"

"So much the more reason for taking our happiness while we can, Mary! How long do you want me to wait?"

"A few days, Joe!"

"That's rather indefinite."

"Three days, then! Take that long to think it over, Joe—before I let meself go, before I want ye so much that I can't help meself!"

And suddenly, before he could answer, she started past him. "Go away, Joe!" she whispered. "Please!" and sprang into the hospital-tent and out of his sight.

[19]

It was love in the midst of arms. On the morning after the passage between the two, before Hal had had time to get the dreams of Mary out of his mind, there came Lieutenant Stangholz to make another "search"—this time of the Greeks in the colony.

The militiamen were especially set against the men of this race, because many of them had seen fighting in the Balkan wars, and were not so easily subdued as Slavs and Hunkies. They were for the most part unmarried men, and lived by themselves in a separate group of the tents. Four of them had been in the party with John Edstrom which had been tortured up at North Valley, and after that a number of them had got guns, and were open in their threats. The enemy had spies among them, of course, and when the spies would report such threats, the militiamen would reply in kind. Half a dozen of the strikers had made affidavit to having heard Lieutenant Carroll, the crony of Stangholz, declare that he meant to kill Louie. These affidavits had been published in the miners' paper—with the result that the militia officer now threatened to kill the makers of the affidavits as well.

Hal was routed out of bed, and ran over to do what he could to avert the threatened trouble. Stangholz had brought about a dozen men, and it was as if they walked in a den of fierce beasts, who crouched and watched for a moment to spring. Louie the Greek endeavored to argue with the officer, and was thrown back with an oath; there came a snarl from a hundred throats, and it really looked as if unarmed men were about to hurl themselves upon naked bayonets.

Hal had evolved a method of dealing with Stangholz. The ex-mine-guard with the face of a bull-dog would curse the rich young man as furiously as he would any striker, but Hal knew that in his heart he was in awe of that "mystic spell" which had been able to throw open the door of General Wrightman's dungeon of torture. So, when the Lieutenant began to bully his victims, Hal would come and stand by in silence—as if he were a statue of public opinion. The officer would go on blustering, cursing; but he never went to the same extremes as when Hal was away.

Others hurried up, peace-makers, helping to control the Greeks: Mary Burke, Mrs. Olson, Kowalewsky, the Polish organizer, Jerry Minetti. Jerry had now had two or three weeks of freedom and fresh air and wholesome food, and was more than ever a power among the strikers. He pleaded and argued and cajoled—and meantime the militiamen were ripping up tent-platforms, slitting up mattresses, breaking open trunks. They did not find a single gun—not because there were none to find, but because the strikers had been warned by previous experiences, and had found safer hiding-places, and not taken too many into their confidence.

The soldiers went away at last, partly because of their non-success, and partly because of the menacing aspect of the crowd. They took money and other property, as usual; and that was the last time, men said. So many said it that Hal was convinced it was true; if Stangholz came again, the Greeks would have their guns ready in the tents, and there would be fighting at the very start.

It had come to be the general conviction of the Horton people that the militia were bent upon wiping out their colony, as had been done in the case of Harvey's Run. The soldiers had been making such threats for months, but now it was noticed that their words had become definite—they would set a time, as if they had heard rumors of the intentions of their officers. "You've only got a few more days now! We are going to fix you all right!" And they would point to the Black Hills, fifteen or twenty miles out on the plains, where the strikers had been herded in the struggle of ten years ago. "That's where you're bound for! The Black Hills for you!" Hearing such remarks, the strikers' women and children had got into the mental state of those sects of religious fanatics who anticipate the coming of the judgment-day, and sell their worldly goods and gather in the fields to pray.

[20]

Different people responded in different fashion to such terrors. Some gave way to them utterly; they would go about white-faced and abject, cowering at every sound. But others had spirit, and fought in their own souls, and if you were a psychologist, you would be interested in the signs of their inner state. Some would brag about what they were going to do, working themselves into belligerency as a defense against their own cowardice. Others went about their affairs unmoved; but you discovered that there were two types of these—some who were merely dull and unimaginative, and others who were highly strung, but determined not to let it show. There is a story of a young officer, going into battle with his teeth chattering, and saying, "Yes, I'm afraid; and if you were half as afraid as I am, you'd have run away long ago!" Hal suspected that Mary Burke belonged to this group, and he was sure that little Mrs. David did.

Old John Edstrom was lifted out of bed by the general excitement. He was feeble, suffering from shock, as the doctor said; he was covered with cruel bruises, and could hardly take any food. Nevertheless, he went about in the tents, talking to the people in his gentle, winning voice, and when you had listened a while you would realize that in this shrunken and twisted and beaten old body dwelt one of the bravest spirits you might ever meet in life. For

Edstrom had been through more than one of labor's agonies, and had sat late on many a lonely night, meditating them, working out his philosophy. It was a high and difficult philosophy; at first when you listened to it you might think the old man was in his dotage, or that he was talking empty words that he had got out of some book.

He would say to people—to rough and ignorant men, to women bent and worn with household drudgery, even to little children—that they were part of a mighty host which had been marching for ages, and would march for ages after they were dead—the host of labor, moving out of slavery into freedom and joy. So they must not think about themselves too much, they must not fear for themselves; whether they won or lost, whether they lived or died, was not the question, but only whether they helped to bring labor nearer to its goal. The old man would go farther yet—he insisted that they would help more if they would have nothing to do with guns, if they would hold themselves quite above the wickedness of their enemies; they must fight, of course, and fight hard—but with their minds, with their souls. And of course most of the people to whom he talked had no idea about such fighting as that; they did not know what he meant, and when he explained it, they could not see what good it would do. It sounded like religion, which most of them despised, because it was preached by hirelings of their oppressors.

But they knew the old Swede was a remarkable man, because he had read so much, he could tell them things about their own countries, which many of them, being ignorant peasants, had never heard before. He could tell the Greeks about episodes in their history, both ancient and modern, he could tell the Bulgars about their long struggles with the Turks, he could tell the Italians about Garibaldi, who had not been afraid to appeal to the utter-most heroism in his followers. He had not offered them higher wages and "prosperity"; he had said, "Men, I bring you wounds and death!"

—And then would come vehement arguments; for Garibaldi had been a fighter, and so had Leonidas, and all the other heroes; they had not told their followers to stay quiet and suffer whatever their enemies did! The old man would try to explain, that the world had moved on, that there might be other methods of gaining one's ends in a democracy. But you could not get Greek and Bulgarian and Italian peasants to see that. Democracy had no meaning to them, they had no votes and no share in government; to them the tyranny of the "G.F.C.", manifested through gunmen and militia, was exactly the same thing as the tyranny of Austrian and Turk.

[21]

The old Swede came into Hal's tent after supper. He had something to say; and Hal, seeing how feeble he was, made him lie down on the cot—on his face, because of the bruises on his back.

What he wanted was to persuade Hal to go away from Horton. He said frankly what he did not say to the others, that he had become convinced that the tent-colony was to be destroyed, and he thought that Hal's life was too valuable to be thrown away. It sounded like harsh doctrine, but it was a fact nevertheless, that ignorant and helpless working-people would have to be sacrificed by thousands in this class-war; but a man who could think, who had the rare gift of leadership—he ought to be saved. How many young men were there who had money and education, who were willing to come and help the workers in their struggles? Such a man was like a watcher on a hill-top, he could spy out the pathway ahead, and speak words of warning and guidance to the marching host of labor.

Hal had not thought much about the possibility of losing his own life in this struggle; he had been so wrought up about what was happening to other people that he had really forgotten himself. But Edstrom insisted that he could no longer count on the "mystic spell" to protect him; the militia were so enraged that they would kill him as quickly as the next man. And besides, if it came to a fight, the bullets would fly promiscuously, and would not be considerate of what they struck. Hal had been on the ground long enough to know everything, and now the thing for him to do was to go up to Western City, and use his power to stir the public conscience.

Mary Burke came in, to get her patient and put him to bed in the hospital-tent. Hearing Edstrom's argument, she backed it up with intense conviction. But Hal would not give in. It was indeed "harsh doctrine", that he was to consider himself too good to share the dangers of his friends! Could he say that he had more right to life than Mary herself? Could he be sure that in the long run he would be of more use to the working-class?

Outside, as they talked, there came to them the voices of some of the children, and suddenly Little Jerry opened the flap of the tent and peered in, to say good-night to his pal and play-fellow. And there was another argument! Who could look into the eager eyes of that child, and watch his quick mind unfolding, and say what possibilities of service to the working-class might be hidden in him?

Little Jerry entered. He had been to the school-tent, where Mrs. Olson had an hour of story-telling before bed-time—thus doing her part to keep up the morale of the community, to banish night-mares from little children's sleep. The Dago mine-urchin had been hearing a thrilling tale about a thing called a "geenee", which came out of a bottle, and spread all over the sky, and might be compelled to do a man's bidding, if only the man knew the spell. "Jesus," said Little Jerry, "how I wisht I had one o' them geenees!"

"What would you do with him?" inquired Hal; and the youngster's imagination was turned loose in Arabian

Nights' fashion. That "geenee" would come out of his bottle and proceed to the militia-camp; he'd go tramping on the tents, and scattering destruction; he'd smash them militiamen like they was cockroaches! You bet when they seen him on the ball-ground, they wouldn't come to bother no games!

"An' I'd send him to get that feller Jesus Christ," declared Little Jerry. "He'd lift him up by the back of his neck, and take him and do him like they done Mr. Edstrom!"—This was not some new and terrifying kind of blasphemy of Little Jerry's invention; it was merely the established manner of reference to Lieutenant Stangholz!

Hal laughed and told his little friend about a new kind of "geenee" which Mr. Edstrom proposed to raise up. It was a slow process, bringing into being this labor genius, but some day he would be big enough to take "Jesus Christ" and all the rest of the gunmen and soldiers—and not smash them or torture them—but put them to doing their share of honest work in coal-mines. And with this new and more modern form of fairy story, Hal broke up the argument, and strolled home with the Dago mine-urchin, and went into the tent to say how-dy-do to Rosa, and to take a peep at the little black-eyed Dago doll.

They could only whisper, for Big Jerry lay asleep on his cot, and sleep was not so easy for him, since his long torturing in prison. Little Jerry had learned to go about on tip-toe, for his father was apt to be cross; it was very distressing, for he had never been that way before. Like many another child, Little Jerry made the discovery that heroism and martyrdom, in their familiar domestic aspects, are not at all the romantic and picturesque things they are portrayed in the books!

[22]

The next day chanced to be Sunday, the Greek Easter. The Greeks were accustomed to make much of this holiday, celebrating it with old ritual and festivities, after the fashion of the German Christmas. They got out their national costumes, which every man had brought with him from the old country; they hoisted Greek and American flags, and in the bright sunshine of an April morning they turned out in the square in the midst of the tent-colony to dance their national dances. It was a grand festivity for everyone. The Greeks had a big new tent which they set up on the edge of the field, and they prepared a banquet to which all the strikers were welcome. There was music and sandwiches and coffee, and a ball-game in the afternoon, and dancing and singing in the evening.

But all that day the militiamen were prowling about, interfering with the fun. The ball-game took place in an open field which the strikers had rented and paid for; they were amusing themselves in a harmless way, and surely they had a right to be let alone. But the militiamen came, as they always did to these ball-games; they stood by the base-lines and jeered at the bad plays, and when no one paid attention to them, they began aiming their rifles, pretending to shoot the players. Finally one of them stuck out his bayonet and tripped a base-runner; and so the game had to be stopped.

Hal went off to appeal to Captain Harding, who happened to be at the Horton station, making arrangements for the train on which his company was to be carried to Western City next morning. Hal got him on the telephone, and was able to have the soldiers ordered away from the ball-ground. They went, jeering. "You play your game today; we'll play ours tomorrow!" That remark spread among the strikers, and at night, while the dance was on, there came a rumor that the militiamen were surrounding the tents, and were about to make an attack. The dance broke up in confusion, and the Greeks took turns mounting guard all night with guns in their hands.

In the midst of events such as this, needless to say Hal had had no time to make love to Mary Burke; he did not even have time to think about the matter as he had promised. He would see her in dreadful agitation—not for herself, but for all these helpless women and children. He would realize afresh the devilishness of what was going on here, and would go over to the Horton station, and with militia brutes standing by, glowering at him, would send telegrams to Wilmerding and Lucy May and Governor Barstow—he even sent one to Peter Harrigan, who read all his mail, and so, no doubt, his telegrams!

Early the next morning came Lieutenant Carroll with a squad of men. There was a strike-breaker being held in the colony against his will, the Lieutenant declared, and he had orders to fetch the man. The Lieutenant gave the name, a Greek name; and Louie was called, and said that the man was not in the colony, had not been for a long time. He was a cripple; why should anybody want to hold him against his will? The Lieutenant shook his fist in Louie's face and called him a liar; the man was there, and they would have him without delay.

The officer went to the telephone and called Major Curran, his superior. The Major spoke to Louie, and told him to come to the militia encampment to see him. The other Greeks gathered about their leader, urging him not to go; it was a trap, they were trying to get him out of the way before they attacked. Hal and Mrs. Olson were summoned, and they also advised Louie not to go—because they saw the excitement under which the Greeks were laboring, and knew that Louie was the only one who could restrain them. But Louie finally said that he would meet the Major at the railway-station, which was half-way to the militia-camp.

Hal walked out part of the way with him, discussing the situation. Even in the midst of the excitement and alarms, some of the Greeks had been able to remember that this was Easter Monday, and they were dancing in the square with music—a violin, a mandolin and a flute. The mournful wailing of the flute followed the two men as they walked, and Louie changed his step to it, noticing the music, even while he was talking about the saloon-keeper Major of Militia, and what he might be meaning to do to the tent-colony.

They stopped on a slight rise of ground and looked about them. They could see over to the militia-encampment, and remarked that the troopers were saddling their horses and fastening on their cartridge-belts. Suddenly came the call of a bugle, drowning out the music. It came from Water-tank Hill, an elevation which commanded the tent-colony; and Hal looked and saw that two machine-guns had been transported to this hill during the night. "Look at that!" he exclaimed. "Can they really be going to attack us?"

Louie hurried over to the railroad-station to see Major Curran; and meantime Hal stood watching. He saw groups of the militiamen moving down the arroyo which led to the steel bridge crossing the railroad-track—another important position if there was to be fighting. In fact, it seemed to him that all the surrounding country was alive with khaki-clad figures. He stood hesitating; should he go over to the railway-station and appeal to the Major and his cousin? Or should he return to the tent-colony and help keep the strikers in hand?

[23]

Five minutes must have passed after Louie had reached the depot, when suddenly Hal was startled to hear a loud explosion from somewhere back of the militia-camp. It was followed by a second explosion, then by a third. He did not know what to make of the sounds; it transpired afterwards that three bombs had been made by Lieutenant Stangholz and set off under his orders, for a signal to members of the new "Troop E", stationed up in the canyons at the coal-camps. The meaning of the signals was that there was trouble, and that the troopers were to come down to Horton. But to the strikers, ignorant people, a prey to terror, these signal-bombs were some form of artillery which the militia had brought up, and were using against them.

Instantly, it seemed, the tent-colony was swarming like a bee-hive; the Greeks poured out with guns in their hands, making for the railroad-cut and the steel bridge. Looking toward the depot, Hal saw Louie running, frantically waving a white handkerchief and shouting. Hal ran towards him. "Major Curran say all mistake!" he cried. "Don't want to fight! Want ever'body come back!"

"Why has he posted those machine-guns?"

"Don't know. He say ever'body stop, don't make no more trouble." But it was too late. At that moment came the whir of the machine-guns on Water-tank Hill, directed upon the colony, and the tents were riddled by a hail of bullets. The frenzied women and children rushed out into the streets, scattering in every direction.

Hal and Louie dashed towards the colony. The first person they met was Kowalewsky, to whom Louie repeated his breathless sentences. Major Curran had promised that there should be no violence. He must search for the missing man, but he would try to do it decently; the strikers might trust him. Louie rushed off, waving a white handkerchief, in pursuit of the strikers, with the Polish organizer following him, and bullets from the machine-gun kicking up spurts of dust about his feet.

Hal plunged into the tents, to get the women and children into the cellars which had been dug for their protection. The first tent he came to was that of the Burke family, where he found Mary gathering the children and getting them down through a hole in the floor. Old Patrick was at the bottom, catching them in his arms. They lifted Mrs. Jonotch down, no easy task, for she was a considerable bulk; then all her children were piled in, one after another, and Tommie Burke and Jennie, and Mrs. Ramirez, a Mexican woman, and her children.

So the cellar was filled, and still there were women and children screaming and crying. Mrs. Zamboni, the Slavish widow, was almost insane with fright, and Hal took her by the arm, and carrying or dragging all her children, they went down the street to the tent of the Minettis, where there was also a cellar. As they entered the tent, a bullet struck a china bowl and sent the fragments flying; the children set up terrified yells.

"Jump down! Jump down!" cried Hal to Rosa, and he helped her into the cellar, and handed her the little baby and the second child. "Where's Big Jerry?" he cried.

"He got gun; he go fight!" answered Little Jerry; and added, "I got gun too, I go fight!" He had a little air-gun which Hal had given him, and was starting out of the tent to follow his father, when Hal caught him by the arm. "I kill them militias!" he cried; but Hal handed him, kicking and screaming, to his mother.

Hal helped down Mrs. Zamboni and her swarm of children as they came. Then he rushed on to the next tent, where he found John Edstrom, weak and ill, but giving what help he could to women and children. Seeing that the trap-doors in some of the tent-platforms were no more than thirty inches square, Hal got an axe and knocked loose a few boards, so that those in the cellars might not be stifled.

While he was busied thus, Mary Burke came running up to him. "There's a crowd out there in the field!" she cried; and Hal, following her down the street, saw a great number of women and children huddled like a flock of frightened sheep.

"They must come in here," he said, "and get under ground."

"But there won't be room enough!" There had been more than twelve hundred people in the tent-colony.

Suddenly Hal thought of the well which was over by the railroad pump-house; twenty or thirty feet wide, and perhaps a hundred feet deep, with rickety stairs leading in a circle to the bottom, and several platforms on the way. He had considered this as a place of shelter in case of trouble. "Take the people there," he said to Mary; and seeing Mrs. Jack David, he called to her to help, and the two women ran out into the field.

[24]

Meantime Hal went back to his task of getting those in the tents out of reach of the flying bullets. He came upon an Italian family, concealed in a packing-case in back of their tent; he dragged them out, and persuaded them to lie in a drainage trench. In another tent he found a Lithuanian woman crouching behind a stove, her eyes staring wildly, her teeth chattering so that she could not speak. She had shut her baby up inside the oven! Fortunately, there was no fire in the stove, but the child would have suffocated in a few minutes. Hal carried the child and dragged the mother, putting them into one of the rifle-pits which the Greeks had dug behind their tents.

Louie and Kowalewsky joined Hal, having run the gauntlet of fire once more. Louie had taken an even greater risk, running out toward the militiamen and waving his white handkerchief; but it had availed nothing—they had shot at him, wounding him in the arm, and putting bullets through his coat. He shouted to Hal that somebody was firing from the tent-colony and this was drawing the fire upon women and children; he and Hal raced about, searching for these men, shouting to them to desist. It was some time before Hal realized the truth—that the sound which they thought was firing from the tents was the bursting of explosive bullets. This use of explosive bullets was vehemently denied by the militia, but it was a point on which the subsequent testimony of witnesses was overwhelming. One of the bullets hit a stove near Hal, and splinters of steel cut his clothing and hands.

Mary Burke came in, breathless and gasping, having run all the way back from the well; she showed Hal where the heel of her shoe had been carried away by a bullet. She told him that sixty or seventy women and children were crowded onto the rickety underground platforms, which trembled when anyone moved. There being no more people left in the tents, Hal and Mary went to the cellar where they had put the Burke family. There was just room for two more to squeeze in, and there they stayed, packed like sardines, for hours, while the firing went on.

The sun beat down upon the tent, and in the course of the afternoon the hole, which was only six feet square, became stifling hot; the children were whimpering and the babies screaming, and it was necessary to get water for them. So Hal and Mary clambered out again. Thinking that the militiamen would not fire upon the tents if they realized that only non-combatants were in them, they ran to the hospital-tent. Hal put a red-cross badge upon his sleeve, and Mary Burke put on the white costume of a nurse. They knew that the militia-officers could see these signs through their field-glasses, but they found that they had only made themselves targets. As Mary went about, taking water to the children, the bullets followed her so that people begged her to stay away.

They went back to the shelter; but late in the afternoon little Jennie Burke fainted, and it was necessary again to get water. Hal started to go, but Tommie, with the eagerness of a boy, climbed out ahead of him. He limped with his lame foot across the tent, and the next instant came an explosion, and he fell upon the floor.

There was a moment's stillness; then the women screamed, and Hal caught hold of the platform and lifted himself out, and saw the horrible thing that had happened. The boy was lying in a pool of blood, the whole back of his head blown away by an explosive bullet.

"Don't come up! Don't come up!" he cried to Mary; and he lifted the body and carried it to one side, so that the blood and brains should not trickle down upon the people. Then he went back into the hole, and told what had happened, and caught Mary in his arms as she swooned away.

[25]

It was late in the afternoon when the firing showed signs of dying away. Hearing the shouts of men in the street, Hal decided to investigate, and climbed out. He saw a sight which struck a chill to his heart. The tents were on fire! And in a moment he saw why they were on fire. A militiaman in uniform was coming down the street, carrying in one hand a pail full of liquid, and in the other hand a broom; he stopped at one of the tents, and dipping three or four times into the pail, splashed the stuff over the canvas. Then he put a match to it, and the tent went up in a roaring blaze.

"What are you doing?" Hal shouted. "There are women and children in those tents!"

The man turned and stared at him. "Who the hell are you?" he demanded; and stepping across the street, he began to sling his broom upon a second tent. When Hal sprang toward him, making to interfere, the man whirled and stuck the broom into his face. Hal recoiled, and the man struck another match. At the same moment a second militiaman ran out of the tent, carrying an armful of clothing.

Hal saw another man approaching the tent of the Burkes, also with a broom and pail. "There are people in there, down in the cellar!" he shouted.

"Well, get them out, and be quick!" said the man, with an oath. And he pulled back the flap of the tent. "Come up out of there!" he cried. "Be quick about it, if you know what's good for you."

"Have you got orders to burn these tents?" Hal demanded.

"Sure thing!" was the answer. And looking inside, the man saw old Patrick Burke. "What the hell you doing in there?"

"I'm trying to get out my children."

Hal turned, and looking down the street, saw Lieutenant Carroll and a group of men, some of them in uniform, some of them not. He knew this officer for a ruffian, but he could not believe that he would permit the burning up of women and children. He ran to him and made a frantic appeal. The Lieutenant's response was to leap into the tent and proceed to throw the people out as if they had been grain-sacks.

"Hold on!" cried Patrick Burke. "I've got a boy killed here!"

"God damn you, you old red-neck!" cried Carroll. "I know who you are—you've done as much shooting as anybody!"

"I never done no shooting!" protested the old man. "I never had a gun."

"I've got a notion to kill you right here!" replied the Lieutenant, with a string of unprintable oaths. Then he saw "Red Mary" climbing out of the hole, followed by her sister Jennie, who had got her picture in the paper when she was kicked in the breast by General Wrightman. Both these people had deserved cursing from the militia, and now they got it. Old Patrick got more cursing because he was slow and clumsy, his arms and hands being slippery with blood. Finally, however, he managed to get the body of his dead boy onto his shoulder, and staggered away through the smoke and flame, followed by the cowering women and children. Then the kerosene was slapped onto the tent, and it went up in a blaze like a dried Christmas tree.

Hal started again to protest, and Lieutenant Carroll whirled upon him. "What the hell have you got to do with this?" he shouted. Then, to one of his men, "Take this fellow out of here."

"Is Captain Harding about?" demanded Hal.

"I don't know whether he is or not. He's got nothing to do with me if he is."

"Look here, man!" shouted Hal, wildly. "Do you want to burn up women and children? Don't you know the cellars are full of people?"

"We're getting them out aren't we?"

"You're not doing anything of the sort! Your men aren't even looking inside!"

These words seemed to bring the Lieutenant to his senses. He turned to some of his men, who were carrying off armfuls of stuff: "Hey, you! Drop that loot, and get these red-necks out!" And he began to curse them, as furiously as he had cursed the Burke family. They were not soldiers, they were a bunch of pan-handlers and bums!

That was really the truth about the membership of this newly organized "Troop E"; it was not a militia-body, but a mob. Its enlisted members had never had a drill, nor even a roll-call—many of them had not yet got their uniforms.

They did not know their officers, and their officers did not know them; now they were turned loose, each man to follow his own impulses. Some of them were dragging out trunks and boxes, prying them open with bayonets, or smashing them with axes. You saw men going down the street, laughing and joking, carrying clothing, cigars and food; you saw others risking their lives to drag women and children out of the burning tents.

[26]

There was still time to save people, for there was no wind, and the flames were not spreading; each tent had to be separately kindled. No one paid any attention to Hal Warner, for the reason that so many others wore civilian clothing. Knowing where the cellars were, he ran from one to another shouting to the people to come out. In some cases the women were so dazed by terror that he had to spring down into the holes and lift them out bodily; then they would stand in the middle of the street, sobbing and moaning, confused by the glare of the flames and the yells of the raiders.

In one of the cellars Hal found the body of a woman on the ground. He did not know whether she had fainted or been suffocated; he lifted her out, and was climbing out himself, when he saw something which made him crouch back. A group of half a dozen militiamen were bringing in two prisoners: one of them Kowalewsky, the Polish organizer, and the other poor old sick John Edstrom. A moment later came others with a third prisoner, Louie the Greek.

Poor Louie had been beaten, but still he was not thinking about himself; he was pointing to the burning tents and shouting, "Women and children in there! Women and children in there!" They struck him, but still he would not be silent.

And then came Lieutenant Stangholz, rushing upon the scene. He caught Louie by the throat, as a terrier might catch a rat, and with a torrent of profanity, accused him of having taken part in the fighting. Louie answered that he did not have a gun, he had never had a gun in his life. He called the men about him to witness that they had not found any weapons on him; and then again he pointed towards the burning tents, crying that there were women and children there. He started towards the tents in his excitement; and Stangholz turned to one of the militiamen, grabbed the rifle from the man's hand, swung it and brought down the stock with a crash upon Louie's head. The Greek went down like a log, his cries about women and children stilled at last.

And Stangholz glanced at the weapon, which had been broken by the blow. "I've spoiled a damned good rifle on that red-neck," said he. The crowd of militiamen laughed.

Old John Edstrom was trying to argue. "Gentlemen, I have never had a gun. I was trying to stop the fighting!"

But the militiamen would not hear him. "Get a rope!" somebody shouted. And Hal, peering over the edge of the hole in the platform, saw a man come running up the street with a coil of rope over his arm. "String them up!" was the cry; and they cut the rope into lengths and made nooses, which they fastened about the necks of their three captives, lifting the half-conscious form of Louie for the purpose. "String them up!" They started towards the telephone pole by the headquarters tent.

Hal knew full well the danger of his own position; nevertheless he began to climb out of the hole. But before he had revealed himself, Lieutenant Stangholz changed his mind. He must have realized that it would be a dangerous matter for the militia to hang their prisoners. Hal heard his voice, dominating the clamor, ordering the men to take the ropes off; he gave other orders in a low tone, which Hal did not hear, and then he turned and walked away.

It was all over before Hal had time to realize what was meant. Stangholz had seen service among the Mexicans, and he knew their custom, called the "law of flight". The militiamen jerked the shuddering and moaning Greek to his feet, and thrust the other two prisoners forward. "Run, damn you, run!" they yelled; and Kowalewsky and Edstrom started down the street, while Louie staggered three or four paces. At the same instant a dozen rifles blazed, and all three of the men went down as if struck by lightning. Hal sank into the hole beneath the platform, overcome with horror. For several minutes afterwards he heard the crack of rifles—when the body of Kowalewsky was examined, there were fifty bullet-holes found in it!

[27]

When Hal looked out again it was night, and the tent-colony was a blazing inferno. The platform which covered him had caught fire, and he was choking with the smoke. Everywhere he looked, the looters were still at work; a group of them were dancing about the blazing ruins, shouting and singing, waving whiskey-bottles in the air. Nearby Hal saw a couple of others, emptying the contents of a trunk into the street.

Then suddenly Hal heard a familiar voice. "What are you doing, you fellows? Drop that, and go about your business!"

Hal clambered out of the hole, and confronted his cousin; and for a minute the two stood staring into each other's eyes. "Well, Appie!" said Hal, at last. "How do you like this?"

When the other spoke, there was no life in his voice. "I couldn't stop it," he said. "I've done everything in my power."

"They're not taking your orders to-night, hey?"

"They're not, indeed," the Captain answered. "As God is my witness, I had nothing to say."

"Do you know they've murdered Louie the Greek? And John Edstrom, and Kowalewsky, the Pole? Shot them down in cold blood!"

The other stared at him. "No, I didn't know it."

"And look at these tents! There were women and chilren in the cellars! There may be women and children in them still!" Hal thought suddenly of all the places into which he had helped to crowd people. He thought of Rosa Minetti and her three babies; of Mrs. Zamboni and her eight. "Come help me!" he cried. "We may save some yet!"

Captain Harding hesitated for a moment, then took Hal by the shoulder. "You mustn't stay here!" he said. "They'll do something to you if they see you."

"What do I care?" cried Hal. "There are women and children under those burning tents!"

But the other held him by force. "You must get out of here!"

"Damn you!" cried Hal. "You stood for this! You let it happen! And now you want me to be a coward too!"

"Old man," replied the other, "you're out of your head. You must go away from here."

And there in the light of the burning tents, the two came near to having a fight. The matter was ended by Captain Harding calling two of his troopers. They took Hal by the arms, and with the officer arguing and commanding, they forced him away from the tent-colony.

When they were out in the darkness, Appie was able to get him to listen to reason. If only he would promise to go away, Appie would return with his men, and do all he could to save the people in the cellars. The militiamen would obey his orders, provided they were not made furious by the sight of Hal. So finally the young man promised, and started away across the plain.

He found himself near the railroad, and suddenly remembered the well in which so many people had sought refuge. He turned towards it; but before he got near he heard bullets whistling past, and realized that he had come into the danger-zone once more. There was still enough light for his figure to be shot at, and so he began to run, and reached the well and called. The people were still inside, upon the rickety platforms and steps.

At the top was Mrs. David, who had taken the throng in hand. While she and Hal were discussing what to do, a freight-train approached, and stopped at the water-tank. The militiamen who were doing the firing were posted on a ridge on the other side of the track, and Hal realized in a flash that this train had cut them off. "Now's the time to get out!" he cried; and the others took up the cry, and the crowd began to pour out of the well.

A couple of hundred yards down the track was an arroyo. If once they could make that point, they were safe. But before they had got half way, the militiamen had found out what was happening, and several of them rushed towards the engine and leaped upon the running-board, leveling their rifles at the head of the driver. They forced him to stop watering his engine and to move his train out of the way; after which a fusillade of bullets poured after the fleeing crowd. Mrs. David, carrying one of her children and dragging the other, found afterwards that she had got a shot through the sleeve of her coat, and another through her skirt. Several of the children were shot.

But the crowd got to the arroyo, and crouched in the shadows, or made their way along, stumbling over rocks and irregularities in the darkness. Some of them traveled for miles in this way. One woman gave birth to a baby during the flight, and was out in the cold and rain unassisted for forty-eight hours. Others came to a ranch, where they were given food and shelter; and next day the militiamen visited this ranch, and completely gutted the house, leaving a note tacked upon the door explaining that this was what happened to those who gave encouragement to "red-necks"!

[28]

Late that evening Hal came into the town of Pedro. The streets were swarming with excited men and women; anyone who had news from Horton was instantly surrounded by a throng. One heard the most hideous tales that night, so that one did not know what to believe. Strikers had been shot, some of them beaten to death, their bodies thrown into the burning cellars; there were witnesses ready to swear that they had seen militiamen using their bayonets to drive people back into the flames. The estimates of the dead ran from twenty-five to a hundred—and there was no way to find out the truth.

Hal went to the union headquarters, where he met Billy Keating, almost beside himself with excitement, the perspiration standing out in beads on his forehead, making streaks down his full-moon face. He had been over to Horton, and had tried to get through the cordon which the militia had thrown about the colony; they had threatened him with arrest, and driven him away, and now he was through with the job of reporter. "It's a gun for me!" he exclaimed; and he asked Hal, "Have you had enough yet?"

"Yes," was the answer, "I've had enough." Hal had had it out with himself during his flight down the arroyo. He was done with talk, with fine sentiments!

That was the mood of nine people out of ten in Pedro, he found. Store-keepers, cab-drivers, bar-tenders—even doctors and lawyers—they were asking for guns. And Johann Hartman was ready for them; he invited Hal and Billy and a party of a dozen others to a blacksmith's shop on the outskirts of the town, and in a shed in back of the shop he disclosed some packing-cases. The covers were knocked off, and there, each neatly wrapped in excelsior and paper, were beautiful new shiny army-rifles. In other cases were full cartridge belts, all ready to be strapped on.

A fine and satisfactory thing was "preparedness", when you really came to need it!

That group marched forth, and other groups came. It was rainy and cold and dark, but no one delayed on that account; some of them set out to march on foot, others started in automobiles. And the same thing was happening in all the towns and villages of this coal-country—the roads were thronged with parties of men marching towards the scene of the "massacre". You were amazed when you heard the roll-call of these parties; they were not merely striking workingmen, but prosperous bankers and merchants of these company-ridden towns!

Hal and Billy and their party left the automobile on the outskirts of Horton, for they did not want to take any chance of being ambushed or captured by militiamen. But their entry into the town was unmolested, the enemy having thought discretion the better part of valor, and retired beyond the tent-colony. In the village one heard new stories of horrors; the members of "Troop E" seemed to have gone quite insane with fury—they had spent the night burning and smashing everything in the colony.

At the railroad-station, which had been turned into an emergency hospital, Hal met Mary Burke, still at her duties. She came to him, her face like a mask of grief. "Have ye heard the news, Joe?—who's dead?"

And she stood with her two hands clasped together, staring at him. "The Minetti children."

"Oh, my God!" he cried.

"All three of them! And Rosa—she may live, they say, but we can't find out for sure."

"What happened?"

"They were in one of the cellars. They were suffocated—some of them burned. And Mrs. Bojanic and her children, and Mrs. Alvarez and hers—all in one cellar."

"But those people weren't in the Minetti cellar!" exclaimed Hal.

"It wasn't the Minetti cellar, Joe. It was Tent Fifty-eight—the maternity-tent."

"But Rosa and her children weren't there!"

"They were, Joe. They were found in it."

"But I put them in their own tent—with Mrs. Ramirez and—"

"I don't know about that, Joe; but they were taken from under the maternity-tent." And Mary named three Italian women who had been near, and had brought the information. "There's no doubt about it. They were burned, but not too much to be recognized."

Hal turned away to hide his emotion. Little Jerry, his play-mate and pal, and Maria, who had been the baby when he had boarded with the Minettis at North Valley, and Gino, the new baby, the pretty little black-eyed Dago doll—burned, but not too much to be recognized! It seemed to Hal that he could still feel Little Jerry kicking and struggling in his arms. "I kill them militias!"

Long afterwards Hal solved the mystery of the Minetti family's presence in the maternity-tent. Rosa, who survived, a distracted and desolate wreck, told the story of that night of horror. Her own tent having been set fire to, she climbed out of the cellar with the children; but then, bewildered by the flames and smoke, the shooting and the yells of the militiamen, she sought another place to hide. She thought of the maternity-tent, which had a big roomy cellar, floored and timbered and fitted with a bed, upon which the babies of the colony had been born. She fled into this, and when the tent and platform above were fired, she and the other women and children covered their heads with blankets and hid from the smoke. They became unconscious, and when the place was opened up, it was found that eleven children were dead, and two women, both of them about to become mothers.

"Where's Big Jerry?" asked Hal.

"I don't know," said Mary. "Out fighting, I guess."

"Does he know yet?"

"I don't know that."

"And where are the bodies?"

Mary answered that the militia had them. They would not let anyone come near the colony. They had sent out for wagon-loads of quick-lime, it was reported; no doubt they wanted to destroy the bodies, so that no one would ever know how many had been killed.

Hal gripped his gun, in sudden fury, and turned to the rest of the party. "Come on, boys," he said. "It's coming on daylight, and we have work to do!"

[29]

In the darkness and rain they marched up into the hills, where the strikers had sought refuge. When they were challenged, they answered, "Friends," and the strikers crowded about to welcome them. Hal questioned them—who was with them and what arms they had, who was wounded, who dead. He told them the news from the tent-colony, and there were cries of rage. There could be no doubt about the mood of these men; they were ready for real war. And Hal was ready too. For seven months he had been watching and enduring—spending all his energies in enduring. It had been a heavier strain than he had realized; and now suddenly his energies were released—now he would have decision, action!

It was daylight, and he looked about. The strikers had dug themselves a regular line of entrenchments; but this did not appeal to his mood. "Boys," he cried, "we don't want to stay hiding out here in the rain! We want to punish them, don't we?"

They answered with a yell; yes, that was what they wanted!

"You fellows that have real guns and know how to shoot! Where are these Greeks and Montynegroes that were in the war?"

So they crowded around him. They would do whatever he said, they would follow him to the death. He should have led them long ago!

Hal picked out the men who could be relied upon, dividing them into two companies, to be commanded by Billy Keating and himself. Billy had borrowed a pair of opera-glasses, and from the top of the ridge they looked over the position and mapped out their plan of campaign. They would move forward in open order, finding shelter where they could, and advancing at regular intervals, spreading out on both sides and taking the militia in the railroad-cut on the flank. "And keep moving, boys, we're none of us going to stop! And remember, don't shoot too often. The noise won't frighten them—we must hit something!"

So they started in; and very soon the militiamen realized that their foes meant business now. They beat a hasty retreat—all but one, who was hiding behind a big rock, shooting deliberately and carefully. Hal marked this man's position, and behind a rock he lay down to watch.

Hal Warner was no wage-slave, who had spent his life digging coal; he was a member of the privileged classes, and one of his privileges had been the use of a rifle. He had hunted mountain-goats in Mexico, an occupation which requires endurance, patience and keen eyesight. So he watched, and when the militiaman raised up to fire again, he fired first, and the militiaman disappeared.

Hal poked up his hat on the end of his gun, and moved it about for a while, but there came no response. So at last he sprang up and rushed the place. There, flat on his face, lay his human target, wth a clot of blood in his hair where the bullet had come out. Hal looked at him, and the realization came over him with a wave of horror, so that he went sick, and had to lean on his rifle. He had killed a man!

It was a big fellow, six feet high, with beefy legs and beefy shoulders, which would never move again. Hal took in every aspect of him—the khaki uniform, soaked with rain and stained with red clay, the red stained puttees and boots. He had learned to hate this militia uniform with such bitterness that he trembled when he saw it; but he did not hate it at that moment. He turned the man over, and saw his weather-beaten face, looking dull and placid, as if he had not had time to get interested in what was happening. The bullet had drilled a clean little hole through the middle of his forehead.

And so this was what Hal had come to! He who had started out with a dream of brotherhood and justice upon earth—here he was, a murderer!

Voices within him cried out in protest. But he did not have time to listen to them. A bullet went by him, singing loud and shrill, like a telephone wire in a wind. A second later came another, and he felt a pain as if a bee had stung him on the top of his head. He took off his hat, and looked, and saw there was a hole in the band. He realized that this was no time for philosophy or ethics, and dropped behind a big rock.

Hal exchanged guns with his victim, because the latter's gun had a bayonet, and he thought he might be needing one before long. Then he made another advance, and reached a position from which he could fire into the arroyo. He saw the militiamen moving back, and with another quick shot he caught one of them as he went over the ridge. It was as exciting as hunting mountain-goats—when once you got started at it!

[30]

That day the strikers drove their enemies back into the tent-colony. When they entrenched that evening, they had the colony surrounded; also they cut the telegraph and telephone wires, to keep Major Curran's forces from getting reinforcements and supplies.

Hal put trustworthy men in command for the night, and with Billy drove in an automobile to Pedro. He went into the town quite openly, carrying the militiaman's gun and cartridges, and wearing the hat with the bullet-hole through the band. There was nothing to be afraid of, for the strikers had taken possession of Pedro, and were patrolling the streets with arms in their hands. For fifty miles up and down the railroad the same thing had happened; the mouths of the canyons leading to the coal-camps were in the hands of the "rebels". They had taken possession of the railroads, and were not permitting trains to move with militiamen on board. In Pedro they raided the American Hotel, in which twenty militiamen were quartered, and turned the tables upon these gentry by searching the place for arms!

Billy Keating turned into a reporter again, and sent off a story for the "Gazette", and long telegrams to other papers over the country which might be willing to publish the news. As for Hal, having taken charge of a military campaign at a few hours' notice, he had no end of business to transact. Not half his men were armed, and some of them had no more than a dozen cartridges. Also food must be ordered and sent to them. Hal got Adelaide Wyatt on the telephone, and learning that the "pianos" had arrived, he ordered them delivered by an automobile truck. Also he sent telegrams to Lucy May and Will Wilmerding, to Governor Barstow, even to Peter Harrigan again. In the course of the evening there came a pathetic message for him: "I implore you to come up here. Must see you. Jessie." He smiled a grim smile, and wrote a grim answer: "Sorry. Am busy."

While he was at union headquarters, arranging for the forwarding of supplies, Mary Burke and Mrs. David came in; they had brought by train a party of fifty women and children for whom no shelter could be found in Horton. They told of the distress of the refugees, and of fresh horror in the blackened and smoking tent-colony. Through the day the militiamen had been acting like maniacs—not being content with fighting the strikers in the hills, but keeping their machine-guns playing upon the ruins of the tents, making targets even of chickens and cats. The union had sent wagons from Pedro to get the bodies of those who had been killed, but the wagons were fired upon and forced to turn back. On the county road which passed along the ridge came an automobile with some tourists from California, who did not even know there was a strike. The machine-gun was turned upon them, and for two miles the people rode through a rain of bullets.

It would be only fair to mention the excuse which is given for the conduct of the militia through this mad time. In the course of the first day's fighting, one of the troopers had been shot in the neck, and his comrades had plugged up the wound; afterwards, being obliged to retreat from the position, they had left him lying under shelter. When they came back, some hours later, they found the man's face beaten in with the stock of a gun, and the top of his head shot off. They were naturally made furious by this deed, and their defenders did not fail to make the most of the circumstance. But in all the various inquiries which were conducted, no one was interested to find out what this trooper might have done, to awaken such a frenzy of hatred in the minds of South European peasants!

[31]

Hal went back to his command after midnight, in an automobile piled up with bread and canned provisions, and some boxes, very heavy, labeled "foundry-type". He brought also the assurance that reinforcements were coming; the union was preparing a call to arms to workingmen all over the state.

One of the first persons Hal met when he reached the hills was Jerry Minetti, who had been up cutting telephone wires in the canyons, and had only just heard the news about his family. He wanted to know about Rosa, if there was any chance of her living—but Hal could not tell him. There was no way to get news; the militia would not even give up the dead bodies. Hal put his arms about his friend, who broke down and wept like a child. There was hardly a dry eye among the men who stood round.

Then suddenly Jerry looked up; he was haggard and grim of aspect. "We get them bodies maybe!" he said.

"Yes," agreed Hal; "we'll get the bodies, I think!"

So at dawn the amateur army got in motion; "General" Jack David commanding the center, Billy Keating the left wing, and Hal Warner the right. They drove the militiamen out of their positions by the steel bridge, and got sharpshooters around on the flank, where they could make trouble for Stangholz and his "babies" on Water-tank Hill. By noon-time Major Curran was desperately hard-pressed, and sent a courier up to North Valley, asking that the mine-guards there should create a diversion, in order to weaken the pressure upon his forces. So a crowd of some fifty guards and strike-breakers came out from the mining-village, and crossing over the ridge, opened fire upon the tents in the next colony, that of Greenough.

The result was such as to bring relief to Major Curran—but at a higher cost than the North Valley crowd would have wished to pay. It was another wanton attack upon women and children, and it set the strikers wild. The men in Greenough on one side, and Hal's forces on the other, charged as one body upon their assailants. The rain of bullets did not stop them, they left their dead and wounded on the way, and rushed and carried the ridges, driving their assailants back in confusion upon the North Valley camp. Nor did they stop with that, but rushed the mine village itself, storming the stockade, breaking down the gates and going through the place like a cyclone. The guards fled up the canyon; the strike-breakers took refuge in cellars or under their beds; while Cartwright, Alec Stone and Bud Adams, who had led the attack on Greenough, sought shelter in the mine-shaft, with their families and the office-force, some thirty-five people in all.

Nearly two years ago, when Hal was driven out of North Valley, he had told the men that he would return; and here he was keeping his pledge—making a tumultuous entry, with crack of rifles and shouts of fighting men. At his side was Jerry Minetti, who had been a "shot-firer" in former days, and had returned to his old occupation—with a repeating rifle and an automatic revolver!

They went through the camp and up the canyon—for their business was with the mine-guards. They came upon a group entrenched behind a barrier of rocks, and as they were charging, Jerry suddenly gave a grunt, dropped his rifle and clutched at his body, and went down in a heap. Hal had time for only one glance—then he took shelter in a depression of the ground, and began working his way around to the side; he got to where he could outflank the hidden enemy, and when he had got back his breath and steadied his hand, he put a bullet through the leg of one of them. The rest sprang up and fled again, pursued by a dozen strikers, firing and yelling.

Hal went back to where Minetti lay, with a couple of his fellow Italians bending over him. There was blood upon his lips, and upon his face a look of anguish which wrung Hal's heart. Jerry could not lift himself, and Hal bent down and caught his gasping words, "Rosa! Rosa!" Hal pressed his hand and gave a promise—he would find Jerry's girl-wife, if she were still alive, and would see that she did not suffer want. A minute or two later poor Jerry passed away.

Tears ran down Hal's cheeks; for he had truly loved this Dago family. He carried a thousand memories of them, beginning with the first meeting—Big Jerry striding down the street with Little Jerry at his heels, trying desperately to keep step, unaware of the smiles of observers. So proud and eager these Dagos had been, so full of the joy of life; and now they were utterly wiped out!

There was little time for tender memories, however. Down the canyon, in the direction of the mine, Hal heard the sound of heavy detonations, and saw a vast cloud of smoke rolling up to the sky. He ran down, and found that the infuriated strikers, of whom there were now a hundred in the village, had set to work to level it with the ground. They had blown up the jail in which Hal had been imprisoned; they had blown up the company-store and the breaker-buildings; they had set fire to Reminitsky's and the other boarding-houses, which were roaring furnaces; they were getting ready to blow up the mine-shaft and everybody inside. Some fifty strike-breakers had been captured, and the hot-heads wanted to treat them as Louie the Greek, Kowalewsky and John Edstrom had been treated two days before.

So Hal had to make a test of his authority as a military leader. He rushed here and there, commanding, exhorting. He set others to work, shouting his orders in half a dozen languages. They would take the lives of no strike-breakers—strike-breakers were workingmen, and had a right to be taken into the union. They would not wreck the stope of the mine-shaft, because they might need it before long—they might decide to run the mine! They would mount a guard about the shaft, and when those inside got ready to surrender, they would be treated as prisoners of war. Before long Hal found "General" Jack and put him in charge. It was too late to save the village of North Valley, but Hal did not particularly care about that; in truth, he rather shared the feelings of the strikers about it. Let it burn!

[32]

Returning to Pedro that night, Hal learned that the union had issued its call to workingmen to rise and defend their homes and families. The answer was prompt and wonderful. Five hundred miners were starting from the metal-mining district over the mountains; hundreds were coming from the coal-mines of other parts of the state, and even from states adjoining; the whole country was up. Meantime, however, there must be more food, more guns and ammunition. Johann Hartman and Tim Rafferty were at the telephone all day, buying supplies and making arrangements for their delivery.

Mary Burke had been in Horton, gathering refugees and feeding them. A party of two clergymen and some women, under the authority of the Red Cross organization, had at last been admitted into the tent-colony, to bring out the bodies of the dead. So there were new horrors to hear about. The bodies of some of the children had been so burned that they had fallen to pieces when lifted. Mrs. Bojanic had given birth to her baby while on a slab in the undertaker's parlor. Hal exclaimed that such a thing was not possible; but Mary answered that the baby was there. Rosa Minetti had regained consciousness, but was out of her mind. She talked about her little ones, and imagined herself in Italy with them; she "babbled o' green fields". Her breasts were swollen, and the doctor had brought some kind of cup with a suction-pump; Rosa would lie and caress this in her arms.

There was another piece of news, of interest to Hal. His brother was in Pedro, looking for him. He had called up union headquarters half a dozen times, and had even humbled his pride sufficiently to ask "Red Mary" to help him find Hal. He had not said what he wanted, but Hal knew only too well; there would be more arguments, more frantic appeals, more tales of his father's distress of mind; possibly there might even be some half-insane effort to force him away. No, he had no time for Edward just then! He wrote a note to his brother, bidding him drop the matter and go home; and having sent this to the hotel, he slipped away to the home of his friend MacKellar. The realization had come to him suddenly that he had forgotten sleep for three days and nights on end!

Next morning came the newspapers, and he learned that Cartwright and Alec Stone had succeeded in getting out a message from North Valley, telling of their peril and calling for help. This caused great excitement, of course, and Hal thought of going back to the camp, and persuading the strikers to let their prisoners go. But something happened which called him even more imperatively—an outbreak of desperate fighting in Sheridan. The sheriff-emperor was in control of this town, with a newly-organized "troop" of mine-guards, and he had conceived the idea of "teaching the strikers a lesson". Early that morning he posted a machine-gun on the porch of the General Fuel Company's store, and opened fire upon a street lined with miners' homes. The bullets went through the frail shacks as if they had been of paper; men and women were struck, and children playing in the street were mown down. It happened that Billy Keating was in Sheridan, and he took command of the strikers, and called for Hal and "General" David, and opened a pitched battle with three hundred militiamen and guards.

Up in Western City, the Governor had ordered more militia to the field, and General Wrightman had been frantically trying to get them together. But one company of eighty-two men mutinied in their armory and refused to go. The General offered them double pay, but they stood by their decision, and he was too much cowed by the popular clamor to attempt to punish them. At this critical moment the newspapers published the fact that during the course of the strike more than fifty officers of the militia had resigned, unwilling to do the work of the coal-companies.

But in spite of this, the General had got some reinforcements in hand, and loaded them onto a train, including several flat-cars with machine-guns and ammunition. He came to Horton, and took charge, and immediately received a visit from the two clergymen and the women who, under the authority of the Red Cross, had gone into the ruins and brought out the bodies. They now pointed out to Wrightman that forty or fifty people were missing, and they asked permission to make a search for more bodies. The permission was granted, and they went in—starting immediately toward one of the places where it was reported some bodies had been covered with quick-lime. Before they had got to the spot they were placed under arrest, and brought back to the General, who cursed them furiously. One of the clergymen was the villain who had been guilty of "besmirching the uniform of the soldier", by complaining of the misconduct of militiamen on the streets. The commanding officer now shook his fist in this villain's face, declaring that "pimps, preachers and prostitutes all look alike to me!" When Hal heard this story, from the clergyman's lips, he tried to send it in a telegram to Wilmerding, and Billy Keating tried also to send it to the "Gazette"; but the telegraph company refused to transmit the General's language, declaring it "obscene"!

[33]

During this crisis there were wild doings in the State Capitol. Hal got only swift glimpses at the time, being too busy even to read through the newspaper accounts; but when the trouble was over, and he met Lucy May and Will Wilmerding, he got the inside story. The first accounts of the "massacre" had come in the morning, and Wilmerding had rushed to see the Governor. Finding that nothing could be hoped for from this distracted wretch, he had proceeded to John Harmon's office, where a conference was held to discuss means of rousing the state. People came from every walk of life, both men and women—among the latter Mrs. Edward S. Warner, Junior. The decision was taken to call a mass-meeting on the State House grounds, and the Reverend Wilmerding undertook to notify the police authorities.

Lucy May told about this scene, which had been so funny that even in the midst of her excitement and distress she had hardly been able to keep her face straight. The clergyman had called up the chief of police, and there in the office of the miners' union, before a large company, had conducted a telephonic debate upon the rights, privileges and immunities of citizens in a free republic. The clergyman's voice had rolled and thundered, just as it was wont to roll and thunder from the pulpit; as the discussion waxed hotter, the reverend orator began to make gestures into the telephone—the gestures which Hal had watched on so many Sunday mornings from the family pew. "What, sir? Do you know whom you are addressing, sir? An ordained clergyman of the church! (Gesture of the right hand) Very well, sir—if you have no respect for my office, then perhaps you will have respect for the aroused citizens of this community. Let me inform you, sir, (Gesture of the admonitory fore-finger) I have not called you up to ask permission for this meeting, I have called you up to inform you that this meeting is to be held. It will be held, (Gesture of the clenched fist) regardless of anything that the debauched police-force of this community may threaten or attempt. You will be well-advised, sir, if you instruct your officers not to show their faces on the Capitol grounds during that meeting! (Gesture of the tossed head)."

And this advice the chief of police took; there was not a blue uniform in sight on the afternoon of the meeting. In spite of a pouring rain five thousand people packed the grounds, and they passed resolutions demanding the immediate impeachment of the Governor, and calling upon the citizens to arm and assemble.

Nor was this mere verbiage. The citizens meant it. It seemed that every third man you met was organizing a military company. The members of some of the unions had got arms, and were publicly drilling; many of them were setting out by automobile and train for the coal-country. It was the thing Hal had foretold to Congressman Simmons—civil war!

Lucy May made an effort to see Peter Harrigan again—this time at his office. When the old Caliban-monster refused to admit her, she set out to carry into effect the threat she had made—to raise up an insurrection of women against him! She and Adelaide Wyatt summoned a gathering of women at one of the big hotels, and it was resolved to march on the State House next morning, to demand an interview with Governor Barstow, and stay with him until they had got what they wanted. The papers published the summons, and next morning there was a throng of several thousand women—and women with votes, accustomed to having their way with politicians! They literally camped out before the Executive office, making speeches, singing hymns, doing everything they could to upset the nerves of a little cowboy Governor. They stayed day and night, and there was no way to get rid of them.

Their demand was that the Federal troops should be called to the scene, to put a stop to the bloodshed. This had been Hal's proposition to the congressional committee, and so Lucy May and Adelaide thought that they were carrying out his wishes. They were amazed when they received a telegram from him, telling them that they were playing into the hands of the enemy. Before long, there he was on the telephone, a faint voice, pleading in desperate excitement. The time for Federal intervention was past—what was wanted now was for people to keep their hands off! There were only some six hundred of the militia ruffians, and the strikers would drive them back into the mountains, or wipe them out altogether, and that would be the end of corporation rule in the coal-country!

"I tell you we've got everything in our hands! We're taking mine after mine—we've got a revolution accomplished!"

"But Hal, you're burning things—you're killing people!"

"We're not killing anybody that deserves to live! Keep your hands off! Let us settle our own affairs!"

And next day there was another telephone controversy, this time between Hal in Pedro and John Harmon in Western City. What madness was this which the newspapers reported? The union leaders were planning to grant a truce! Throwing away everything they had gained, going back into their old slavery! They had brought the operators to their knees at last—and now they proposed to lie down, to let Federal troops come in and deprive them of all they had fought and bled for!

It was the difference between a young man of the leisure class, high-spirited, accustomed to having his own way, and a representative of the toilers, a self-made and self-taught man, schooled in patience and obedience. John Harmon was not a revolutionist, he had never thought about a revolution, and did not know what to do with the advantage that had so suddenly come to him. What he was asking was a living wage and decent working conditions for his people; it had never occurred to him in his wildest dream that it might be his destiny to seize the coal-mines and run them co-operatively!

[34]

Jim Moylan and the young men were with Hal, but the older men had their way, and the truce was declared. But it proved that the truce could not be kept; both strikers and militiamen were like wild animals turned loose—they would fight, and nobody could stop them, and anybody that tried would get hurt. The mine-guards fired into the Oak Ridge tent-colony, and as a result the strikers charged the Oak Ridge mine, razing everything inside it.

And here was more work for Hal. Word came to Pedro over the phone that General Wrightman was sending a force of militiamen to retake the mine-property, and the strikers swarmed from every direction to defend it. Hal gathered a party of twenty picked men, and posted them near the entrance to the Oak Ridge canyon, where they would be hidden from anyone coming up the road. He himself took post a few yards in advance, giving orders that they were to wait for his signal, then to shoot, and shoot to kill. He lay for a couple of hours, until he made out an automobile winding its way up the road. As it came nearer, he saw that it carried some of the hated figures in khaki, and he gripped his rifle and made ready. They would wipe out that bunch at one volley, and then they would have an extra automobile for moving ammunition!

The car passed from sight for a minute or two, then suddenly it came round a turn directly in front of Hal. He looked, and his heart stood still with dismay. Sitting in the front seat, alongside the driver, was Captain Harding!

Hal saw the terrible plight he was in. Could he kill his cousin? Could he give the signal and let others kill him? No, he could not!

It was necessary to act instantly, for at any moment the men behind him might take matters into their own hands; and of course, if one shot were fired, it would be too late. Hal dropped his rifle, and springing up, rushed forward, pulling his handkerchief from his pocket and waving it. "Stop! Stop!"

The driver of the car obeyed. He was a militiaman in uniform, and there were four other militiamen in the back seat, all with rifles in their hands. They craned their necks and stared at Hal, who waited until he was close, and then said, "Appie, you must turn back."

"What do you mean?" demanded the Captain.

"Don't ask me," said Hal. "Don't stop to talk about it. Turn back and get out of here."

"But I have business up the canyon."

"Don't you know that Oak Ridge has been taken by the strikers?"

"No, I didn't know it."

"Well, it has. So you can't go forward."

"I have orders," declared the Captain.

"Your orders aren't to get yourself killed!" And then, with swift intensity, "For God's sake, Appie, don't be a fool! I'm trying to save your life. Every second may be too late."

Captain Harding answered, coldly, "I don't think you know the man you're dealing with." And he took one glance about the landscape, to see where the enemy might be hidden. Then came his command, quick and sharp: "Get out, and get under cover!"

The five militiamen, clutching their rifles, leaped out of the automobile. Before Hal had time to realize what was happening, they had run across the road and flung themselves down behind the rocks. Captain Harding stepped behind the automobile and stood waiting with his revolver drawn.

Hal was beside himself with indignation. "You dog!" he cried. "When I was trying to save you!"

"Did I ask you to save me?" demanded the other. He was almost as angry as Hal, but controlled himself better.

"Have you no honor at all?" Hal exclaimed.

"My honor is not in your keeping. My duty is not to you, but to the state."

Hal looked about him, in desperation. He could not see any possible escape from this predicament. He had betrayed his fellows, put them in a trap. "If you must have a fight," he cried, suddenly, "let's settle it between us two!" And he drew the revolver which was at his belt.

But Captain Harding would not even look at him. "Rubbish!" he said. "It will take more than that to settle this matter."

"All right!" exclaimed Hal. "But let's begin with this anyhow!" He caught his cousin by the arm and jerked him round.

Appie, however, was not to be moved. "Forget it!" he said, contemptuously. "I'll do no fighting with you."

"Whom do you think you'll fight with? Don't you know that if you fire on my men, I'll kill you?"

"Very well," said the Captain. "As you please about that." He was looking in the direction of the strikers, trying to estimate where and how many they were. Some of them were now showing their heads, staring at the perplexing scene; but none of them made a move to fire.

As for Hal, he thought of walking out to rejoin his men, leaving it for his cousin to shoot him, or to order the militiamen to shoot him, as he saw fit. His hesitation was because he realized that such an action would not help his men in the least. He had put them in a plight from which they could hardly escape without loss of some of their lives.

The young man stood motionless, heartsick, numb. The strikers, now probably realizing his predicament, and their own, lay back out of sight and waited; the militiamen also waited—and so this strange, almost ridiculous situation continued for a couple of minutes.

At last Hal turned to his cousin again. "Can't you see there's no sense going on with this? You're outnumbered three to one. You're only throwing away the lives of your men."

"What do you propose?" demanded Harding.

"I propose that you let my men go up the canyon without firing on them."

The other considered the matter. He had had time to realize the military aspects of the affair; perhaps also he felt that he had vindicated his reputation for courage. "I'll not fire on them if they don't fire on me," he replied, at last.

"Will you give that order to your men?"

Captain Harding gave it; and then Hal turned towards the strikers, and took a few steps in their direction. "Boys," he called, "I've made a mess of it for you, but I couldn't help it. I couldn't shoot at my cousin. Call it off and go up the canyon; please. Don't shoot, and the militiamen won't shoot at you."

After a few moments' hesitation, several of the strikers stood up and started to back away. Others, less trustful, began crawling away on the ground. But as the militiamen did not fire, they were reassured, and stood up and went off. Farther up the canyon they came together on the road and passed from sight.

[35]

Captain Harding gave an order, and the car was turned about. Then he went up to Hal and touched him on the shoulder. "Come," said he.

Hal was standing in an attitude of utter dejection. "Why should I?" he asked.

"Come," repeated the other, firmly.

"Let me alone. I'll not travel in your car."

The other grasped him by the arm. "You are my prisoner," he said.

Hal looked at him for a moment. He had no more heart for disputing; he answered, dully, "Oh, very well. I don't care." He had betrayed the strikers and disgraced himself! What did it matter where he went?

His cousin put him in the seat beside the driver, and stood on the running-board beside him; so they went down the canyon, Hal staring before him, but seeing nothing, his shoulders sinking lower and lower. He had a sudden terrible reaction from the strain and excitement of the past week. He was sick, sick.

He hardly noted the remark which his cousin made to the five militiamen as they were coming into the village of Horton. "Men, I think we'd best say nothing about this. I don't know just what I'm going to do about it, and I don't want anything known until I make up my mind."

"All right, sir," replied the men, promptly. They were loyal to their young officer, and could appreciate his plight.

Captain Harding took his cousin to the little hotel, and went upstairs with him to one of the rooms. There was a big upholstered chair in the room, and Hal sank into it wearily, and closed his eyes; for a considerable while he paid no attention whatever to the conversation of Captain Harding.

But after a while this conversation began to filter into his mind. He heard Appie say that he, Appie, was in the devil of a predicament. And gradually Hal made out why. Hal had been taken with arms in his hands, in insurrection against the authority of the state; he was liable to pay the penalty with his life, and obviously it was any militia officer's duty to give him up.

"Well, why don't you?" exclaimed Hal, at last. "I failed in my duty, but that's no reason you should fail in yours."

"*You* failed in *your* duty?" echoed Harding, puzzled.

"Of course. Wasn't it my duty to shoot you? But I was a coward, and didn't do it."

That point of view had evidently not occurred to the other. "May I ask why it was your duty to shoot me?"

"Because," replied Hal, "I found you with arms in your hands, engaged in maintaining a régime of infamy which I had sworn to exterminate. But you were my cousin, and I hadn't the courage of my convictions."

Captain Harding did not seem to know how to deal with such an argument. There was a pause, and then Hal added, "There are exponents of the class-war who say that only proletarians should be trusted in the movement. It would seem that I'm a proof of their contention."

There was nothing to be gained by discussion with a lunatic. The officer began to pace up and down the floor, consumed by his own thoughts. Could he give Hal up to imprisonment and possible execution, with the frightful scandal it would involve? But on the other hand, if he let him go, what would become of discipline? How could he face his men? What would he say to his superiors?

After a while Hal looked at him, and realizing the torment he was inflicting upon himself, remarked, "Cut it out, Appie, and give me up. It won't make any difference. Nothing will come of it."

"They would hang you!" cried the other.

Hal laughed. "Hang a millionaire's son in this state? You know they couldn't keep me in jail three days!"

"You're guilty of murder!" exclaimed the other.

"I am that, of course—guilty of several murders. I shot one of the men of your own company, I think—a big beefy animal in khaki—"

"*Hush!*" cried Captain Harding. He looked about him as if he thought the walls might have ears.

"I drilled a hole clean through his forehead. And I shot another one as he was running down the railroad-cut. I got two mine-guards at North Valley, and I think I winged a third at Sheridan. I did all that—and in spite of it, they couldn't keep me in jail three days!"

Captain Harding had resumed his pacing of the room. He was in a terrible condition of agitation.

"Make a test of it," persisted Hal, defiantly. "You believe in the law—you're going to practice it, make your living out of it. Go tell Wrightman what I've just told you, and see what he'll do about it!"

"Hal," protested the other, "you may please yourself by flouting the law, but surely you must realize that I am one of its officers—I have taken my oath to maintain it—"

And Hal laughed, a wild, half-hysterical laugh. "Poor fellow! He's a lover, and he believes in his mistress, and he wants to fight anyone who doubts her virtue! But I—I know her—the bedraggled old harlot! I have followed her about the streets at night—I have tracked her to the filthy dens where she makes her bed! I know that she will lie with any man that puts gold into her palm; so I pity the poor fool who believes in her and won't listen to the truth!"

So they had it back and forth; until little by little it became clear to Hal that his cousin's nerve had failed, like his own. He would not, *could* not do his duty!

He wanted Hal to go away, to go home; but Hal answered that he had enlisted for the war. Probably he had destroyed his influence with the strikers, but still, in a fresh emergency, he would have to do what he could to help them.

"Then you'll have to stay here!" exclaimed Captain Harding. "I'll hold you myself, since you make it necessary."

"Your private prisoner?" laughed Hal.

"Yes, my private prisoner."

"Well, you've exactly as much right as the General has to hold *his* prisoners. But be careful I don't fall out of the window."

"Hal," pleaded the other, "won't you give me your promise and quit fighting?"

The other considered, and then answered, "Suppose I asked *you* to promise and quit fighting?"

"As a matter of fact, I don't expect to do any more."

"Indeed! What were you doing in that automobile?"

"I was on my way to see some members of the guard who are wanted as witnesses. You evidently haven't heard that I've been appointed on a committee to investigate the events at Horton."

"No, I hadn't heard that." Hal was interested, and his cousin told what had happened since that dreadful night of destruction. The sights he had seen had been too much for Harding; he had made up his mind that the murder of the three prisoners was an intolerable crime, and he had gone up to Western City with the intention of preparing a statement concerning the conduct of the militia, and giving it to the newspapers. But as fate would have it, on the train he met a fellow-officer, whom he told of his intention; this officer argued and pleaded with him, and when they got to Western City, he called in Major Cassels to help. They finally persuaded Harding to accept as a compromise the appointing of the three of them as a commission to go down and make an investigation into the conduct of the guard.

Hal was first thrilled with this story, and then made heartsick. If only he could have been on hand at the critical moment, to hold his cousin to his bold resolve! Now, of course, it was too late; they had got their nets about Appie, they would soon have a ring in his nose, and be leading him where they pleased. —And so in the event it proved. Before they got far in their "investigation", Major Cassels was objecting to some of the witnesses Captain Harding produced, and to some of the questions he asked them; finally he was using his authority as Captain Harding's superior officer, to forbid him to summon certain witnesses at all!

[36]

The question of Hal's immediate fate was decided by a compromise. Captain Harding saw that he lay back in the chair with his eyes closed, and desperate weariness in his face. "Boy," he said, in a different tone, "you're pretty nearly done up!"

"I know I am."

"Don't you want to rest?"

"I don't know what I want."

"Why not make an agreement to stay in this room for a few hours? So we'll both have a chance to think it over—"

"And you have a chance to get hold of Edward and Dad. Is that your plan?"

"Edward was here for two days, Hal—looking for you."

"Where's he gone?"

"He went back home. What else could he do?"

There was a pause; then again the other began to press his proposition. Finally Hal gave his promise—he would stay in the room until six o'clock in the evening. And so Appie, relieved of his anxiety for the moment, became human and solicitous. Could he get a doctor for his cousin? Could he get him some food? Hal answered that he wanted nothing but to be let alone, and so the other went out, closing the door behind him.

During the past week Hal had had an experience which falls to men only in great crises of history. He had known the soul-shaking emotions of martyrdom. He had known what it was to be able and willing to throw his life away as he would a withered flower; to go with the clashing of cymbals and the blare of trumpets in his ears, to be blind, dizzy, walking upon air, transported out of himself, so possessed with rage that he might have been torn limb from limb without feeling it. He had forgotten that he had a body; he had gone on and on, living upon his nerve, consuming his own substance—

And now suddenly came reaction, as violent and extreme as the former excitement. Exhaustion possessed his body, despair possessed his mind. He who had set out to make people happier, to make the world better—he had killed several men, he had cruelly wounded others—and he had accomplished nothing, absolutely nothing! He saw his week's proceeding as an insane delirium, a drunken debauch; he saw all his two years of strife and pain from the point of view of his brother and his cousin—a thing of utter futility.

He flung himself down on the bed, where in the end the claims of a worn-out body took precedence, and he fell into a sleep. But it was a sleep tormented by nightmares. He was back in the burning tent-colony, and women and children were shrieking and rushing to him for help. He was holding Little Jerry in his arms, and as the child struggled, his burned flesh came off, and he fell to pieces in

Hal's hands. And then came the sounds of cannon-firing; the reports beat upon his brain, there was a crash of shells about him, he struggled to run, but his legs would not move—

So, with beads of sweat on his forehead, he opened his eyes and realized that someone was knocking persistently on the door of the room. Before he could answer, the door began to open, and a face appeared. For a moment he thought he must still be dreaming. The face was that of Jessie Arthur!

[37]

She spoke his name, her voice a whisper. And so he realized that it was no dream, and started up—because he had his coat off, and his aspect and circumstances were not such as befitted the receiving of a young lady. Jessie, realizing this, stepped back, half closing the door; but it was only for a moment—when he had slipped into his coat, she appeared again. "Hal," she whispered, "I must talk with you."

"Come in," he said; and she came, and summoning all her resolution, closed the door behind her. Then, still holding the knob, she stood gazing at him, intense excitement in her aspect.

"What is it?" he cried.

She answered, her voice still faint and trembling, "I have done what you asked me to."

"What I asked you to?" Perhaps he should have known what she meant—but so much had happened since they had had their talk.

"I have given up everything, Hal! I have run away from home!"

"*Jessie!*"

There was dismay in his tone, rather than welcome, and she, with her woman's intuition, recognized it instantly, and turned even paler, and seemed to sway against the door. "You told me to do it, Hal! You said I must be willing to give up everything and follow you. So I did it!"

There was a silence. Then, "Sit down," he said, and made a move as if to help her. But she went to a chair alone, and sat down, gazing at him out of frightened eyes.

"Hal!" she whispered. "You don't want me?"

"You don't understand, Jessie. I've been through such horrible things since you saw me! I can't think about—about us."

"I know, Hal; and I've been so frightened. I was afraid you'd be killed. I've suffered agonies—simply agonies. I couldn't stand it any more, I had to come to you!"

Again there was a silence, a long, trying silence. The tears came suddenly into the girl's eyes, and she clasped her hands together. "Oh! You don't love me any more!"

"But Jessie, listen—I've ruined my life. I've been fighting, and I've killed men. I may be punished—sent to prison—"

"I know, Hal. I met Captain Harding—that's how I knew you were here. But I don't care—I can't do without you. I will share whatever happens."

"They might hang me!" he exclaimed.

It was an attempt at evasion—and it availed not at all. "I don't care what they do, Hal! Don't you understand? I've run away! From Mamma and Papa—everyone and everything!"

There was desperation in her voice—but no more than in Hal's mind. "Jessie, I had no idea—" And he stopped.

"What is it, Hal? You don't trust me now? Answer me!"

"It isn't that, Jessie. But you see—something has happened. When you asked me if there was anybody else, I told you there was not—and that was true. But there has come to be since. When I thought you had given me up, I asked Mary Burke to marry me."

She started from her chair, and horror came into her eyes. "Oh, Hal!" And she took a step towards him, her hands stretched out. "*Oh!* You can't mean it!"

"I mean it, Jessie."

"Oh, surely, surely you couldn't marry that woman! It's too dreadful!"

"I don't know that I shall marry her, Jessie; she has not given me her answer."

"Not given you your answer?"

"She told me to wait and think it over."

Jesssie had to have time to grasp the meaning of this last incredible statement. Some instinct told her that this was not a situation to be handled by hysterics; she became suddenly calm, mistress of herself. "Hal," she said, "sit down, won't you? There is so much I have to say to you." And she came to him, and when he had seated himself in a chair, she knelt at his feet, gazing up into his eyes.

"Hal, I know that you've lost your faith in me; and I don't blame you—only you can have no idea how cruelly I've suffered, how hard it's been for me, without a soul to understand or help. I kept hoping and praying that this trouble would be over, and that you'd come back to me. I had no idea it could last so long, or be so serious. But Hal, don't you realize, I love you with all my heart, and I want to stand by you, I want to understand what you're doing, and why. You must know that I love you, Hal! You must know that you love me; it's just anger that has seized you, because I didn't come, because I left you alone in your trouble! And you were here with this other woman, you felt sorry for her—and I feel sorry for her, too, Hal, I know she's had a hard life—but oh, you can't love her as you love me! You must know that, surely, Hal! Think what it would mean if you married her—a common woman, that couldn't understand—"

He started to interrupt, but she would not let him; she caught his hand in hers, and rushed on, with desperate pleading in her voice: "Oh, maybe I'm mistaken, maybe I'm doing her an injustice, I don't really know her. But Hal, I know that *I* love you, and I knew you first—you were promised to me, Hal! And it isn't as if I didn't know that you love me. You do, and it's just a momentary misunderstanding—you're disgusted with me, and I don't blame you—I'm ashamed of the way I've behaved. But I'm going to be different now, Hal; I'm yours, and I'm going to do what you ask me to—anything, anything! Don't you understand, Hal—I'm yours, yours! I love you, I love you!"

And there she was clinging to him, with her beautiful, sensitive face upturned, quivering with emotion; her trembling hands were clasping him. In that desperate emergency she cast away those reserves which are supposed to be essential to young ladyhood; he was hers, and she would have him—no vile common woman should take him away!

So Hal felt again the power of that magic spell, that thrill of young love which is almost too intense to pass as happiness. Yes, he had known her first, she had the first place in his heart. He took her in his arms—how could he refuse to take her? And there she was, sobbing on his shoulder, clinging to him, quivering in his embrace. He was amazed by the storm of emotion which seized upon her, a terrible, soul and body-shaking tempest, incredible in a girl so delicate, so hitherto reserved. "I love you! I love you!" she whispered, again and again; and so how could he mention Mary Burke, how could he even think of her again? How could he do anything but respond to her embraces, and drink this cup of madness with her? He kissed her, and her lips met his, and seemed to cling to him and hold him, while a current of fire ran through his being, melting his resolution, burning up his fears and hesitations. Yes, she was his! She loved him in a way that was not to be denied, she sent into the deeps of his being a call that no man could leave unanswered!

[38]

When this soul and body-shaking tempest was past, Jessie was weak, almost swooning; he stood holding her in his arms, and her head was thrown back and her eyes closed—but even then her instinct did not fail to guide her.

"What are we going to do, Hal?"

"How do you mean, sweetheart?"

"We've got to go away, or something. See the dreadful position I'm in! I've thrown away everything for you, I've ruined myself! I've come here alone, and when people find it out, they'll never have anything to do with me any more."

"When did you leave home?" he asked.

"Late last night. I said I had a headache and went to bed, and when everything was quiet I got up and stole away and took the night train. I locked the door of my room, and put a note on the outside, saying that I hadn't been able to sleep all night, so please not to disturb me. That'll put them off a while, but some time today they'll break into the room, and they'll know exactly where I've gone, and Papa will telephone down here, and they'll come and arrest me, and Papa will take me and lock me up where I'll never see you again. That's what I'm so frightened about, Hal—they may be after me any second!"

"You say Appie knows where you are?"

"Yes; I couldn't find you, and I had to ask for him. I thought he might help me."

"What did he say?"

"Why—he said he thought you needed me."

Even in the midst of many bewilderments, Hal could not help smiling at this. But Jessie could see no particle of humor in the situation. "You *do* need me!" she exclained. "You look so ill, so perfectly dreadful! And don't you see, Hal, how it was—I thought I could come to you and take care of you—I thought you would welcome me, that you'd be glad of what I'd done—"

"So I would, Jessie—"

"I thought—I thought we could be married right away, as you had said before; then they couldn't carry me away, they couldn't separate us. But now—now I'm disgraced forever! I've come to this horrible place—to your room! And if you cast me off, no one will ever speak to me again—"

"No, dear," he answered, "we'll go and get married."

There was nothing else he could say; he had invited her, and now he would have to go through with it, and silence once for all the voices of protest in his soul. Obviously, every moment that she stayed here in this room, she was more deeply compromised; he would have to take her out. He had promised Appie not to leave the room, but Appie would not count this a breach of his word, but rather an intervention of providence to relieve him of an impossible burden.

So Jessie wiped away her tears, and made a stab at the straightening of her hair, which had been blown awry in the tempest of emotion. They went down stairs—not too intensely preoccupied to note the significant smiles of loungers in this third-rate village hotel. Hal did not have to ask where to go, for he knew the justice of the peace, as it happened—the village coal-merchant, to whom, in the early days of government by gunmen, the strikers had many times made futile appeals both for justice and for peace. Now, however, the official was more cordial—this was the sort of work a justice of the peace could do gallantly, even in the midst of civil war! Captain Harding, having been found and summoned by telephone, came swiftly in an automobile; and with him for one witness and the village constable for the other, the irrevocable step was taken—just about ten minutes before the constable received a telephone message from police-headquarters in Western City, instructing him to seek out and hold under guard a runaway young lady named Jessie Arthur—age twenty, height five feet seven, hair golden brown, eyes brown, wearing a grey tailored suit and a small grey hat with a golden pheasant's wing.

[39]

It was a strange and troubled honeymoon they had: the bridegroom being so much concerned about a strike-war that it was only by starts and flashes that he would realize that he was married; while as for the bride—well, she would think about her father and her family, and be seized with terror; and then she would realize that she had got the man she wanted, and she would be dizzy with rapture; and then again she would look about, and find herself completely surrounded by hordes of incendiary foreigners, and she would have fits of shivering again. She had come with her mind made up to be wholly and completely Hal's, to subordinate herself and her ideas and wishes to him; she would not once ask him to come away, or to refrain from any of his revolutionary proceedings—and by this heroic and loving resolution she held fast for not less than four and twenty hours.

Fortunately there was no more fighting. The truce which had been declared was made permanent by the news that Federal troops were on the way to take command of the situation. It was a striking commentary upon the generally accepted idea of the strikers as outlaws, that from the moment they knew that the militia had been ordered away, and that they were to be actually protected, not a single shot was fired in that entire coal-country!

Meantime, Hal had one duty that could not be shirked; he must tell Mary Burke about his marriage. "Can't you write to her?" Jessie ventured, timidly; but he answered that decency required him to see her, and Jessie said no more. Early next morning they took the train to Pedro, where Mary was helping in the care of a couple of hundred refugees.

She had already heard the news, of course; for the reporters had been "on the job"—here at last was an occasion when a "millionaire Socialist" might take his rightful place upon the front page, even of Tony Lacking's "moral paper"! Here were all the elements of "romance"—a truly and inescapably and essentially American "romance", with two of the city's wealthiest and therefore most influential names, and a desperate class struggle to serve as spice, and a real physical running away, and a real serious appeal to the police! A young "millionaire Socialist" might hand a ten dollar bill to a village constable and bid him keep still, and the constable might in all sincerity intend to do it, and think that he had done it; but of course he could not help telling the story to his wife, and his wife could not help telling the baker's wife and blacksmith's wife—and so it would be easy for a reporter, coming along an hour or two later, to get as many highly colored versions of the matter as he might wish. And obviously, of course, any competent reporter would realize that one of the first things the eloping young lady would do would be to send a telegram to her father, telling the news and begging for forgiveness; and any American reporter would know ways of getting a glimpse of that telegram—or, failing the glimpse, would feel perfectly safe in "faking" it!

There was nothing said about "Red Mary", of course; there was no romance about her, none for her. "Red Mary" was out begging old clothing among the towns-people of Pedro, for women and children who had been driven out of their tent-homes and caught in the rain, and had nothing dry to put on, and no money to buy things. Having buried her young brother only two days before, the girl had plenty of reason for regarding herself as badly treated, and for staying at home and being overwhelmed by grief; but instead, here she was with young Rovetta and a hand-cart. When Hal met her, she was helping to push this vehicle, loaded to top-heaviness with cast off clothing of every size and color and shape.

Needless to say, the young "millionaire Socialist" had been looking forward to this interview with no little perturbation. It is one of the disadvantages incidental to being too eligible, too attractive—this sinking at the heart as you walk down the street and see a girl approaching; this sense of unutterable self-abasement, of desire to sink through the side-walk, to disappear entirely from sight and even from memory. In vain the too-eligible and too-attractive one tells himself that the law does not allow him to marry two women, and that in such a matter it is a question of first come, first served. The uncomfortable sensations do not leave him—especially when they are accompanied by a sense of pity so strong as to make his heart bleed.

Mary had a band of black about her arm, the only reminder of poor crippled Tommie with the back of his head shot off. She had no trace of that Irish complexion which had caused Hal to say that she must dine on rose-leaves; on the contrary, her face was wan and deeply lined, and when she saw Hal, her effort to smile was infinitely tragic.

Rovetta tactfully went on with his cart, and Mary put out her hand. "I've heard the news, Joe," she said. "I wish ye luck."

"Mary," he began—and then he did not know what to say. "I'm ashamed to meet you, Mary."

"You mean about us, Joe? Sure, don't think of it! 'Twas only a dream, Joe. Such things don't happen!"

"I want to tell you, Mary, Jessie came to me—she had run away from home, and I couldn't—" He stopped; it did not seem quite right to apologize for having got married; but on the other hand it did not seem quite right not to apologize!

"I know how it was, Joe—at least I read in the paper, and I could guess. And I know ye were promised to her, and I know 'tis better. She's your own kind. Ye know how it is, I like to be honest, and I'll not try to deny I'd have liked to had ye, Joe Smith; but I knew I could not have made ye happy. 'Twas a dream, as I say; such things don't happen to the likes of me."

There was more pain for "Joe Smith" in these brave words than there would have been in either tears or reproaches. He stood looking at the girl, with her wan features and her lips twisted into another attempt at a smile. She had spoken the true word in that last—such things did not happen "to the likes of her". Freedom and joy did not come to the likes of her; "romance" did not touch her with its golden wings. For the likes of her were poverty and frustration, humiliation and loneliness, at the best the bitter mockery which the world and its preachers call "duty". Pushing a hand-cart down the street, for example, and collecting old clothes for naked and shivering outcasts—while reading in the papers about the triumphs and delights of those who have been born to power and ease!

It was the way of the world, summed up in one picture; and as Hal contemplated it, he was suddenly seized with shame for those raptures of young love to which he had abandoned himself. Those charms which had seduced him, those allurements of the senses, those sweetnesses of the "honeymoon"—stolen sweets, extracted from the sweat and blood of people like this girl before him!

"Joe," she said, reading his every thought, "ye must not be takin' it like that! Ye must give her a chance, and maybe ye can teach her, and both of ye be happy. Only—I must say this, Joe—don't forget us—I mean the miners. They need ye so bad!"

"I know it, Mary; I'll never forget!"

"We been beaten, Joe! We got nothin' to show for this long fight! Ye must realize that—"

"Mary, please believe this—I've told it to Jessie, and I'll never change. I'm going on helping the working-people, all the rest of my life. You and I will find ways—some new plan—"

"You and I can never work together like we been doin', Joe." And she smiled sadly at his protests. "There's no use thinkin' about that—ye'll find it out in time. But there'll be many ways ye can help the miners, and if I can be sure that ye won't lose interest, then I'll not feel so bad about losin' ye, Joe—I'll be able to tell meself it was a help to ye to think ye were in love with a workin' girl!"

[40]

The fighting being over, there was nothing very pressing for Hal to do in Pedro, and Jessie was pleading with him to come away. He was so pale and thin, so utterly worn in body and mind; anyone could see that he was going to be ill if he did not rest. Surely they were entitled to a *little* happiness on their honeymoon! —So she lured him away, to a part of the mountains where there was no coal, but instead a beautifully appointed "camp", with ladies and gentlemen in picturesque outing-costumes, seeking rainbow trout in crystal-clear torrents.

But there came each day a newspaper, which could not be kept from Hal; so in a short while he was chafing, and they went back to Western City, where a struggle was developing about the Horton "massacre"—a struggle of publicity and politics. The killing of unarmed prisoners, the burning of helpless women and children, had accomplished this much at least—it had broken the conspiracy of silence of the great press association. There were meetings and demonstrations all over the country; telegrams of protest were pouring in upon the little cowboy Governor, and the big magazines were sending investigators—trained men of independence and conscience, who wanted to know, not merely about the events of that night of horror, but about the grievances which had caused a civil war. So here at last Hal had a chance to be heard; his friends the Minettis and John Edstrom and Louie the Greek had accomplished that much by their deaths!

The young married couple went to stay with Hal's father, who welcomed them with open arms; and the first thing Jessie did was to post off to fling herself into the crater of Vesuvius. That volcanic mountain had answered her telegram, casting her out forever. She had chosen her bed, let her lie in it; she had brought disgrace and shame upon her parents, she had brought their grey hairs in sorrow to the grave—and all the rest of the things that fathers say when they are disobeyed. Now Jessie stormed the house, rushing past poor helpless Horridge, who had orders not to admit her, and flinging her arms about her father's neck and clinging there, sobbing as she had learned to sob for her purposes. And what could poor "Mr. Otter" do? What can any father do, who has brought up his favorite daughter to have everything in the world she wants, and discovers to his dismay that the thing she wants is a man whom *he* does *not* want?

"Mr. Otter" did what all volcanos do—he rumbled and grumbled and emitted fumes. She had got her way, but she would live to rue it! That fellow would go on, he would drag her with him—next thing she would be in jail herself, and expecting her father to bail her out! Well, she might understand right away that he would not do it! He would have nothing to do with that young puppy! He had trouble enough already, with everybody he knew whispering and joking about his Socialist son-in-law! Every time the young scape-grace chose to get into the paper, it would be mentioned that his wife was the daughter of Robert Arthur, head of the hither-to respected banking-house!

—Not more than a month later they were telling all over town a joke about this head of the hither-to respected banking-house. Perry White, president of the Red Mountain Coal Company, had come unannounced into his office, as a boyhood friend had a right to do, and had observed the old gentleman making a quick move to hide behind his coat-tails a book he had been reading. It would be a jolly lark to catch a leading citizen, father of half a dozen grown sons and daughters, indulging in the surreptitious reading of a French novel or a volume of *risqué* jokes; with this in mind Perry White seized the book and dragged it from its concealment. He could hardly believe his eyesight—a book about Socialism! For weeks thereafter old Mr. Arthur had no peace when he went to lunch at his club. In vain his protests that he had merely been trying to see what the young scape-grace was up to, to be able to answer his arguments! No, the subtle contagion would infect him, he would be convinced before he knew it; the city must prepare itself for the news that the head of the banking-house of Robert Arthur and Sons was wearing a red button, and distributing his dividends among "comrades" and "fellow-workers"!

[41]

The struggle of publicity and politics continued without abatement for months, and poor Jessie found to her despair that her husband was making himself as much a nuisance to the bankers and coal-operators of the city as if he were down in the strike-country with a gun. They were trying with all their resources to hush things up, and here was a young man of their own class, with their own prestige and sense of power, talking to magazine-writers and causing them to spread broad-cast his wild slanders against the state; speaking at mass-meetings, and working up hysterical women, so that they formed committees and plagued the lives out of politicians and newspaper editors! And this, just when a session of the legislature had to be held, to reimburse those who had paid the expenses of putting down an insurrection!

This legislature was a smooth machine, carefully put together and thoroughly oiled by Peter Harrigan and his associates. There were a few men who could not be controlled, and these would make speeches; but then someone would move the "previous question", and the machine would roll over the orators. There would be a bond issue of a million dollars, to reimburse the Clearing Association—thus saddling the state with an interest charge of fifty thousand dollars a year for all time. Also there would be a bill authorizing the state authorities to search homes and disarm citizens at any time—in flat violation of the constitutional guarantee of the right to bear arms. As a sop to the insurgents, there would be a committee to investigate the causes of the strike—which committee would never do anything. And then the Coal King's legislature was ready to adjourn.

The leaders of the militia had also a bit of important work cut out for them. They realized that the happenings at Horton left them in a dubious position in the eyes of the law. Might it not some day occur to a politician, anxious for the applause of the mob, to prosecute militia officers for the killing of unarmed prisoners? So the Judge-Advocate evolved a brilliant scheme. The law provided that no man should be twice placed in jeopardy of his life for the same offense; now let the militia proceed to court-martial its own officers and acquit them—thus making them forever immune!

From which it may be realized that Major Cassels was a first-class lawyer, who earned the salary paid to him by the metalliferous mine-owners! The court-martial met with all military pomp and ceremony; it tried thirty-seven murderers in a bunch, putting through the proceedings with machine-like celerity. The only witnesses were officers and troopers—for of course the strikers refused to recognize this solemn farce. With one exception, the verdict was an absolute acquittal. In the case of Lieutenant Stangholz, who was charged with both assault and murder, he was acquitted on the latter charge, and on the former, that of assault, he was found guilty but innocent. Some lawyers might have thought this difficult, but not so a Judge-Advocate of Militia! The gun which Stangholz had broken over the head of Louie the Greek having been exhibited at the trial, the court-martial found the defendant "guilty of the facts as charged", but declared that the court, by reason of the justification, "attached no criminality thereto"!

Meantime in the Kingdom of Raymond a hand-picked grand jury of deputy-sheriffs and company employes was proceeding to vindicate the majesty of the law by bringing four hundred indictments against strikers and their leaders, nearly all of them for murder. These cases were to be prosecuted with vigor and were to drag on for a couple of years, costing the union hundreds of thousands of dollars. The most important case was that of John Harmon, who was accused of killing a mine-guard—not because anybody had ever seen John Harmon with a gun in his hand, or had ever heard him threaten this mine-guard, but because he had been in control of the strike, and therefore responsible for all that happened. Under such a theory of law, it was obvious that Peter Harrigan and his associates were guilty of the murder of the Horton victims; but you saw no grand juries proceeding to bring indictments against Peter Harrigan! To employ Hal's metaphor, it was the bedraggled old harlot flaunting her shame upon the streets.

[42]

She flaunted it again in the case of Hal Warner—this time not upon the street, but more decorously, in the private office of his brother. She came in one of her numerous disguises—the form of Judge Vagleman, grey-haired and impressive, yet genial, shrewd as an old fox—an advocate of rich men's causes who had held his own in a world of fiercest competition. He had come now as a friend, a fellow member of St. George's, moved by consideration and deep anxiety. Edward, being a man of the world, would understand that the chief counsel of the "G.F.C." of necessity knew something of what was doing and planning in connection with the strike prosecutions; he knew and would say frankly that the case of Edward's brother was giving great concern. The boy had killed several people; he made no bones about it, he went about telling everybody, and that put the authorities in a most awkward dilemma. It would be a scandal to indict for murder a member of one of the leading families; but on the other hand it was almost as much of a scandal not to indict him.

Edward answered, as one man of the world to another—the Judge must realize that he had done everything in his power to restrain his brother, but in vain. What could the Judge suggest?

The Judge answered that the boy ought to disappear for a while, until things blew over. He would be quite honest and admit that he had two motives in making the suggestion. The disturbance Hal was making was disagreeable to those in authority, and they naturally wanted to keep him quiet; but also it would be for Hal's good, and the good of the family.

Of course, Edward answered; he had been urging his brother to quit the state. Appie Harding had been urging it also. But the boy was courting martyrdom.

The consequences might be very serious, said the Judge; and pretty soon Edward, as a man of the world, began to realize that there was more then friendship and church fellowship in this errand of Old Peter's legal adviser. He had come to convey to the Warner family the definite threat that if Hal did not shut up, he was to be indicted for murder in the first degree, and that once the proceedings had started, they would go through to the bitter end.

So it was necessary for Hal's brother to stiffen his back somewhat. With the utmost tact, and in his best man of the world manner, he reminded the Judge that he, Edward, had come to know a good deal of the inside story of the strike, and it was a pretty rotten story; the boy had not a little cause for his indignation.

"Of course," said Vagleman, "I admit that a lot of the business has been raw as hell. But what can you do—with a bunch of anarchist labor leaders—"

"I know," replied Edward, "but none the less, there are laws, and if your side pays no attention to them, you can't blame the other side for following suit. My point is that the boy has a case, and he's managed to win over my father. You must understand that if he is indicted, the old gentleman will come to his defense; he'll engage counsel and he'll fight the case, not merely in the courts, but before the public. You know, he's old and feeble, but he's capable of being aroused, and he won't consider his health, any more than Hal did. You must realize too that he knows an awful lot about Peter Harrigan's affairs."

There was a still more awkward aspect of the matter which Edward had to mention—his wife. She had been completely convinced by the boy's story, and it was all Edward could do to hold her back from day to day. If Hal was indicted and in danger of prison, she would go completely "off". As the Judge knew, a woman could make the devil of a lot of talk.

Of course, admitted Vagleman, tactfully; and he knew that Mrs. Warner was a very determined lady, not to be easily diverted from her course. —There was a gleam of a smile about the Judge's lips, from which Edward divined that he had heard about the visit to the lair of the old Caliban-monster.

It was not his wife alone, Edward made haste to add. There were a number of women who could make trouble. There had recently been organized a so-called "Fair Play Society", and while its membership was not large as yet, there were some women among them whom even the "G.F.C." must think twice before attacking: for example, the widow of the former Chief Justice of the state, the man who had written that grim prediction about the dragons' teeth.

The Judge admitted all that. He did not want any trouble; he was willing to go to any extreme to avoid it; but—and here he spoke with firm decision—the defiance of the law must cease, the authorities had made up their minds to it!

Edward answered, as firmly as the Judge. If the defiance of law was to cease, he would suggest the first step. He had just seen a secret communication from the secret organization of the coal-operators of the state, pointing out that there was to be a general election in the fall, and that the one issue would be the coal-strike and its suppression. It was of vital importance that this election should result in a vindication of "Law and Order", and so it had been voted that the membership should assess themselves the sum of five cents a ton for all coal mined during the year, so that the state might be made safe for the Republican ticket. Now, said Edward, no one knew better than Judge Vagleman that there existed a law against contributions to campaign funds by corporations, and the coal-operators who complied with that program of the association would each and every one be liable to five years in jail. It must further be pointed out that this was a Federal statute, not so easy for local interests to evade; the operators were in somewhat the same position as strikers who could fight the state militia, but had to give up when Federal troops came on the scene!

—Edward said all this with his smoothest manner of man of the world; but he looked the old legal fox in the eye, and he made sure that the old legal fox was getting exactly what he meant. For that assessment fund of the coal-operators to do its work, it must be kept secret; if the public were to hear about it in advance, ten times the amount would not be enough to carry the election. So far, neither Hal nor Edward's wife had got wind of the matter; but if it were a question of saving Hal from prison, why then the Judge must understand that some of Hal's friends might give him a "tip", and the criminal secret might be spread upon the front page of the Western City "Gazette"!

The two friends and fellow church-members parted; and such was Hal Warner's trial for murder! Hal never even knew about it. Edward, still in his role of man of the world, said nothing to his brother, but instead went to see Garrett Arthur, the "bond worm", and caused him to convey to his youngest sister a harrowing picture of the imminence of hanging for her husband. So Jessie fled to her husband in hysterics, and made desperate attempts to persuade him to go back to Europe, to continue his studies of the Socialist and Syndicalist movements!

[43]

The Horton tent-colony was rebuilt, and the strikers settled down to a life of waiting. But Hal did not go back to live among them, for the work he had now to do could be better done in the city. Mary Burke and Mrs. Olson, Rovetta and Jack David were in charge, and kept him informed as to events. The Federal troops had been welcomed with music and flowers, just like the militia; but this time the confidence of the strikers was not so cruelly misplaced. Under the new regime there were no outrages, and hence no violence; the strike became a test of endurance between the coal-operators and their slaves—but with the conditions so arranged that only one outcome was possible. The strikers were deprived of their right to picket, and were not allowed near the depots; on the other hand the operators were allowed to bring in strike-breakers, provided they came "of their own initiative"—and of course it was a simple matter for such "initiative" to be provided.

What this meant was that the power of the United States government was being used to hold down Peter Harrigan's slaves while he starved them into submission. The President sent ambassadors and would-be mediators, to try to persuade Old Peter to make at least a pretense of concession; but Old Peter stood firm as a rock, he would make no pretense. So there began a struggle for the public opinion of the country—into which struggle Hal threw himself with fiery ardor. He wrote articles and leaflets, he addressed meetings of all sorts, he started a citizens' league and a free speech association. One thing at least all his clamor accomplished—it broke down the indifference to public sentiment which up to now had been the most conspicuous fact about the Harrigan regime. For the first time in history, Old Peter issued statements to the newspapers; as things got hotter yet, he was forced to set up a regular publicity-bureau—a sort of journalistic fire-department, to put out the flames of popular indignation. He had congressmen making speeches in Washington, he had judges and society-ladies running to plead his cause with the President. He sent General Wrightman to tour the state and defend the militia; he sent Major Cassels to tour the whole country on the same errand.

Hal happened to be in Washington, having gone to see the President himself, when he heard that the Judge-Advocate was to speak in a church, and he went there and confronted his eminent legal friend, and gave him a most miserable half-hour. The kind of thing that Cassels was doing may be judged from his story of little Jennie Burke, who had testified under oath how she had laughed when General Wrightman fell from his horse, and how the General had mounted again and ridden her down and kicked her in the breast. Now, all over the country, at meetings of chambers of commerce and women's clubs and church congregations, Major Cassels was giving a sample of the falsehoods told about the militia—a girl had sworn that General Wrightman had jumped from his horse and kicked her in the breast! Was it likely that a person as elderly as an Adjutant-General could kick that high? And the audiences would laugh—being impressed by the Judge-Advocate's genial manner, as well as by his imposing title. What chance had they to find out about poor little Jennie, who was crippled for life, and might die of cancer in the end?

Confronting such things, it was hard for Hal not to become rabid. To see this elaborate conspiracy of falsehood and suppression—and back in the tent-colonies, the slow, relentless strangling of the hopes and lives of men! In the early days at North Valley, arguing with Jeff Cotton, the camp-marshal, Hal had pointed out the fundamental issue of this struggle—the question whether a ton of coal was to consist of two thousand pounds or three! The miners said two, Old Peter said three—and that his will might prevail, all the forces of a state had been set into motion; a thousand gunmen had gathered, the militia had been called out, a million dollars had been spent, men, women and children had been tortured and murdered! And now, to clinch the victory—here was the regular army of the United States, and a publicity-campaign to poison the mind of the entire country!

Old Peter had got a press agent, a highly trained person who was paid a thousand dollars a month. His name was Oakes, and as the strikers came to understand the character of his work, they gave him the surname of "Poison". One of the first things he did was to get out a bulletin headed, "Why the Strike was Forced on the Miners". He quoted from the report of the secretary-treasurer of the union, which showed that the national vice-president of the union, in charge of the strike, had received a yearly salary of $2,395.72, and a year's expenses of $1,667.20. "Poison" Oakes put these two together, calling it all salary, $4,062.92; then he added the expenses again, making a total of $5,720.12; finally he said that all this had been paid to the national vice-president for nine weeks' work on the strike—thus showing that he was paid over ninety dollars a day, or at the rate of $32,000 a year! By the same method "Poison" Oakes showed that another official was paid sixty-six dollars a day; that John Harmon had received $1,773.40 in nine weeks! Mother Mary was listed at forty-two dollars a day; the actual fact being that for her work as an organizer she had been paid $2.57 a day—and this not including any of the time that she had spent in jail! The bulletin of "Poison" Oakes, containing these falsehoods, was mailed over the country to the extent of hundreds of thousands of copies, and the union leaders received many letters of inquiry and denunciation. They exposed the false statements, and demanded that the operators correct them; but this was a detail to which "Poison" Oakes never got round.

[44]

In his efforts to rouse the public to the meaning of these events, Hal invented a phrase which covered the situation—that Peter Harrigan had murdered labor, and now was proceeding to loot the corpse! Everybody else had suffered loss—Old Peter alone had profited from the struggle! The union was out of pocket several millions of dollars, the consuming public was out several millions more, some of the smaller operators were in bankruptcy—but Old Peter had made a fortune! He had laid up enormous stocks of coal before the strike, and these he had sold to the public at a big advance, alleging scarcity. Now that the stocks were reduced, he was starting his mines again, with a fresh supply of slaves; and for this procedure he had not merely made the state pay the bill—he had made the country think of him as a philanthropist!

"Poison" Oakes gave out another statement, detailed and explicit, showing how for many years the common stock of the General Fuel Company had not paid a cent in dividends. Peter Harrigan was running his mines as a matter of charity, to keep ten or twelve thousand men employed! This statement was sent everywhere by the great press association, which served as an aqueduct for Old Peter's ideas, and as a concrete dam to the ideas of his opponents. How was anyone to know about the bonds of the General Fuel Company, which had earned regular interest right along? About the preferred stock, which had earned regular dividends? About the sums which had been spent in improving the property—so that the Coal King had doubled the value of his holdings within a few years! Or about the surplus which the company had accumulated—sufficient to have paid six per cent a year on its common stock if it had chosen! Old Peter was keeping his money in an inside pocket, so to speak!

All along the line was defeat and ruin for Hal's friends. Billy Keating had to flee the state in order to escape arrest; he went into Mexico, and the "Gazette" very kindly invented a report that he had been killed by bandits! Professor Purdue, who had acted as counsel for the strikers, was thrown out of Harrigan College; Will Wilmerding was driven to resign from St. George's. How could he remain in a church whose rector appeared at a meeting of the "Law and Order League" and declared the militia justified in all it had done, and that if he could have had his way, every miner's home would have been blown up with dynamite? Such being the mood of the ruling classes of the state, it was easy to get a jury to find John Harmon guilty of murder, and a judge to sentence him to prison for life. Jim Moylan was under indictment, expecting soon to be tried; so also were Johann Hartman, Jack David, Rovetta, Klowowski—everyone who had been active in the defense of the tent-colonies. Many were in jail—and those who were out on bail were in no better circumstance, for they were not allowed to leave the state, and yet, being blacklisted, could get no work within the state!

Some few Hal was able to help: poor old Mike Sikoria, with his damaged arm; Jennie Burke, with her injured breast; Rosa Minetti, dazed and hysterical by turns—these he could save from the full consequences of defeat. But he could not help all who were under indictment, he could not help all who were scattered to the four winds, impoverished and marked for persecution. They would pay for their effort after freedom a penalty proportioned to the courage they had displayed. In this war, as in all others, it was "Woe to the conquered"!

The time came when the inevitable admission of defeat had to be made; a convention of the strikers gathered, and the fourteen months' struggle was formally called off. The slaves would go back to their galleys—until they were driven by unendurable torment to another revolt. These great strikes came at regular periods of ten years; and the thought which goaded Hal to madness was of those next ten years. Ten years of life such as he had lived for three months at North Valley! Ten years of starvation and despair for thirty or forty thousand human beings!

[45]

There was no way Hal could prevent it; it was life and he must face it. He must fight the battle in his own soul, to make wisdom out of his humiliation, resolution out of his despair. He must go over what had happened, and organize and order it in his own mind; he must work out a new program of action, testing it by new experiments, revising it to fit new facts; he must study the ideas of other men, weighing them and judging them, bringing order out of the chaos of their contentions. In other words, he must learn to think; and this is a slow and tedious process, which does not lend itself to picturesque narrative nor afford stirring and dramatic climaxes.

One thing Hal had come to see quite clearly: that beautifully simple formula of syndicalism which he had brought back from Europe did not fit the situation. The Syndicalist might wish ever so hard to ignore the state—but the state would not let itself be ignored. It would come in and smash your labor organization, no matter how strong you might be; it would smash you before you had a chance to become strong! So inevitably, by automatic reaction, your labor organization was driven into politics, the strike-war became a war of ballots.

Then—because some men have room for only one idea in their heads—you would have another beautifully simple formula, that of the pure and simple politician, the orthodox Socialist, who preached that salvation was found in the ballot alone. What would happen if you followed that formula was obvious enough; the labor men had seen it happen in this very state, having duly elected a radical governor, and stood by helpless while he was barred from office! When that happened to Socialist candidates, the pure and simple politicians would be the first to come to the unions, to ask for "direct action".

You needed the strike to back up the ballot, and the ballot to reinforce the strike. Thus Syndicalism and Socialism were two feet, and the wise man did not hop on either foot, he walked on both—and he kept his eyes open in addition! The worst thing any man could do was to adopt a formula, making it as a bandage over his eyes, so that he fell into the traps which his cunning enemy planted. And here was the service which a young man of the leisure class could do for groping labor; he had time to think, he was independent of all ties; so, if he kept true and steadfast, he could be, as old John Edstrom had said, a watcher on a hill-top, spying out the pathway ahead, and giving signals to the marching host of labor.

In fulfilment of this function, Hal would travel about the state, addressing meetings of whoever would come to hear him. And Mary Burke would go along, and speak at these meetings. It seemed to Jessie a shocking thing that Mary should travel about with Hal and appear on the same platform with him—considering what had been said about them, and might continue to be said. Surely one must pay some attention to what other people thought! But Jessie found that it was not wise to argue about this with

Hal, who was so wrapped up in his propaganda that he did not care what even his wife thought; so, in order to keep things respectable, the wife would make it a point always to accompany her husband on these pilgrimages, no matter what the sacrifice of convenience and comfort. In order that all the world should know that the wife was present, Jessie would sit upon the platform while the speech-making was going on; in order that people might understand that she had a proper wifely sympathy with her husband's activities, she would smile at the interesting places in his speech, and when the meeting was over she would shake hands with all sorts of strange and uncomfortable people; workingmen who wore no collars, or worse yet, celluloid collars; ranchmen with big calloused hands; angular young women with prominent teeth and short hair; queer old ladies with fervent convictions; poetical young men with Windsor ties; pale, harassed-looking agitators who had been in jail, and must be treated with especial deference on that account. All this was a part of the process which the Russians term "going to the people"; and Hal would explain to Jessie that the process was unavoidable—it was a burden upon progress caused by the blocking of the ordinary channels of information, the corrupting of newspapers and magazines. There were things that the people must know, and there was no way to teach them save to tell it by word of mouth.

So both Hal and Mary developed into orators: quite as a matter of course, and without thinking about it especially. The first time that Mary walked out on a platform and faced an audience deliberately and in cold blood, she was frightened so that her knees shook visibly; but presently her theme took hold of her—she was telling the story of life at Horton, and she would be back in the tents, taking care of the wounded, or on the street in Pedro, arguing with General Wrightman's troopers, or shut up in jail, singing the union song from the window, or leading the women and children over the plain, with machine-gun bullets knocking off the heel of her shoe and cutting holes in her dress. It was a story of varied adventure, with humor and pathos and terror; and Mary soon found out what parts of it moved her audiences, made them laugh or shed tears or cry out with horror.

Old Patrick's daughter became "Red Mary" in a new sense of the word; a sinister figure, a name of terror and reprobation throughout the state. She worked incessantly, with ceaseless energy; for she could see that she was accomplishing something, and so hope came back into her heart, and purpose into her life. It was a different thing, meeting human beings face to face in this way—quite different from sitting helpless through the agony of a strike, suffering wrongs heaped upon wrongs, and crying to that blank and deaf and lifeless thing which was called "the public".

This difference was a circumstance which puzzled you at first; it took you some time to work out the explanation. The thing which was called "the public" was not really the public at all; it was not a human, a real thing, but an artificial creation, an institution; it was newspapers, political machines, churches, colleges—organizations maintained and controlled by privilege. The mass of the people were warm-hearted, they loved justice, they believed in democracy and the American tradition; the only trouble was that the truth was withheld from them. They simply did not *know* what was being done to workingmen all over their great country, from Lawrence, Massachusetts, to Wheatlands, California, from the lumber-camps of Louisiana to the copper-mines of Michigan.

So Hal and Mary traveled about, meeting the people face to face, telling them what they had seen with their own eyes; and they saw men's hands clenched in anger, they saw tears running down the cheeks of women. When the meeting was over, the people crowded about to pledge their aid; they passed resolutions, they gave money, they went home to tell their friends and neighbors. And they were all men and women who had votes, and were accustomed to make use of them; they did not forget overnight what they had heard; they would organize, and keep in touch with you; so, if you had patience and determination, you would find yourself in the end with the means of punishing politicians who had betrayed the public interest, and sold the state to predatory corporations.

There was an English poet of the Chartist times, whom his enemies in derision dubbed "the Corn-law rhymer". One does not read about him in text-books of literature, because his work is not to the taste of those classes which tend the shrines of culture. But in an old book which Hal loaned her, Mary came upon some verses called the People's Anthem—as noble in utterance and as perfect in form as any hymn in English. Because they voiced so perfectly the longing in her heart, Mary always closed her meetings with a stanza from this Anthem. The heart of Ebenezer Elliott has long since been dust, but in its time it suffered imprisonment and ignominy, and its cry is as moving to an audience in the Rocky Mountains as in an English factory-town:

"When wilt thou save the people?
O, God of Mercy! when?
Not kings and lords, but nations!
Not thrones and crowns, but men!
Flowers of thy heart, O God, are they!
Let them not pass, like weeds, away!
Their heritage a sunless day!
God save the people!"

POSTSCRIPT

The reader of this book will wish to know what relation the story bears to fact. "The Coal War" is a historical novel, dealing with the Colorado strike of 1913-14. The writer has availed himself of the fictionist's privilege, only so far as concerns matters of no social significance. He has given imaginary names to towns, coal-camps and characters, for the sake of convenience, and in the interest of good taste; but all the characters having social significance are real persons, and every detail of the events of the strike is both true and typical.

The reader who wishes to investigate the evidence for these statements will find in another volume, "King Coal", a list of documents, comprising some eight million words, three-quarters of it affidavit or sworn testimony. It hardly seems worth while to reprint this list here; it will be more to the point if I give a few definite references. Thus, the reader who is interested in the character of Mike Sikoria, as he appears in this volume and in "King Coal", may find him on pages 679-694 of the Congressional Committee report, entitled "Hearings before a Sub-Committee of the Committee on Mines and Mining, House of Representatives, Sixty-third Congress, Second Session, pursuant to House Resolution 387." The story of "George Tareski", who was made to dig his own grave, has been taken bodily from the same report, pages 2052-58. In the same manner, episode after episode has been made out of the sworn testimony of actual participants or witnesses. Every case of injustice, every practice named among the causes of the strike, has been sworn to by witness after witness. The speeches made at the convention which called the strike are from stenographic records. The statements as to political conditions were sworn to by actual participants in these practices. The statements as to violations of the state mining-laws, and the maintaining of peonage during the strike, have been sworn to by state officials.

As to the picture drawn of the conduct of the militia, it has been made from the testimony of several hundred witnesses, a large number of them having no connection with the strike or with the strikers. Professor Purdue is a real person, who has told his own story—and part of the story of Will Wilmerding as well. Captain Harding is also a real person—though perhaps I should state that I never met the gentleman, but have merely imagined a character upon the basis of his testimony and report. I have taken the liberty of having him present at the Ludlow massacre, although he was not actually there. Billy Keating is a real person; so are MacKellar, Louie the Greek, Jim Moylan, John Harmon, Mrs. David, Rosa Minetti—the testimony of all these people having been given under oath.

To give even a summary of the evidence would require a volume. Therefore I will select one representative episode, and give a detailed summary of the evidence upon it. For this purpose the "Ludlow massacre" will serve best—it being the most conspicuous single event, and likewise the one about which there has been the most controversy. I remember, after the publication of the militia testimony, reading an editorial remark in a leading New York newspaper, to the effect that the editors were glad not to have to believe that the militia of the state had deliberately set fire to the tent-colony. Now, when a conservative and cautious-minded newspaper editor does not wish to believe a thing, he can find reasons for having his way; but I assume that the readers of this book wish to believe the truth, and I think one can fairly say that the evidence upon this point is sufficient to have hung a number of officers of the militia if there had been such a thing as justice in the State of Colorado.

In my story of the events at "Horton", I have for the most part followed the militia's own account, the report made by three militia officers, Major Boughton, Captain Danks and Captain Van Cise, under circumstances described in this novel. I have taken the liberty of departing from their narrative on only two important points. Let me first state these points, and my reasons for disputing their opinions upon them.

It is their claim that rifle-firing was begun by the strikers, and had continued for some minutes, before the three signal-bombs were set off by the militia. Against this, the testimony of the strikers is unanimous that not a shot had been fired anywhere when the signal-bombs were heard. But my reason for rejecting the statement of the three militia officers is not the testimony of the strikers; it is the evidence brought before the United States Commission on Industrial Relations, impugning the good faith of the officers upon this point. There was a disinterested witness available, Mr. M.G. Low, pump-man of the Colorado and Southern Railroad water-tank at Ludlow. This witness, a decent and fair-minded man, had watched the events from close at hand, and he testified that first Major Boughton and then Captain Van Cise came to him and questioned him as to what he had seen; he told them that the bombs had been fired first, and that the first sounds of rifle firing had come from the militia position on Water-tank Hill. Yet having had this statement, the officers failed to call him before their body or to mention his testimony in their report!

The other contention of the militia report to which I have not considered it necessary to pay attention is the one that the first fire in the tent-colony "was due either to an overturned stove, an explosion of some sort, or the concentrated fire directed at one time against some of the tents." The stoves at Ludlow were of the kitchen range variety, not to be upset by rifle bullets. As for the concentrated fire of rifle bullets setting fire to tent-canvas, I have not been able to find any military man who would consider such a theory. The statement seems the more superfluous, because the militia officers admit that shortly afterwards other tents were deliberately set fire to by the uniformed militiamen, in the manner described in the affidavits of the strikers and their wives. Here are four paragraphs from the report bearing upon this point:

> "We find that the tents were not all of them destroyed by accidental fire. Men and soldiers swarmed into the tent-colony and deliberately assisted the conflagration by spreading the fire from tent to tent.
>
> "Beyond a doubt, it was seen to intentionally that the fire should destroy the whole of the colony. This, too, was accompanied by the usual loot.
>
> "Men and soldiers seized and took from the tents whatever appealed to their fancy of the moment. In this way, clothes, bedding, articles of jewelry, bicycles, tools and utensils were taken from the tents and conveyed away.
>
> "So deliberately was this burning and looting done that we find that cans of oil found in the tents were poured upon them and the tents lit with matches."

Now I have carefully investigated all the evidence bearing upon the Ludlow massacre, and believe it may safely be declared an established fact that the destruction of the tent-colony was deliberately carried out by militiamen; also that it had been planned in advance and was directed throughout by militia officers. There is, in the first place, the affidavit of Mrs. Susan D. Hollearin, postmistress at Ludlow, who witnessed a great deal of the fighting from her office, which was used as headquarters by the officers. She testified that before the fighting began she heard Major

Hamrock say, "Line up, boys, and get ready." She testifies that the bombs went off before any shots had been fired. She testifies that she heard Lieutenant Linderfelt, in command of the machine-guns, say, "Keep working on the tent-colony. Shoot everything that moves." She testifies "I looked down and saw that one tent was on fire at that time. Later another tent caught fire. There was no wind, just a little breeze. The tents seemed to catch fire from the outside, and caught in spots. About the time the train came in, the women and children in the colony began to scream, and could be heard in the depot, and I heard the soldiers giving the women and children orders to 'Get a move on you, and get the hell out of here,' and such statements as that." She testifies that in the morning, hearing that Mary Petrucci's children were still inside the tents, she took a flag of truce. "When I got nearly there I saw three or four soldiers leaving. They had been firing the few remaining tents. I saw this."

There is the testimony of W.J. Hall before the Coroner's inquest. Mr. Hall was the driver of an automobile, having nothing to do with the strike; he testified that he was stopped on the road by the militiamen, who took his car and used it for the moving of a machine-gun. He heard the orders given "to go in and clean out that colony. For them to drive out and then to burn the colony."

The testimony of Doctor Aca Harvey, of Aguilar, before the Coroner's inquest. He was trying to attend a dying man who had been shot through the head. He carried a white flag, but was fired upon. The militia used explosive bullets. He saw two fires started separately. Mr. Hayes, his companion, climbed up onto the water-tower with his field-glasses, and described to him men in uniform throwing oil upon the tents, setting fire to them; he could see the blaze.

R.J. McDonald, before the Coroner's inquest, a military stenographer employed by the troops, went out with them to see the fighting, and at six P.M. he heard one of two militia officers (whom he named) say, "There is thirty minutes yet before dark, and we have to take and burn the tent-colony."

A.J. Reilly, freight-brakeman on the Colorado and Southern freight-train, stopped to let a passenger-train pass. He saw the tents blazing. "Saw a man in military uniform touch a match to the third tent. Ten or fifteen more stuck their guns up to our faces and told us to 'move on and be damn quick about it,' or they would shoot us."

"Did you say anything to those men?"

"No, sir, not a word."

"Who were those men who stuck their guns up to your faces?"

"Uniformed men."

The testimony of William Snyder, store-keeper at the Ludlow tent-colony, whose twelve-year-old boy had his brains shot out by an explosive bullet.

"Did anybody come to your tent while you were there?"

"Yes, the militia came there."

"What did they say to you?"

"They set fire to the tent, and opened it and came in."

"They set fire to your tent?"

"Yes, sir."

(The remainder of the remarks of the militia, as testified to by Mr. Snyder, are given in the text of this novel—except that a number of unprintable oaths have been omitted.)

The affidavit of Mrs. Alcarita Pedregon. She swears, among other things, "I seen a militiaman come over there and look inside the tent and strike a match and set fire to the tent. I stayed in the tent until it was all burned up. There were eleven children and two women suffocated with the smoke where I was. I lost two children in this cave when the tent burned. I don't know where my husband was at this time. I looked up out of the hole and saw the soldier set fire to the tent with a match. I lost everything I had in this fire."

The affidavit of Mrs. Ed. Tonner. "My tent was so full of holes like it was like lace, pretty near. It could have been about four when little Frank got his head hurt, and a little while after this they tried to set the tents on fire. I kept bobbling my head up and down, and Mr. Fyler said, 'For God's sake keep your head down, or you will get it blown off.' About six o'clock they turned around and tore the tent between the two tents, and they set the broom on fire with coal oil, and they set the tent on fire, with me right underneath with my five little children. Then Gusta Retlich, she helped me out with the children, grabbed them up, and then we run to a Mexican lady's tent farther down, and then Louie the Greek helped me, he helped me down into a hole and threw water in my face as I was fainting with all the children."

The affidavit of Mrs. Gloria Padilla: "At sometime late in the afternoon they started to burn the tents. When the tents were first fired they did not burn my tent, later in the evening the soldiers came back to fire the rest of the tents and they heard my children crying, and they said there is a family in there, and they helped me out and took me and the children to the depot. While at the depot three Mexican guards got mad at the women and said they ought to be burned in the tents."

The testimony of Mrs. Margaret Dominiske, before the United States Commission on Industrial Relations, in New York City. "About three o'clock was when they started to shoot so awfully hard. And about a quarter to six one of the men from the arroyo came up to where we were. He said we had better get out of here because he said there is about fifty militia right close to the camp, and he said they are burning up the tents and if you crawl out here you can see it. So I crawled out and I looked down and I saw about six or eight tents burning. And then I saw five militiamen cross from the tents that was burning over to those that was not burning, and three of them had torches, and two had cans. I don't know what was in the cans, but I think it was oil. They went into one of these other tents, and I got back into where my children were. Pretty soon some said it looks like there is a train, that that will be our only chance of escape. So I went and crawled back out and looked out again to the tents and I saw the militia going into them, they was all on fire, so I judge from that they had set it on fire, and when this train came–"

"Chairman Walsh: You say that they had torches and that they were lighted?

Mrs. Dominiske: Well it looked like a broom, to me, that is what it looked like from where I was at, looked like they were brooms lit. Then when the train came, why we all got out of the well and out of the barn and went to the arroyo. And on the way there, as she stooped to get under a fence, one of the ladies had a big apron on and she stooped to get under the fence and there was a bullet passed right through her apron, and another passed over my head and exploded. It was an explosive bullet and exploded right in front of another lady, and she had a baby in her arms, and she fainted. We got into the arroyo and we went down to a ranch about five or six miles from Ludlow."

The above is only a part of the evidence bearing upon one episode of the novel; I do not exaggerate in saying that I could prove fifty other incidents in as much detail. Before writing the story I studied eight million words of printed or typewritten evidence, in addition to many millions which I gathered directly from the lips of witnesses. I invite anyone who suspects me of exaggeration to examine at least a part of this evidence himself.

When I wrote "The Jungle" it was my purpose to call attention, not so much to evils connected with the country's meat supply, as to injustices suffered by the working people who prepare it. As I said afterwards, I aimed at the public's heart, and by accident I hit it in the stomach. Now once more I am aiming at the heart. This book goes out as an appeal to the conscience of the American people: an appeal for millions of men, women and children who are practically voiceless—not merely in Colorado, but in West Virginia, Pennsylvania, Michigan, Alabama, a score of states in which miners and steel-workers have been unable to organize and protect themselves.

It is an appeal against the hideous nightmare of Government by Gunmen: a new form of sovereignty which has grown up in America, a system nation-wide in scope, having many millions of dollars invested in it and employing tens of thousands of men. The public is told that its purpose is the preservation of order; but let the reader of this book carry away one definite certainty—the purpose of Government by Gunmen is not order, but oppression, not peace but slavery. These "guards" and "detectives" come, not to prevent violence, but to provoke it; not to protect industrial property, but to crush labor organization. They are private mercenaries, fully equipped for military campaigning, as well as for secret diplomacy; trained in every phase of their peculiar kind of warfare, and as merciless and irresponsible as the *condottieri* of Italy. If they are permitted to go on developing their power in this republic, they will bring upon us a slave rebellion as bloody and cruel as those of Spartacus and Eunus against Rome.

APPENDICES

EXPLANATION OF EDITORIAL PROCEDURES

Three drafts of *The Coal War* survive at the Lilly Library, Indiana University, typescripts which may be designated A, B, and C, early to late. In the original purchase of the Sinclair Papers in 1957 the Lilly acquired the early A typescript, consecutively paginated 1-290, the first 41 pages of C, and several sections of *The Coal War* designated here as B. Not until the autumn of 1968, in its fourth acquisition of Sinclair Papers, did the Lilly obtain the remaining 376 pages of C.

B and C typescripts were initially identical in content, C the ribbon copy and B the carbon. Since Craig Sinclair's hand is distinct from Upton's, an examination reveals that she was responsible for all B emendations, while he made every revision in C. A comparison of both versions indicates that, in all probability, Sinclair and his wife worked through them simultaneously at first, each holding a copy, and then Sinclair made further changes at a later date. These final revisions are almost uniformly of a minor nature, generally a matter of word substitution. Although no differences exist between B and C drafts in the vast majority of pages, there can be no doubt that C is a later version than B. C incorporates most B emendations but deliberately rejects others. An example will make this process apparent and, at the same time, illustrate the nature of Sinclair's revisions.

In Book III of *The Coal War*, prior to final emendation, Hal Warner is engaged in a dialogue with Congressman Simmons.

> "You see," explained Mr. Simmons, "ours is a government of divided powers. The task of this committee was to investigate a possible need of new legislation."

"Will you recommend any new legislation?"

"It's difficult to think of any that would remedy this present situation."

"Let me suggest something then, Mr. Simmons. You would end this struggle if you made membership in a union compulsory in coal-mines."

It was evident from the look on Mr. Simmons' face that he was not going to recommend anything like that! "It would seem to me," he said, "that the problem is to get the present laws enforced."

"Well," said Hal, "I'll compromise on that. How are we to do it?"

"It's the duty of the Governor of your state."

"But he won't do his duty. We've spent five miserable months proving that! So what next?"

In B, Craig made a single emendation, the substitution of "would make": "You would end this struggle if you [would make] membership in a union compulsory in coal-mines."

As may be seen from the text (pp. 260-261), when Sinclair made his final revision of C typescript, he wished to make Hal appear more direct, forceful, and authoritative. Sinclair initially added "would make" as Craig suggested, but then excised it. He saw that his earlier use of "something" and "anything" was vague, and he substituted the much more concrete "law" whenever possible, four times in all. In place of the wordy sentence, "You would end this struggle if you [would make] membership in a union compulsory in coal-mines," Sinclair emended to: "Let me suggest a law, Mr. Simmons–a law that would end this struggle at once. Make membership in a union compulsory in all coal-mines." Given this tone, Hal's later talk of compromise and his weak interrogatories are inconsistent. Therefore, when Simmons states "that the problem is to get the present laws enforced," Sinclair removes altogether Hal's uncertain rejoinders: "I'll compromise on that. How are we to do it?" It is readily apparent that Sinclair's final emendations in C draft are direct, emphatic, and more effective than his earlier version.

The differences between Sinclair's hand and his wife's, evident after a close inspection of manuscripts and handwritten letters, make it a relatively straightforward matter to join the two C fragments together and establish the final text of *The Coal War* with certainty. The fact that C is the ribbon copy, and B the carbon, introduces yet another safeguard in the identification and choice of C version as copy-text. This text of *The Coal War* is a critical, unmodernized reconstruction: critical in the sense that it is not an exact reprint of the copy-text; unmodernized in that, apart from established editorial procedures, every effort has been made to present the text in as close a form to Sinclair's finally revised typescript C as possible.

Editorial alteration is of two basic types. First, since Sinclair did not catch every typographical error, they are editorially corrected. Second, inconsistencies in spelling, capitalization, word-division, and punctuation are normalized by utilizing the principle of most frequent usage. Sinclair adhered to a virtually uniform practice of placing punctuation outside quotation marks when the quoted matter was a phrase or term, as opposed to conversation. His few departures from this custom may all be attributed to his own or his typist's carelessness. Sinclair's practice is clear and, since it was a quite common system of punctuation in America at least through the 1930s, his customary pattern has been adopted.

In matters of doubt, the first edition of *King Coal* has been turned to for authority. The copy-text is everywhere legible, and Sinclair's revisions are clear. Editorial alteration has been conservative and always weighed against the values of authorial purpose and possible stylistic idiosyncrasy. But for the exceptions previously noted, this text of *The Coal War* represents a faithful transcription of typescript C as finally revised and submitted to the Macmillan Company for publication.

NOTES TO INTRODUCTION

1 John D. Rockefeller, Jr., *The Personal Relation in Industry* (New York: Boni and Liveright, 1923), p. 99.

2 William Dean Howells, *A Traveler from Altruria* (New York: Hill and Wang, 1968), p. 65.

3 Quoted in Gabriel Kolko, *The Triumph of Conservatism* (Chicago: Quadrangle, 1967), p. 13. Two other discussions of the misnamed Progressive Era are particularly helpful: William Appleman Williams, *The Contours of American History* (Cleveland: World, 1961), pp. 343-424; and James Weinstein, *The Corporate Ideal in the Liberal State, 1900-1918* (Boston: Beacon, 1968).

4 Williams, p. 351.

5 David Shannon, *The Socialist Party in America* (Chicago: Quadrangle, 1967), p. 5; Weinstein, pp. 17, 120.

6 *Rebel Voices,* ed. Joyce L. Kornbluh (Ann Arbor: University of Michigan Press, 1968), pp. 12-13.

7 *Conditions in the Coal Mines of Colorado: Hearings before a Subcommittee of the Committee on Mines and Mining. Pursuant to H. Res. 387*, 63rd Cong., 2d Sess. (Washington: GPO, 1914), II, 2841-42. Hereafter cited as *Hearings*.

8 *The Colorado Coal Miners' Strike: Industrial Relations: Final Report and Testimony Submitted to Congress by the Commission on Industrial Relations, Created by the Act of August 23, 1912*, 64th Cong., 1st Sess., Doc. 415 (Washington: GPO, 1916), IX, 8638, 8642, 8881. Hereafter cited as *IRC*.

9 Former U. S. Senator Thomas H. Patterson testified that the introduction of non-English speaking immigrants into the mines was

the deliberate policy of the operators. A labor contractor had informed him, "Twelve or fourteen years ago I was furnishing them [CF&I] men . . . and when I would get the orders for men frequently I would get them in writing, and they specified the number of one nationality and the number of another nationality—all speaking different languages—and no English-speaking nationality." This policy, Patterson concluded, was intended to provide the operators with men who "would cause them as little trouble as possible." *IRC*, VII, 6499.

10 *Biennial Report of the Secretary of State of Colorado* (Denver: Smith-Brooks, 1912), p. 37.

11 *Hearings*, II, 2658-59.

12 *IRC*, VII, 6557,; IX, 8492-93.

13 Ibid., IX, 8499.

14 Ibid, VII, 6771; IX, 8910, 8498.

15 Ibid., VII, 6777; IX, 8547, 8540, 8543, 8545.

16 Ibid., IX, 8548-50, 8500.

17 Ibid., VII, 6434, 6554; IX, 8501; VII, 6360; IX, 8494; *Twelfth Biennial Report of the Bureau of Labor Statistics of the State of Colorado* (Denver: Smith-Brooks, 1911), p. 27.

18 *Hearings, Brief for the Striking Miners*, 28.

19 *Hearings*, II, 2007; Donald J. McClurg, "Employer Policy in Colorado Coal to 1915," pp. 193-94; *IRC*, IX, 8437.

20 *Hearings*, I, 21; *Fourteenth Biennial Report of the Bureau of Labor Statistics of the State of Colorado*, p. 212; *Hearings*, I, 27-8, 30; *Twelfth Biennial Report of the Bureau of Labor Statistics of the State of Colorado*, pp. 21, 25-26; also cited by George S. McGovern, "The Colorado Coal Strike, 1913-1914," pp. 28-29.

21 George P. West, *Report on the Colorado Strike* (Washington: GPO, 1915), p. 81; see jury verdicts, *IRC*, VIII, 7265-96; *Hearings*, II, 2383; *IRC*, VII, 6786.

22 *Fourteenth Biennial Report of the Bureau of Labor Statistics of the State of Colorado*, p. 164; *Thirteenth Biennial Report of the Bureau of Labor Statistics of the State of Colorado*, p. 57; *IRC*, VIII, 7234; IX, 8414; *Hearings*, I, 76; II, 2038.

23 *IRC*, IX, 8411.

24 *IRC*, VIII, 7138-39.

25 *IRC*, IX, 8411, 8773.

26 *IRC*, VII, 6513.

27 Colorado Senate, *Testimony Taken before the Committee on Privileges and Elections of the Nineteenth General Assembly, Wednesday, February 5, 1913*, 1-2, 843-44, as cited by McGovern, p. 58.

28 Barron B. Beshoar, *Out of the Depths* (Denver: Golden Bell, n.d.), p. 7, also cited by McGovern, p. 58.

29 *IRC*, IX, 8764, 8781-82.

30 Beshoar, pp. 341-44. Like the McClurg and McGovern unpublished dissertations, Beshoar's study is also a fine piece of work. A student of the strike is indebted to all three.

31 Ibid., pp. 49-50.

32 *Hearings*, II, 2539.

33 *IRC*, VIII, 7297-7300; VII, 6791, 6801-2.

34 *Fourteenth Biennial Report of the Bureau of Labor Statistics of the State of Colorado*, p. 150.

35 *IRC*, IX, 8422, 8416.

[36] Ibid., 8419-20.

[37] Beshoar, p. 58.

[38] The *Denver Express*, September 24, 1913, cited by Beshoar, pp. 63-64.

[39] West, p. 106.

[40] *IRC*, IX, 8421.

[41] McClurg, pp. 222-24.

[42] Beshoar, p. 93.

[43] *The Military Occupation of the Coal Strike Zone of Colorado by the Colorado National Guard, 1913-1914* (Denver: Smith-Brooks, 1914), p. 14; George S. McGovern and Leonard F. Guttridge, *The Great Coalfield War* (Boston: Houghton Mifflin, 1972), p. 45; Beshoar, pp. 92-95.

[44] Beshoar, pp. 102-5.

[45] McGovern, pp. 221-25.

[46] *IRC*, IX, 8422.

[47] Ibid., VIII, 7118.

[48] Ibid., IX, 8427.

[49] Ibid., VII, 6833, 6707.

[50] *Hearings*, II, 2879.

[51] Ibid., 2873.

[52] Ibid., 2874.

[53] West, pp. 141-42; Raymond B. Fosdick, *John D. Rockefeller, Jr.: A Portrait* (New York: Harpers, 1956), p. 150.

[54] *IRC*, IX, 8429-30.

[55] West, p. 125; Helen Ring Robinson, "The War in Colorado," *Independent*, 78 (May 11, 1914), 246; West, p. 126.

[56] *IRC*, VII, 6349. This account of the Ludlow Massacre is largely based on Beshoar, pp. 166-80; McClurg, pp. 276-312; McGovern, pp. 270-308; and Colorado National Guard's report, *IRC*, VIII, 7311-23.

[57] McClurg, p. 280.

[58] *IRC*, VII, 6350-51.

[59] Ibid., VIII, 7321.

[60] *Fourteenth Biennial Report of the Bureau of Labor Statistics of the State of Colorado*, p. 188.

[61] *IRC*, IX, 8430.

[62] The *Rocky Mountain News*, April 22, 1914, p. 1, cited by Beshoar, pp. 187-89.

[63] The *Denver Express*, April 22, 1914, p. 1.

[64] Beshoar, p. 190.

[65] West, p. 132.

[66] Beshoar, p. 230, *IRC*, John A. Fitch, "Law and Order: The Issue in Colorado," The *Survey*, 33 (December 5, 1914), 242; McGovern, p. 360.

[67] "The Struggle in Colorado for Industrial Freedom," No. 14 (August 25, 1914), p. 2.

[68] *IRC*, IX, 8609-10, 8869, 8882-83.

[69] "President Wilson's Plan for Peace in Colorado," The *Survey*, 33 (September 19, 1914), 608.

70 *IRC*, VII, 6688, 6690.

71 Ibid., 6691-92.

72 Ibid., 6693.

73 Ibid., 7766.

74 Ibid., 7845.

75 West, p. 186.

76 Rockefeller, pp. 91, 93-95.

77 *IRC*, VIII, 7847-48.

78 Ben M. Selekman and Mary Van Kleeck, *Employees' Representation in Coal Mines* (New York: Russell Sage, 1924), pp. 185, 178, 188, 186.

79 Ibid., pp. 384-87.

80 Mary Craig Sinclair, *Southern Belle* (New York: Crown,1957), p. 154.

81 The *Appeal to Reason*, May 16, 1914, p. 2.

82 The *New York Times*, April 30, 1914, p. 5.

83 The *Appeal to Reason*, May 16, 1914, p. 2.

84 The *New York Times*, May 1, 1914, p. 4.

85 Ibid., May 4, 1914, p. 3.

86 The *Sun*, May 1, 1914, p. 1.

87 Upton Sinclair to John D. Rockefeller, Jr., May 3, 1914. Sinclair Papers, L file. Lilly Library, Indiana University. Unless otherwise specified, my use of Rockefeller refers throughout to John D. Rockefeller, Jr. Sinclair's letters went unanswered.

88 The *Evening Post* (New York), May 4, 1914, p. 2.

89 The *New York Times*, May 5, 1914, p. 3.

90 *Southern Belle*, p. 161.

91 The *Denver Express*, May 14, 1914, p. 8.

92 The *Arapahoe Republican and the Englewood Tribune* (Englewood. Colorado), May 22, 1914.

93 The *Appeal to Reason*, May 30, 1914, p. 1.

94 Ibid.

95 Ibid. Since this resolution was to cause a great deal of controversy, it is useful to consider it here:

> Resolved, that a joint committee of six members, three selected by the Senate and three selected by the House, said members to be selected by the body of each house, shall be appointed and directed to confer and advise with the Governor and other executive officers of the State to the end that the legislative department may render all assistance in its power to the executive department in the enforcement of law and the maintenance of order, and to consider ways and means of restoring and maintaining peace and good order throughout the State; and to investigate and make report at the next session of the legislature upon the following matters and subjects.

There follows a list of subjects to be investigated, such as wages paid the miners, number of lives lost during the coal strike, and whether large numbers of gunmen were imported and employed by the operators. Neither the word "mediation" nor the phrase "to assist in settling the strike" appears in the resolution. *Senate Journal of the Nineteenth General Assembly of the State of Colorado: Extraordinary Session* (Denver: Smith-Brooks, 1914), pp. 177-70. Hereafter cited as *Senate Journal*.

96 The *Appeal to Reason*, May 30, 1914, p. 2.

97 The *Denver Post*, May 18, 1914, p. 1. Governor Ammons was under no illusions about the purpose of the resolution. In a message to the Senate on May 16, he wrote: "I have the honor to inform you that I have approved and filed with the Secretary of State the following resolutions and bills: May 16, 1914, S.B. No. 2, relating to intoxicating liquors. May 16, 1914, H.J.R. No. 6, re joint committee to *investigate* [my emphasis] the strike and report to the General Assembly; if it is not in session to report to the Governor and Attorney General." *Senate Journal*, p. 182.

98 The *Appeal to Reason*, May 30, 1914, p. 2.

99 Upton Sinclair, *The Brass Check* (California: Published by the Author, 1928), p. 165. The Associated Press did send the following story out of Washington. "President Wilson expressed satisfaction with the situation after he had received Governor Ammons' reply late tonight. It was said by officials in close touch with the president that Wilson was greatly pleased with what had been done after he had been informed by Governor Ammons of the work of the Colorado legislature, and that he hoped the State would assume control of the situation in the near future so the Federal troops might be withdrawn." This story can only be regarded as an Associated Press fabrication. The next day the Washington correspondent of the *Rocky Mountain News* wired: "At the White House it was stated that nothing has been given out which would justify the statement printed in some of the morning papers that the President is entirely satisfied with the telegram received yesterday from Governor Ammons." The *Denver Post*, May 17, 1914, p. 1; the *Rocky Mountain News*, May 18, 1914, p. 1.

100 Upton Sinclair to John D. Rockefeller, Jr., May 26, 1914. Sinclair Papers, L file, Lilly Library, Indiana University.

101 The *New York Herald*, May 11, 1914, pp. 3-4; Leonard Abbott, "The Fight for Free Speech in Tarrytown," *Mother Earth*, 8, No. 4 (June 1914), 107-11.

102 The *Sun*, June 1, 1914, p. 1; *New York Times*, June 1, 1914, p. 1; Leonard Abbott, "The Fight for Free Speech in Tarrytown," *Mother Earth*, 8, No. 4 (June 1914), 108-9.

103 The *New York Times*, June 14, 1914, p. 11. The *Times* report reads: "Mrs. Gould invited all Tarrytown residents, including the [village] Trustees, to the meeting, the only restriction placed upon the free speech agitators being that they do not allow I.W.W. men to share in the speechmaking." In this newspaper account, as in numerous others, the term "I.W.W." is inaccurately used and refers, in point of fact, to the anarchist *Mother Earth* group. Although a small number of IWW members and sympathizers participated in the demonstrations, the IWW was not formally engaged in the Tarrytown free speech fight.

104 The *New York Express*, June 15, 1914, p. 3.

105 The *Appeal to Reason*, June 27, 1914, p. 3.

106 The *Sun*, June 23, 1914, p. 1; *New York Times*, June 23, 1914, p. 1; *The Brass Check*, p. 192.

107 The *New York Times*, June 26, 1914, p. 8.

108 Ibid., July 6, 1914, p. 2.

109 The *Appeal to Reason*, July 18, 1914, p. 3.

110 "The Lexington Explosion," *Mother Earth*, 9, No. 5 (July 1914), 132.

111 The *Call* (New York), July 7, 1914, p. 3.

112 Alexander Berkman, "A Gauge of Change," *Mother Earth*, 9, No. 5 (July 1914), 167-68.

113 The *Sun*, July 14, 1914, p. 6; *New York Times*, July 12, 1914, p. 3; "The Lexington Explosion," *Mother Earth*, 9, No. 5 (July 1914), 137-39, 141, 143, 145-51. The most comprehensive account of this memorial demonstration is found in this issue of *Mother Earth*.

Because of its advocacy of violence, this same July number was to prove instrumental both in the successful prosecution of Emma Goldman and Alexander Berkman for violating the Draft Act of May 18, 1917, and in their deportation on the *Buford* in 1919.

114 The *New York Times*, February 16, 1915. Reports published in the *Denver Post*, February 16, 1915, p. 14 and in the *Denver Times*, February 15, 1915, p. 5 are fully consistent with Berkman's statement. With the exception of this explanation by Berkman, however, those who presumably would have known the purpose of the bomb kept it to themselves.

115 Upton Sinclair to Frederick van Eeden, June 17, 1915. Sinclair Papers, L file, Lilly Library, Indiana University.

116 A Proposition For the First Serial Rights of a New Novel "King Coal" by Upton Sinclair, E. L. Doyle Papers, Denver Public Library Western Collection.

117 George P. Brett to Upton Sinclair, December 7, 1915. Sinclair Papers, L file, Lilly Library, Indiana University.

118 George P. Brett to Upton Sinclair, December 31, 1915. Sinclair Papers, L file, Lilly Library, Indiana University.

119 Mary Craig Sinclair to George P. Brett, May 1, 1916. Upton Sinclair's letter to Brett of the same date does not survive. Sinclair Papers, L file, Lilly Library, Indiana University.

120 George P. Brett to Upton Sinclair, May 9, 1916. Sinclair Papers, L file, Lilly Library, Indiana University.

121 Upton Sinclair to Frederick van Eeden, October 29, 1916. Sinclair Papers, L file, Lilly Library, Indiana University.

122 "KING COAL seems to me to be very much better than it was and I am very happy to be able to make you an offer for its publication. . . ." George P. Brett to Upton Sinclair, November 20, 1916. Sinclair Papers, L file, Lilly Library, Indiana University.

123 Edward C. Marsh to Upton Sinclair, November 7, 1917. Sinclair Papers, L file, Lilly Library, Indiana University.

124 *The Autobiography of Upton Sinclair* (New York: Harcourt, Brace World, 1962), p. 327.

125 *Mammonart* (California: Published by the Author, 1925), p. 10.

126 *The Great Tradition* (New York: Macmillan, 1933), p. 197.

127 *Politics and the Novel* (New York: Horizon Press, 1957), p. 20.

128 *King Coal* (New York: Macmillan, 1917), p. 385.

129 *Autobiography*, p. 297.

130 *On Native Grounds* (New York: Harcourt Brace, 1942), p. 121.

131 "Lincoln Steffens and Upton Sinclair," *The New Republic*, 72 (September 28, 1932), 174.